MATHEMATICS RESEARCH DEVELOPMENTS

MATHEMATICAL MODELING IN SOCIAL SCIENCES AND ENGINEERING

MATHEMATICS RESEARCH DEVELOPMENTS

MATHEMATICS RESEARCH DEVELOPMENTS

MATHEMATICAL MODELING IN SOCIAL SCIENCES AND ENGINEERING

JUAN CARLOS CORTÉS LÓPEZ,
LUCAS ANTONIO JÓDAR SÁNCHEZ
AND
RAFAEL JACINTO VILLANUEVA MICÓ
EDITORS

nova publishers
New York

LIBRARY OF CONGRESS CATALOGING-IN-PUBLICATION DATA

ISBN: 978-1-63117-335-6

Published by Nova Science Publishers, Inc. † New York

CONTENTS

PREFACE

This book titled "*Mathematical Modeling in Social Sciences and Engineering*" is devoted to show the power of mathematical modeling to give an answer to a broad diversity of real problems including medicine, finance, social behavioral problems and many engineering problems.

Mathematical modeling in social sciences is very recent and comes up special challenges such as the difficulty to manage the human behaviour, the role of the model hypothesis with the objectivity/subjectivity and the proper understanding of the conclusions.

In this book the reader will find several behavioral mathematical models, that in fact may be understood as the so-called epidemiological models in the sense that they deal with populations instead of individuals.

Fortunately for the readers, they are not going to find in this book questionable mechanical approaches to model collective behavior. Social phenomena unlike the mechanical ones are not driven by unchanging and repeatable without exceptions, but they are driven by trends stated with characteristics, aspects and irregularities that in fact are not reproducible. This means that modeling social behavior one needs to put clear hypothesis about the cultural and particular circumstances (geography, time, cultural values) under which the social phenomena are considered. Individual behavior may be erratic due to emotional influences, but aggregate behavior can be predictable. In fact, the humans we are mimetic (R. Girard), beings of usual behavior (L. Castellani) and the human herding (R.M. Raafat, N. Chater and C. Fritz) and social contagion (N. Christakis and J. Fowler) is frequent and powerful. With respect to the issue of subjectivity/objectivity, and in agreement with M. Weber ideas, social mathematical models are not objective because the hypothesis are linked to only some particular values expressed in the hypothesis (with potential disagreement of a reader who prefers other possibly opposite values). However, subjective does not mean arbitrary, and accordingly the conclusions must be regarded as recommendations.

This book is organized as follows:

- **Part I** contains engineering models with a broad variety of problems from water management, to combustion engine issues, or the Android malware infection propagation.

- **Part II** is devoted to social mathematical models including social addictions such as plastic surgery or fitness activity, or modeling the interregional migration, Spanish religious behavior, for instance.
- **Part III** is mainly addressed to the study some part of the modeling process, such as solving theoretical problems for constructing a model, or just to perform a better way to solve a previous mathematical model including the analysis of numerical aspects.
- **Part IV** focuses its interest on mathematical models in medicine, including a bladder cancer model, diagnosis of cardiac allograft vasculopathy, or modeling patterns of functional dependency in schizophrenia.
- **Part V** contains chapters where the main interest are finance, containing option pricing problems, the estimation of the cost of academic underachievement or modeling the composition of a portfolio to replicate a stock market index.

We thank to all contributors for their participation in this book.

Juan Carlos Cortés, Lucas Jódar and Rafael Villanueva
Fall 2013
Valencia, Spain

In: Mathematical Modeling in Social Sciences ...
Editors: J. C. Cortés López et al.
ISBN: 978-1-63117-335-6

Chapter 1

Second-order Perturbations in Encke's Method for Spacecraft Flybys

*L. Acedo**
Instituto Universitario de Matemática Multidisciplinar,
Universitat Politècnica de València, Valencia, Spain

Abstract

In this work we consider a generalization of traditional Encke's method for the computation of perturbed orbits by including second order effects arising by the interaction of the main body with the first order perturbed orbit of a spacecraft in hyperbolic orbit.

Keywords: Encke's method, Flyby orbits, Solar system perturbations

1. Introduction

Flybys are a common maneuver in spacecraft missions which allows the spacecraft to gain or lose of heliocentric energy with the purpose of reaching their objectives [1]. Doppler analysis of this orbit allows a very precise monitoring of the trajectories and, consequently, it serves the purpose of a very stringent test on celestial mechanics methods. Many conventional effects have been considered in relation with the high-precision calculation of these orbits: atmospheric drag, ocean or solid Earth tides, charge and magnetic moment of the spacecraft, Earth albedo, Solar wind and spin-rotation coupling [2].

The role of these orbits in celestial mechanics theory is particularly interesting nowadays because of the recently announced flyby anomaly. Anderson et al. have analyzed the data for six Earth flybys of five deep-space missions[3]: Galileo, NEAR, Cassini, Rosetta and Messenger that took place between December 1990 and September 2005. Flybys are a common maneuver in spacecraft missions which allows the spacecraft to gain or lose of heliocentric energy with the purpose of reaching their objective [1].

*E-mail address: luiacrod@imm.upv.es

An analysis of the data for these flybys have shown X-band Doppler residuals that are interpreted in terms of a change of the hyperbolic excess velocity, V_∞, of a few mm/s. Anderson et al. have proposed the phenomenological formula:

$$\frac{\Delta V_\infty}{V_\infty} = K(\cos\delta_i - \cos\delta_o) \ , \tag{1}$$

where δ_i, δ_o are the declinations for the incoming and outgoing osculating velocity vectors and K is a constant. The value of K seems to be close to $2\omega_E R_E/c$, where ω_E is the angular rotational velocity of the Earth, R_E is the Earth radius and c is the speed of light. Although this formula works reasonably well for the six flybys studied in the paper, the proposal for the relation of K with the Earth's tangential velocity at the Equator is a daring hypothesis, taking into account that the flybys of other planets with different rotational velocities and radii have not been considered.

In this context, it seems valuable to careful analyze the numerical methods involved in the calculation of perturbed orbits and their sources of error.

In this paper we study second-order effects in perturbative celestial mechanics [4, 5, 6]. Flyby trajectories necessarily deviate from ideal hyperbolic motion as a consequence of the perturbations of the rest of bodies in the Solar system, mainly the Sun and the Moon for their size and proximity to the Earth-spacecraft two-body system. However, the Earth itself cannot be excluded from the perturbation analysis because once the orbit has been distorted by the Sun and the Moon, additional but smaller perturbations are induced by the gravitational field of the Earth. Consequently, the spacecraft orbit should be calculated by taking into account the series of smaller and smaller perturbations that are produced by the Sun and the Moon (or the Earth) and the reaction of the Earth (Sun and Moon) in an iterative way. We show that the second order perturbations in these formalism correspond to an average variation of the total energy along the trajectory with the same order of magnitude observed in the anomaly.

This could not be an explanation for the anomalous result because spacecraft tracking relies upon direct integration methods. However, it is an interesting and advisable fact to peruse careful our assumptions about conventional finite-difference and integration methods in astronomy because they could disclose important differences to be measure by modern spacecraft navigation techniques.

The structure of the chapter is as follows: in Section 2 we develop the application of Cowell's method for the calculation of the lunar-solar perturbations of the flyby trajectory. Also we consider the displacement of the Sun and the Moon in the sky during the duration of the flyby. In Section 3 we calculate the second order corrections of the orbit by the effect of the gravitational effect of the Earth, Sun and Moon on the first approximation. Comparison with observations for several flybys are given. Conclusions are presented in Section 4.

2. Orbit Parameters and First Order Perturbations

Positions of the planets, the Sun and the spacecraft on the sky are usually expressed in terms of the declination angle, δ, which is the angle of the line of sight of the object with the equatorial celestial plane. Similarly, the right ascension angle, α, is the angle between

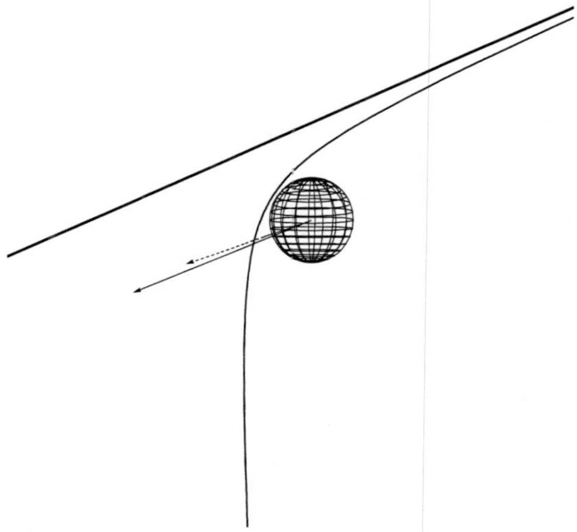

Figure 1. Plot of the NEAR flyby orbit (January 23, 1998). The solid vector points towards the Sun, the dashed vector points towards the location of the Moon at the instant of the closest approach.

the projection of the position vector of the object upon the celestial equator and the first point of Aries (the point where the Sun crosses the Celestial equator at the Vernal equinox). In the following we will use the celestial polar angle $\theta = \pi/2 - \delta$ instead of the declination.

In order to analyze the flyby orbit and its subsequent perturbations, it is highly convenient to define a system of coordinates anchored to that orbit. As such a system we choose a unit vector along the periapsis direction corresponding to the point of closest approach, $\hat{s}$, a second unit vector pointing along the direction of the inclination vector of the orbit, $\hat{w}$, and a third one perpendicular to those two, $\hat{n}$. This third unit vector is defined in such a way that the scalar product with the initial radiovector of the spacecraft, r_{in}, is positive. These vectors are given as follows:

$$\hat{s} = \cos\theta_p\ \hat{k} + \sin\theta_p \cos\alpha_p\ \hat{\imath} + \sin\theta_p \sin\alpha_p\ \hat{\jmath} \tag{2}$$

$$\hat{w} = \cos I\ \hat{k} + \sin I \cos\alpha_I\ \hat{\imath} + \sin I \sin\alpha_I\ \hat{\jmath} \tag{3}$$

$$\hat{n} = \pm\hat{w} \times \hat{s}\ , \tag{4}$$

where θ_p, α_p are the celestial polar angle and right ascension of the periapsis, I and α_I are the inclination and the right ascension of the inclination vector and the sign in the last expression for $\hat{n}$ depends on the orientation of the orbit. The orthogonal system $\hat{\imath}$, $\hat{\jmath}$, $\hat{k}$ is, obviously, the celestial coordinate system.

The NEAR flyby orbit (January 23, 1998) is plotted in Fig. 1. In this case the parameters were (all angles in degrees): $\theta_p = 57$, $\alpha_p = 280.43$, $I = 108.0$, $\alpha_I = \alpha_p + \arccos(-\cot I \cot\theta_p) = 358.24$. The incoming direction is given by $\theta_i = 69.24$ and $\alpha_i = 81.17$. In this particular case, it can be shown that $\hat{n} = \hat{w} \times \hat{s}$. Instead of using time or the true anomaly (the angle formed by the radiovector of the spacecraft and the periapsis vector) to parametrize the orbit, we can use, more conveniently, the eccentric anomaly

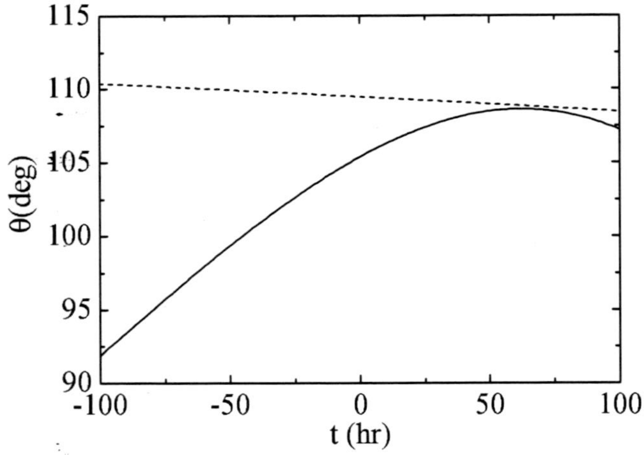

Figure 2. Polar celestial angle of the Sun (dotted line) and the Moon (solid line) before and after 100 hours of the closest approach to Earth of the NEAR spacecraft (January 23, 1998).

defined as follows:

$$\cosh H = \frac{\epsilon + \cos \nu}{1 + \epsilon \cos \nu} , \tag{5}$$

ν being the true anomaly and $\epsilon > 1$ the eccentricity of the hyperbolic orbit. The time of flight can be given in terms of the eccentric anomaly by

$$t = T(\epsilon \sinh H - H) , \tag{6}$$

where the time-scale $T = \sqrt{(-a)^3/\mu}$, a is the semi-major axis of the orbit and μ is the product of the gravitational constant and the mass of the Earth, $\mu = 398600.4\,\mathrm{km}^3/\mathrm{s}^2$. The equations for the radiovector and the velocity of the spacecraft in the ideal hyperbolic orbit are then given by

$$\mathbf{r}(H) = a(\cosh H - \epsilon)\hat{s} - a\sqrt{\epsilon^2 - 1}\sinh H \hat{w} \tag{7}$$

$$\mathbf{v}(H) = \frac{a}{T(\epsilon \cosh H - 1)}(\sinh H \hat{s} - \sqrt{\epsilon^2 - 1}\cosh H \hat{w}) , \tag{8}$$

In order to determine the perturbations of the orbit of the spacecraft in the geocentrical system of reference caused by other bodies (Sun or Moon) the tidal force generated by the difference of forces exerted upon the Earth and the spacecraft must be calculated. In general, this tidal force is given as follows:

$$\mathbf{F}_{\mathrm{tidal}} = -\frac{\mathbf{R}}{R^3} + \frac{\mathbf{R} - \mathbf{r}}{(r^2 + R^2 - 2\mathbf{r}\cdot\mathbf{R})^{3/2}} , \tag{9}$$

where $\mathbf{R}$ is the radiovector from the center of the Earth towards the perturbing body and R its modulus. We must take into account that $\mathbf{R}$ has a significant change during the flyby maneuver which it is considered to last about 200 hours. In Figs. 2 and 3 we have plotted the variation of the polar celestial angle and the right ascension for the Sun and the Moon during the time span of the NEAR flyby. The unit radiovectors of the Sun and the Moon

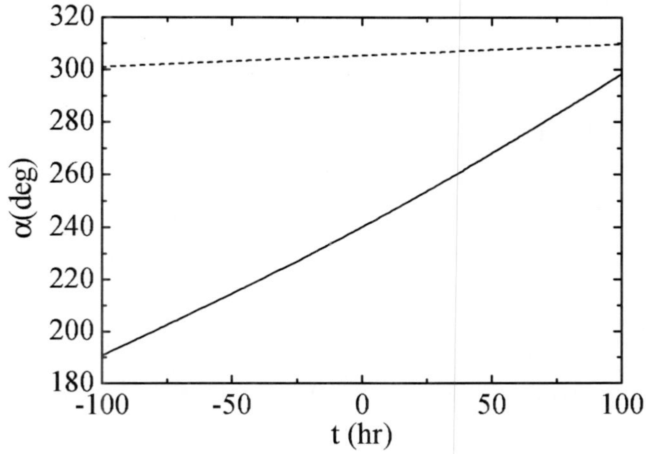

Figure 3. The same as Fig. 2 but for the right ascension of the Sun (dotted line) and the Moon (solid line).

in the geocentrical coordinate system corresponding to the spacecraft orbit are defined as $\hat{R}_S = \alpha(H)\hat{s} + \beta(H)\hat{w} + \gamma(H)\hat{n}$ and $\hat{R}_M = \eta(H)\hat{s} + \chi(H)\hat{w} + \kappa(H)\hat{n}$. From Eqs. (7) and (9) we can now give the components of the tidal force generated by the Sun in the orthogonal system of reference $\hat{s}$, $\hat{w}$, $\hat{n}$ as follows

$$\begin{aligned}
\mathbf{F}_S(H) &= \mu_S\left(-\frac{\alpha(H)}{R_S(H)^2} + \frac{a(\epsilon - \cosh H) + \alpha(H)R_S(H)}{\rho(H)^{3/2}}\right)\hat{s} \\
&+ \mu_S\left(-\frac{\beta(H)}{R_S(H)^2} + \frac{a(\sqrt{\epsilon^2-1}\sinh H + \beta(H)R_S(H)}{\rho(H)^{3/2}}\right)\hat{w} \qquad (10) \\
&+ \mu_S\left(-\frac{\gamma(H)}{R_S(H)^2} + \frac{\gamma(H)R_S(H)}{\rho(H)^{3/2}}\right)\hat{n}\,,
\end{aligned}$$

where $R_S(H)$ is the distance between the Sun and the Earth, $\mu_S = 1.3271244 \times 10^{11}$ km^3/s^2 is the mass of the Sun times the gravitational constant and $\rho(H)$ is the square of the distance from the spacecraft to the Sun:

$$\begin{aligned}
\rho(H) &= a^2(\epsilon\cosh H - 1)^2 + R_S(H)^2 - 2a\alpha(H)R_S(H)(\cosh H - \epsilon) \\
&+2a\sqrt{\epsilon^2-1}\beta(H)R_S(H)\sinh H\,. \qquad (11)
\end{aligned}$$

Once the osculating orbit parameters (the ideal hyperbolic orbit corresponding to the velocity and position at the periapsis) are known, the perturbation tidal force generated by the Sun gravitational field is only a function of the eccentric anomaly, H. A similar expression can be written for the tidal force exerted by the Moon $\mathbf{F}_M(H)$ (in this case, $\mu_M = 4902.8$ km^3/s^2).

Using the relation among the time of flight, t, and the eccentric anomaly, H, in Eq. (6) we can now write the perturbations in the velocity and position of the spacecraft as integrals

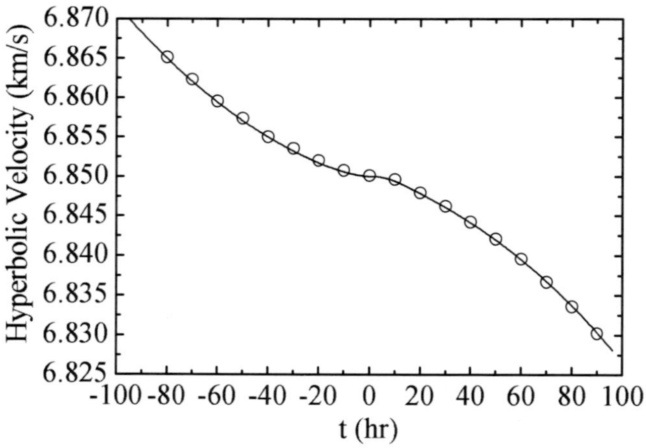

Figure 4. The prediction for the osculating hyperbolic asymptotic velocity (solid line) compared with observations for the NEAR flyby (circles) as a function of time. The instant of time corresponding to the closest approach was taken as $t = 0$.

over H as follows:

$$\Delta\mathbf{v}(H) = T\int_0^H du(\epsilon\cosh u - 1)\left(\mathbf{F}_S(u) + \mathbf{F}_M(u)\right) \tag{12}$$

$$\Delta\mathbf{r}(H) = T^2\int_0^H du(\epsilon\cosh u - 1)\int_0^u dv(\epsilon\cosh v - 1)\left(\mathbf{F}_S(v) + \mathbf{F}_M(v)\right) . \tag{13}$$

In the case of the NEAR flyby we have an orbital eccentricity $\epsilon = 1.81352$, the minimum altitude over the Earth geoid at periapsis is $h_p = 539$ km which, from Eq. (7), implies a semi-major axis $a = (h_p + R_E)/(1 - \epsilon) = -8494.97$ km, where $R_E = 6371$ km is the mean radius of Earth. Consequently, the time scale, T, appearing in Eq. (6) is $T = 1240.13$ sec.

Finally, we can measure the effect of the perturbation in terms of the asymptotic hyperbolic velocity of the osculating orbit at every instant, t. The asymptotic hyperbolic velocity of the osculating orbit at every point of the real trajectory is given by

$$V_\infty^2(H) = \left|\mathbf{v}(H) + \Delta\mathbf{v}(H)\right|^2 - 2\mu_E/\left|\mathbf{r}(H) + \Delta\mathbf{r}(H)\right| , \tag{14}$$

where $|\ldots|$ denote the vector modulus. In Fig. (4) we compare the prediction of Eq. (14) for the hyperbolic velocity at a function of time by performing the integrals in Eq. (12) numerically with the observational results [3]. The agreement is very good and this proves the effectiveness of the perturbation approach in its first order approximation. Nevertheless, the reported flyby anomaly points out towards a smaller correction to these results that should be studied. In the next section, we consider second-order contributions arising from the Earth gravitational field and also to the Sun and the Moon in a second iteration.

3. Back-Reaction of the Earth and Second-Order Approximation

Our objective in this section is to compute the average change of energy for the spacecraft as a consequence of the second-order interactions with the Earth, the Sun and the Moon. As the Sun and the Moon change the spacecraft ideal orbit, a second-order perturbation is superimposed because of the effect of the Earth's gravity. The contribution to the total energy variation can be calculated by taking into account two facts: (i) The force of the Earth at the positions of the first-order orbit is slightly different than that at the ideal orbit:

$$\delta \mathbf{F}_E = -\mu_E \frac{\Delta \mathbf{r}}{r^3} + 3\mu_E \mathbf{r} \cdot \Delta \mathbf{r} \frac{\mathbf{r}}{r^5} , \tag{15}$$

where $\Delta \mathbf{r}$ is the first-order perturbation for the position of the spacecraft as given in Eq. (13). This generates a contribution $\delta \mathbf{F}_E \cdot \mathbf{v}$ to the change in the total energy. (ii) Another second order contribution to the change per unit time of the total energy comes from the first order perturbation in the velocity as follows $\mathbf{F}_E \cdot \Delta \mathbf{v}$. So, Earth's gravity acting upon the first-order perturbed orbit generates an additional contribution to the change in the total energy of the spacecraft as follows

$$\frac{dE}{dt} = \delta \mathbf{F}_E \cdot \mathbf{v} + \mathbf{F}_E \cdot \Delta \mathbf{v} . \tag{16}$$

Similar expressions are obtained for the second-order contributions arising from the Sun and the Moon. For any celestial body other than the Earth the perturbed tidal force is given by

$$\delta \mathbf{F}_{\text{tidal}} = -\mu \frac{\Delta \mathbf{r}}{R^3} + 3\mu \frac{\Delta \mathbf{r} \cdot (\mathbf{R} - \mathbf{r})}{R^5} (\mathbf{R} - \mathbf{r}) , \tag{17}$$

where $\mathbf{r}$ is the spacecraft position vector and $\mathbf{R}$ is the position vector of the celestial body in the geocentrical system of reference. The values of μ for the Sun and the Moon were given in the previous section. The total energy of the ideal osculating hyperbolic orbit at the periapsis is given by $E = -\mu_E/(2a)$. From Eqs. (16), (15) and (12) we can numerically calculate the second-order perturbation for the total energy of the spacecraft as a function of the eccentric anomaly taking as reference the energy of the ideal osculating hyperbolic orbit at periapsis, $E = -\mu_E/(2a)$. The quotient $\Delta E/E$ for the NEAR flyby is plotted in Fig. (5). A similar calculation for the second-order perturbation induced by the Sun and the Moon leads to a small difference in comparison with the Earth's contribution to $\Delta E/E$. In this case, the initial value of the eccentric anomaly was $H_i = -5.667$ corresponding to a time $t = -88.4$ hours before the periapsis. The final point for the integration was taken as $H_f = -H_i = 5.667$. In Table 1 we have listed the relevant parameters for several flybys and the result for the second-order contribution to the energy of the spacecraft. We notice that a steady increase in the total energy of the spacecraft measured in the geocentrical frame of reference is expected. The temporal average can be defined in terms of an integral over the eccentric anomaly as follows:

$$\left\langle \frac{\Delta E}{E} \right\rangle = \frac{1}{\epsilon(\sinh H_f - \sinh H_i) + (H_f - H_i)} \int_{H_i}^{H_f} dH(\epsilon \cosh H - 1) \frac{\Delta E}{E} , \tag{18}$$

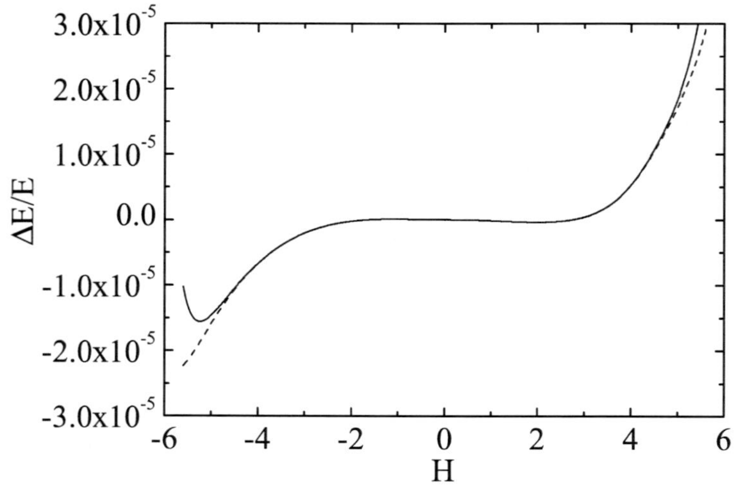

Figure 5. The ratio between the second-order perturbation in the energy for the NEAR flyby and the total energy at periapsis versus the eccentric anomaly of the osculating orbit. The solid line correspond to the total perturbation (Earth, Sun and Moon) whereas the dashed line correspond to the effect of the Earth alone.

where we have used Eq. (6) to compute the total time from the point in the orbit corresponding to the initial eccentric anomaly, H_i, until the final point, H_f. In the case of the flyby anomaly we obtain an average, $\langle \Delta E/E \rangle \simeq 3.82 \times 10^{-6}$ in good agreement with the observed value, 3.93×10^{-6} and the prediction of the heuristic formula in Eq. (1). From Table 1 we infer that second-order perturbations, in particular those arising from the Earth's back-reaction on first order perturbations by the Sun and the Moon, are of the same order of magnitude that the reported anomaly. This suggest that this anomaly is a second order perturbation effect of unknown origin at the present.

Concluding Remarks

Astronomy and astrodynamics are nowadays becoming a high-precision science thanks to the careful monitoring of spacecrafts via Doppler ranging and radiometric techniques. This advances in space navigation also require a parallel improvement in numerical analysis in celestial mechanics in order to fit the reported trajectories properly.

In particular, an anomalous increase of asymptotic hyperbolic velocity has been reported in several flybys of the Earth in recent years [3]. After careful analysis, some anomalies perhaps cannot be attributed to observational error, data processing or to conventional effects within the domain of physical theory. These should be called true anomalies. True anomalies play a fundamental role in the development of physics because they demand the creation of new concepts or even new theories going beyond our present understanding of phenomena.

Table 1. Parameters for spacecraft flybys of the Earth and result for the average increase of total energy: orbital eccentricity, ϵ, semi-major axis, a, in km, celestial coordinates (polar angle and right ascension) for the periapsis and the initial point, θ_p, α_p, θ_i, α_i. Orbital plane inclination, I, and right ascension for the corresponding vector perpendicular to the orbital plane, initial eccentric anomaly coordinate, H_i. The average energy increase, $\langle \Delta E/E \rangle$ is compared with the derivation from Doppler residuals and the phenomenological formula in Eq. (1)

Parameter	NEAR	Galileo-I	Galileo-II	Cassini
ϵ	1.81352	2.4729	2.31941	5.85246
a (km)	-8494.87	-4977.24	-5058.31	-1555.09
θ_p (deg)	57	64.8	123.8	113.5
α_p (deg)	280.43	319.96	302.72	245.59
θ_i (deg)	69.24	77.48	55.74	102.92
α_i (deg)	81.17	266.76	219.35	334.3
I (deg)	108	142.9	138.7	25.4
α_I (deg)	358.24	268.43	163.1	269.28
H_i	-5.667	-5.5	-5	-5.5
$\Delta E/E$ (observed)	3.93×10^{-6}	8.76×10^{-7}	-1.04×10^{-6}	-2.5×10^{-7}
$\Delta E/E$ (Eq. (1))	3.88×10^{-6}	9.21×10^{-7}	-1.05×10^{-6}	-1.33×10^{-7}
$\langle \Delta E/E \rangle$	3.82×10^{-6}	9.23×10^{-7}	-8.1×10^{-7}	-5×10^{-8}

A series of possible conventional explanations of the anomaly including the effect of the atmosphere, ocean and crust tides and other were studied by Lämmerzahl et al. [2] and dismissed as very small to explain the observations. Of particular interest it is the fact that the anomaly is observed both in Doppler as well as ranging data which implies that explanations based only on Doppler effect or spin-rotation coupling (coupling of the helicity of radio waves with the rotation of the Earth and the spacecraft) are not viable.

The search for non-conventional explanations of the flyby anomaly already started a year after the original work by the team of Anderson et al. [3]. For example, the possibility that this anomaly arise as a consequence of the interaction of the spacecraft with a dark matter halo around the Earth have been considered [7].

Anderson et al. suggested a formula given in Eq. (1) which give us a good fit of the anomalous energy change in several flybys during the last twenty years [3]. In this formula the coefficient K was related to the quotient of the Earth linear velocity at the Equator and the speed of light. This identification is rather arbitrary and it is clear from the start that it could be merely coincidental. According to the authors of this work a relation with Earth rotation was possible through a novel mechanism beyond our present understanding of physics.

In this paper we have pursued a conventional model based upon a detailed implementation of classical perturbation theory. The nature of flyby maneuvers require the monitoring of trajectories from the Earth and, consequently, they are naturally followed in a geocentrical coordinate system. From the point of view of this system, any other celestial body will be the source of a tidal force due to the different positions of the Earth and the spacecraft relative to the third body. Moreover, celestial bodies indeed change their positions in the sky during the flyby maneuver which lasts several days and this change must be taken into account. The most important contributions to perturbations in the case of Earth flybys come from the Sun and the Moon. The resulting first-order perturbation trajectory deviates from the ideal hyperbolic orbit and, consequently, the Earth itself must also be considered as a source of perturbations of the first-order trajectory. We have found that these second-order perturbations are comparable to the order of magnitude of the flyby anomaly.

Acknowledgments

The author gratefully acknowledges R. M. Shoucri for many useful discussions and a critical reading of the manuscript. The NASA's Jet Propulsion Laboratory is also acknowledged for their HORIZONs web-based system which was used to compute the ephemeris used in this work.

References

[1] P. H. Borcherds and G. P. McCauley, The gravitational three-body problem: optimising the slingshot, *Eur. J. Phys.* 15 (1994), pp. 162-129.

[2] C. Lämmerzahl, O. Preuss and H. Dittus, Is the physics of the Solar system really understood ?. Laser, Clocks and Drag-Free Control. *Astrophysics and Space Science Library* 349, pp. 75. arXiv: gr-qc/0604052.

[3] J. D. Anderson, J. K. Campbell, J. E. Ekelund, J. Ellis and J. F. Jordan, Anomalous Orbital-Energy Changes Observed during Spacecraft Flybys of Earth, *Phys. Rev. Lett.* 100, 091102 (2008).

[4] H. Pollard, *Mathematical Introduction to Celestial Mechanics*, (Prentice-Hall Inc., Englewood Cliffs, New Jersey, 1966).

[5] J. M. A. Danby, *Fundamentals of Celestial Mechanics*, (Willmann-Bell, Inc., 1988), 2nd ed.

[6] J. A. Burns, Elementary derivation of the perturbation equations of celestial mechanics, *American Journal of Physics* 44, 10 (1976), pp. 944-949.

[7] S. L. Adler, Can the flyby anomaly be attributed to earth-bound dark matter?, *Phys. Rev. D* 79, 023505 (2009).

In: Mathematical Modeling in Social Sciences ...
Editors: J. C. Cortés López et al.
ISBN: 978-1-63117-335-6

Chapter 2

COMMON-RAIL DIESEL INJECTORS BOND GRAPH MODELLING THROUGH THE AMESIM PLATFORM

F. J. Salvador[1,*], M. Carreres[1], J. V. Romero[2] and M. D. Roselló[2]
[1]CMT-Motores Térmicos, Universitat Politècnica de València, Spain
[2]Instituto de Matemática Multidisciplinar,
Universitat Politècnica de València, Spain

Abstract

A methodology to mathematically model a common-rail injection system by using a Bond Graph approach is described in the present chapter. All the physical components of a last generation injector have been graphically represented by block diagrams whose union makes it possible to exchange information between them when the associated equations are solved. The implementation of the model has been carried out through the AMESim commercial software, which contains several libraries that make it possible to easily establish the causality relations between elements. Modelled elements include mechanical components (such as masses, springs, dampers, pistons, etc.) and hydraulic lines and orifices. In this chapter, special attention is given to the equations that represent the physics behind these elements and how they interact with each other. On the other hand, each of these elements requires a complete characterization in order to obtain the relevant parameters that need to be included as the coefficients of the equations. An extensive validation against mass flow rate experimental data is finally performed in order to ensure the model capabilities.

Keywords: 1D-modelling, injection system, diesel, Amesim

* E-mail address: fsalvado@mot.upv.es, Address: CMT-Motores Térmicos, Universitat Politècnica de València, Camino de Vera s/n, E-46022 Spain, Tel: +34-963879659, Fax: +34-963877659 (Corresponding author)

Notation

A	area
b	damping coefficient
C_d	orifice discharge coefficient
C_{wall}	wall compliance
E	material Young modulus
e	eccentricity
F	force
f	friction factor
G	material shear modulus
g	gravity acceleration
k	spring stiffness
L	length
m	mass
n	number of spires in a spring
P	pressure
Q	volumetric flow rate
t	time
U	fluid velocity
V	volume
x	displacement
α	line angle with the horizontal
β	fluid bulk modulus
γ	ball seat valve semi-angle
δ	ball seat geometry angle
ε	material Poisson coefficient
λ	flow number
ν	fluid kinematic viscosity
μ	fluid dynamic viscosity
ϕ	diameter
ρ	fluid density
Θ	mass angle with the horizontal

Introduction

Internal flow in diesel engines injection systems plays a key role in the air-fuel mixture that takes place in the cylinder, affecting the combustion phenomenon and thus having a strong influence on fuel consumption, emissions and noise [1,2].

It is therefore important to develop computational tools that enable to predict the behaviour of the system under different operating conditions, in order to optimise its performance and detect any potential problem. In this chapter, a modelling methodology for common-rail diesel injectors is proposed by using a one-dimensional approach based on the Bond Graph technique and implemented in the commercial software AMESim [3]. The capabilities of this kind of models have already been proved in several works published by the authors [4,5,6].

This chapter is structured as follows. First, a general description of the Bond Graph technique for modelling engineering systems is given and the commercial code AMESim is introduced. Next, a description of the different elements that comprise a common-rail injector model is shown, introducing the equations that represent the physical processes being involved. These elements need to be fed with certain parameters and thus a few remarks are given on how to determine them through a thorough characterization of the hydraulic and mechanical elements of the actual system. Then, the calculation scheme and the numerical resolution of the system state equations are treated. Finally, an example of validation against experimental data regarding injected mass flow rate is shown, in order to demonstrate the capabilities of the model to represent the behaviour of the real system.

The Bond Graph Technique

Bond graphs are a method to graphically represent physical systems. In a bond graph, a set of elements are connected together in a structure that is somewhat representative of the modelled system. Each element has a certain number of ports from which it can be connected to other elements. These connections are represented by arrows and referred to as bonds. The bonds identify power flow paths, through which two kinds of variables are transmitted among elements: effort and flow. In each physical domain, effort and flow correspond to different specific variables. As an example, in mechanical systems, force and velocity correspond to effort and flow, respectively, whereas in electrical systems these variables are voltage and current. However, the fact that there is an analogy between the different physical domains in terms of effort and flow makes it possible for a bond graph to represent multi-domain systems. The multi-domain analogy also makes it possible to classify the different elements of a bond graph as resistive (which dissipate energy), capacitive (store energy), inertial (where the integral of the effort is directly related to the flow by a constitutive law) and effort or flow sources. In addition, the bonds departing from different elements join at a junction represented either by a number 0 or a number 1. In a 0-junction, the flow sums to zero and the efforts are equal. In a 1-junction, the efforts sum to zero and the flows are equal. Half arrows in a bond may indicate the direction the power flows. Bond graphs also give a notion of causality, indicating the side of a bond that defines the instantaneous effort and the side that defines the instantaneous flow. For this purpose, a stroke at one end of the power bond indicates that that end is defining the flow, whereas the opposite end is defining the effort. Finally, it is possible from a bond graph to directly formulate the dynamic equations that describe the system. Causality relations then make it possible to identify the independent and the dependent variable for each element. Thus, some rules apply to causality in order for the model to be realizable.

As an example to illustrate the aforementioned concepts, consider the simple mass-spring-damper system represented in Figure 1(a). Its equivalent bond graph representation is depicted in Figure 1(b). The four elements that constitute the system (namely a mass, a spring, a damper and a force) are represented with a bond graph notation (inertial, capacitive, resistive and effort source, respectively). The bonds depart from each system joining at a 1-junction, since the effort (forces) sum to zero at the junction and the flow (velocities) are equal for all the elements. The causality relations are also indicated, and thus it is immediate to derive the dynamic equations of the system and simulate its behaviour.

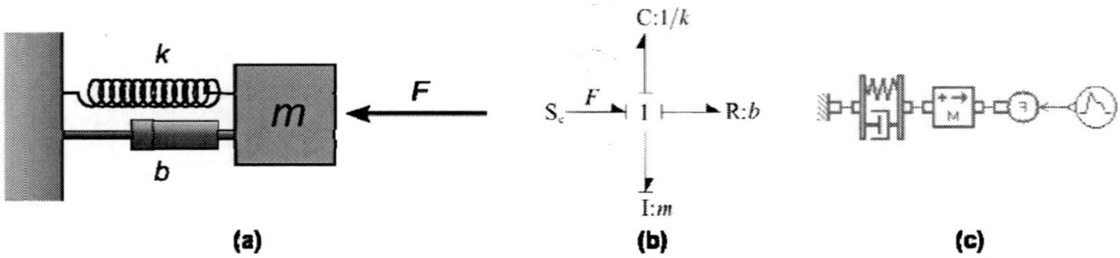

Figure 1. Mass-Spring-Damper system. (a) Physical representation. (b) Bond graph representation. (c) AMESim representation.

The modelling methodology discussed in this chapter is implemented in the AMESim commercial platform, which is based on the bond graph technique. AMESim offers predefined libraries for the different physical domains, containing elements (referred to as components) whose constitutive equations have already been defined. These components include several ports in order to connect them to other components. The bonds are omitted in general so that the representation is more visual, whereas the causality relations can be specified by the user depending on how the different components are connected. The AMESim representation of the mass-spring-damper system of the example is shown in Figure 1(c).

The methodology described in this chapter is based on the AMESim representation of systems. For further reference on bond graph modelling itself, the reader may refer to the work by Karnopp et al. [7].

Modelling Common-Rail Diesel Systems with Amesim

In this section, a description of the most typical elements that comprise a common-rail injector and their modelling in AMESim is given, stating the equations that constitute the model. Since the injectors are mechanical systems conformed by movable elements whose dynamics is hydraulically determined, most of the AMESim components used are either hydraulic or mechanical. A few remarks on the characterization of the actual elements of the system will be given, due to its importance when obtaining the different parameters of the equations that need to be introduced to the model. A complete sketch of an example model together with its real equivalent is shown in the Appendix.

Hydraulic Lines

Modelling high pressure lines (up to 240 MPa in the case of common-rail injectors) is complex since it is needed to reduce the partial differential equations to ordinary differential equations or differential algebraic equations. It is quite common to assume that the variations of the parameters along the line are negligible and thus calculate only this variations with respect to the time. However, this assumption is not valid when the lines are long, since pressure variations may be important due to several factors: pressure waves effects, compressibility variations of the fluid, fluid inertia, friction, etc. It is quite common then to consider a series of nodes along the line where the pressure is calculated by using finite differences or finite elements methods (see Figure 2). It is important to note that the modelling here described assumes isothermal flow along the injector, which means that no temperature changes are considered and the fuel physical properties remain constant.

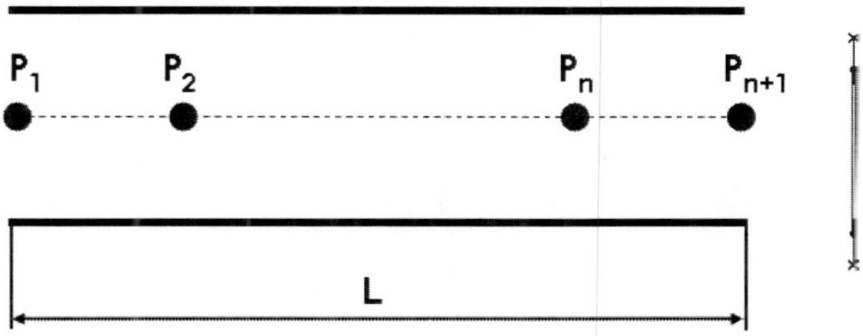

Figure 2. Node distribution of a hydraulic line with its AMESim component representation (in vertical).

Compressibility effects are taken into account by applying Equation 1:

$$\frac{dP}{dt} = \frac{\beta_{eff} \cdot Q}{V} \tag{1}$$

The effective bulk modulus (which takes into account the influence of the air contained within the fluid and the elastic deformation of the hydraulic line) is defined in the following way:

$$\frac{1}{\beta_{eff}} = \frac{1}{\beta} + C_{wall} \tag{2}$$

where C_{wall} can be associated to the total elasticity of the line wall and is defined as:

$$C_{wall} = \frac{2}{E} \cdot \left(\frac{\phi_{ext}^2 + \phi_{int}^2}{\phi_{ext}^2 - \phi_{int}^2} + \varepsilon \right) - \frac{2 \cdot \varepsilon}{E} \cdot \left(\frac{2 \cdot \phi_{ext}^2}{\phi_{ext}^2 - \phi_{int}^2} \right) \tag{3}$$

In order to consider the friction in the lines, the mean velocity in the line is calculated from:

$$\Delta P + \rho \cdot L \cdot g \cdot \sin\alpha = \frac{f \cdot \rho \cdot L \cdot U^2}{2 \cdot \phi_{int}} \tag{4}$$

where the friction factor f is determined through empirical relations that relate it to the pressure loss ΔP, as derived from the Moody diagram [8].

In addition, if the fluid inertia is considered, the volumetric flow rate needs to be calculated from:

$$\frac{dQ}{dt} = \frac{A \cdot \Delta P}{\rho \cdot L} - A \cdot g \cdot \sin\alpha - \frac{f \cdot U^2 \cdot A}{2 \cdot \phi_{int}} \cdot sign(Q) \tag{5}$$

AMESim has different line subcomponents in order to enable or disable friction and inertia. For each of this subcomponents, there are three possibilities in order to define the causality relations with other elements. Thus, it is possible to define if the line receives the pressure as an input and computes the volumetric flow rate as an output or if it proceeds on the opposite way.

Orifices

Hydraulic orifices are resistive elements where the pressure is an input at each port and a flow rate is computed as an output. From the mass continuity equation:

$$\frac{dm}{dt} = \rho \cdot A \cdot U \cdot C_d \tag{6}$$

If the velocity is computed as a theoretical velocity from the Bernoulli equation:

$$U = \sqrt{\frac{2 \cdot \Delta P}{\rho}} \tag{7}$$

Combining Equations (6) and (7), the volumetric flow rate computed from AMESim is then:

$$Q = C_d \cdot A \cdot \sqrt{\frac{2 \cdot \Delta P}{\rho}} \tag{8}$$

The discharge coefficient C_d depends on the Reynolds number. AMESim introduces a non-dimensional parameter that can be interpreted as the theoretical Reynolds number, since it includes Bernoulli's theoretical velocity in its definition:

$$\lambda = \frac{\phi_{hydraulic}}{\nu} \cdot \sqrt{\frac{2 \cdot \Delta P}{\rho}} \tag{9}$$

The discharge coefficient can be expressed as a function of the flow number λ:

$$C_d = C_{d_{max}} \cdot tanh\left(\frac{2 \cdot \lambda}{\lambda_{crit}}\right) \tag{10}$$

This expression means that the discharge coefficient C_d approaches the maximum discharge coefficient C_{dmax} as the pressure drop ΔP increases. The critical flow number λ_{crit} is defined as the flow number that leads to a discharge coefficient that equals 95% of C_{dmax}. An example of this behaviour is illustrated in Figure 3.

Cavitation can also be considered when modelling orifices. However, it is beyond the scope of this chapter. For further reference, AMESim description of the cavitation is based on the one-dimensional flow theory established by Nurick [9].

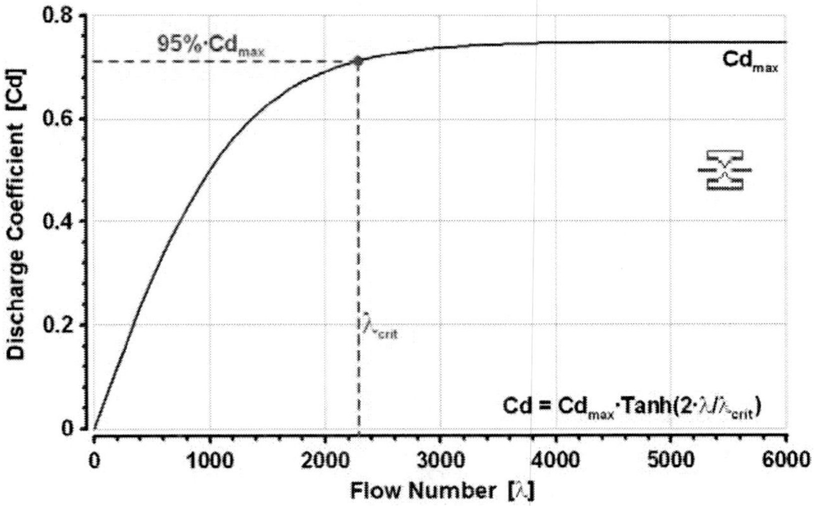

Figure 3. Discharge coefficient of an orifice as a function of the flow number. The AMESim orifice component is also shown.

Hydraulic orifices play a key role in common-rail systems since they are present in the nozzle (which determines the stationary value of the fuel injection rate) and the control volume (whose evacuation and replenishment through orifices controls the dynamic behaviour of the injector). Their small dimensions make it difficult to accurately determine their characteristics. A thorough experimental characterization needs to be performed in order to be able to explain their behaviour. On one hand, a dimensional characterization can be carried out by obtaining silicon moulds of the orifices and visualizing them in an optical or a scanning electron microscope, as described by Macián et al. [10]. This technique is also used to determine the geometry of the internal lines of the injector. On the other hand, a hydraulic characterization is performed in order to obtain the maximum discharge coefficient of the orifices and the critical flow number so that the hydraulic behaviour of the orifice can be explained by Equation 10. This hydraulic characterization is performed by designing test rigs that make it possible to make a fluid flow through the orifice while controlling the pressure drop through it and measuring the corresponding flow rate [5,6].

Masses

Masses of the movable elements and their friction forces are taken into account. Figure 4 shows a mass on an inclined plane subject to an external force and friction. The total force exerted on that mass on the direction of the plane will be:

$$F = F_{ext} - F_{friction} - m \cdot g \cdot \sin\theta \tag{11}$$

The friction force is defined as:

$$F_{friction} = F_{dynamic} + \left(F_{static} - F_{dynamic}\right) e^{\frac{|U|}{3 \cdot U_0}} \tag{12}$$

where $F_{dynamic}$ is the Coulomb's friction, U is the velocity the body moves at and U_0 is the critical velocity, which defines the velocity threshold where there is only dynamic friction. This is known as the Karnopp friction model [11].

From Newton's second law, the acceleration is then:

$$a = \frac{F_{ext} - F_{dynamic} - (F_{static} - F_{dynamic}) e^{\frac{|U|}{3 \cdot U_0}}}{m} \tag{13}$$

The velocity and the position are obtained by successively integrating the acceleration.

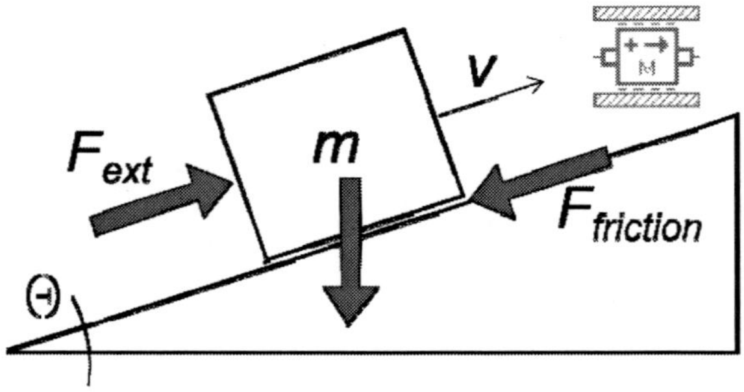

Figure 4. Mass subject to an external force and friction. The AMESim mass component is also shown.

Springs and Dampers

For a physical representation of these elements and the corresponding AMESim subcomponent, refer to Figure 1(a) and Figure 1(c). In this case, the total force is:

$$F = k \cdot x + b \cdot (U_1 + U_2) \tag{14}$$

where 1 and 2 denote the two ports of the subcomponent. The spring compression is computed as:

$$x = x_0 + x_1 + x_2 \tag{15}$$

where x_0 is the initial deformation of the spring (precompression). The spring stiffness can be computed from the geometrical dimensions of the spring assuming it has an helicoidal shape:

$$k = \frac{G \cdot \phi_{spire}^4}{8 \cdot \phi_{spring}^3 \cdot n} \tag{16}$$

Due to the high operating pressures in common-rail injectors, the elastic deformations of the internal elements may reach the same order of magnitude of their displacements. Thus, these deformations must be taken into account. The method defined by Desantes et al. [12] is used, which consists on representing these deformations by a spring with an equivalent stiffness defined from the geometry (longitude and cross-section) of the parts of the element with changes in cross-section (see Figure 5(a) for an example of common-rail injector movable element):

$$\frac{1}{k_{eq}} = \frac{1}{E} \cdot \sum_{i=1}^{n} \frac{L_i}{A_i} \tag{17}$$

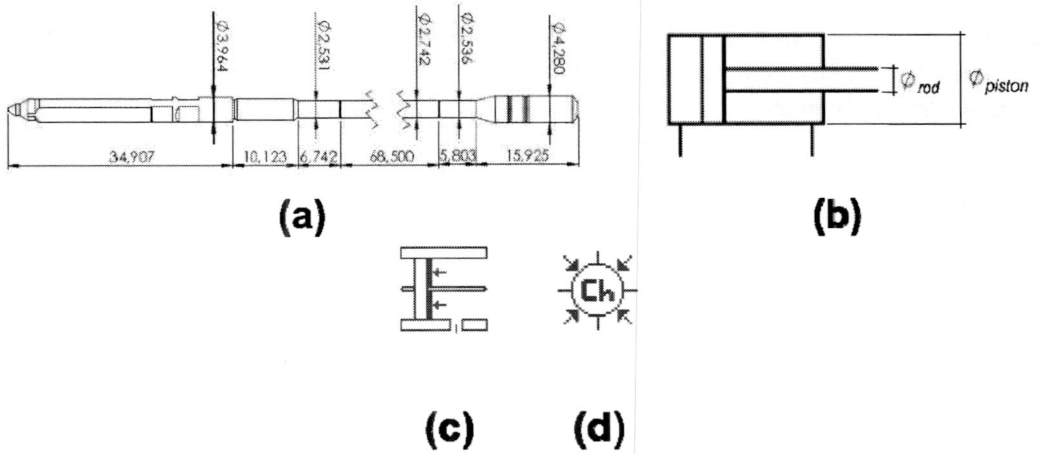

Figure 5. (a) Example of needle of a common-rail injector. (b) Piston diagram. (c) Piston component in AMESim. (d) Hydraulic chamber (volume) component in AMESim.

Pistons and Volumes

Movable elements in common-rail injectors are subject to high pressures. When a change in section occurs in these elements, there is a surface on which the pressure forces act (an annulus section in the case of circular cross-section elements). Figure 5(a) shows a common-rail injector needle where these changes in cross-section may be appreciated. These changes in section are modelled as pistons (see Figure 5(b) and its AMESim equivalent component in Figure 5(c)) where a certain pressure acts. In AMESim, this pressure is an input from an adjacent component, usually modelled as a hydraulic chamber (Figure 5(d)). In an AMESim hydraulic chamber, each port receives a flow rate and a fluid volume as inputs and pressure is computed as an output in the following way:

$$\frac{dP}{dt} = \frac{\beta(P) \cdot \sum_{i=1}^{nports} Q_i(P)}{V_0 + \sum_{i=1}^{nports} V_i} \,. \tag{18}$$

The flow rate coming out from the piston element is:

$$Q = -U \cdot \frac{\pi}{4} \cdot \left(\phi_{piston}^2 - \phi_{rod}^2\right) \tag{19}$$

and the force acting on the piston:

$$F = F_{ext} - P \cdot \frac{\pi}{4} \cdot \left(\phi_{piston}^2 - \phi_{rod}^2\right) \tag{20}$$

Internal Leakages and Friction

An AMESim component computes a laminar hydraulic leakage between a cylindrical piston and its sleeve, together with the corresponding viscous friction. Flow and friction due to Poiseuille phenomenon is taken into account. Figure 6 shows an example of geometry. If the sleeve was also moving, Couette phenomenon could also be taken into account, but this is out of the scope of the present chapter. According to Blackburn et al. [13], assuming laminar flow and a small clearance compared to the piston diameter, the volumetric flow rate through the leakage can be computed as:

$$Q = -\frac{\Delta P}{12\cdot\mu\cdot L}\cdot\left(\frac{\phi_{clearance}}{2}\right)^3\cdot\pi\cdot\phi_{piston}\cdot\left[1+\frac{3}{2}\left(\frac{2\cdot e}{\phi_{clearance}}\right)^2\right]-\frac{1}{2}\cdot U\cdot\frac{\phi_{clearance}}{2}\cdot\pi\cdot\phi_{piston} \quad (21)$$

The viscous friction is:

$$F_{friction} = \Delta P\cdot\pi\cdot\frac{\phi_{clearance}}{2}\cdot\frac{\phi_{piston}-\phi_{clearance}}{2} - 4\cdot\pi\cdot\mu UL\cdot\left(\frac{\phi_{piston}}{\phi_{clearance}}-1\right) \quad (22)$$

Thus, the total forces on the piston are computed as:

$$F = F_{ext} - F_{friction} \quad (23)$$

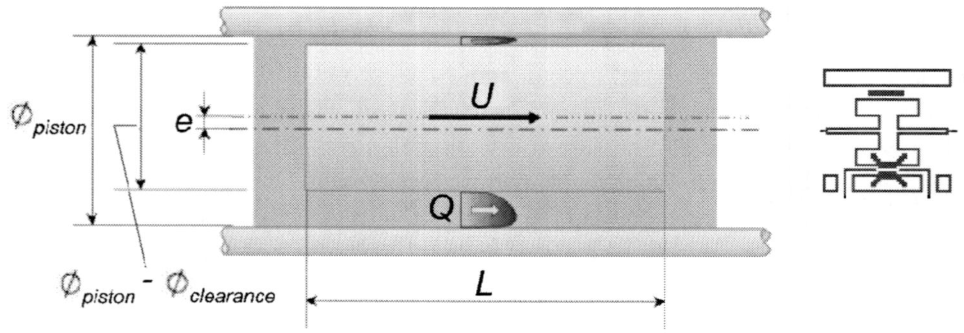

Figure 6. Geometry between a cylindrical piston and its sleeve for leakage calculation purposes. The AMESim component for these calculations is also represented.

Variable Section Orifices

The hydraulic orifices treated previously had a constant geometrical area. However, in hydraulic systems, there are some elements where the area is a function of the pressure and/or the external forces, such as the valves. In the case of common-rail systems, these elements are the nozzle seat, a command piston and a ball valve, as exemplified in Figure 7. Several AMESim components represent different geometries in order to accurately model the area variations in those orifices. Figure 7 shows the AMESim components for the illustrated variable section orifices.

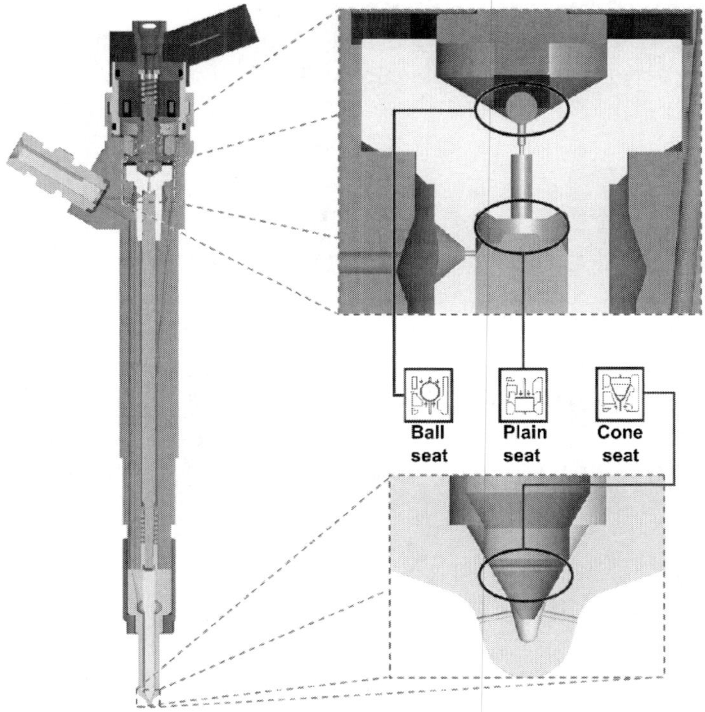

Figure 7. Variable section orifices in a common-rail diesel injector.

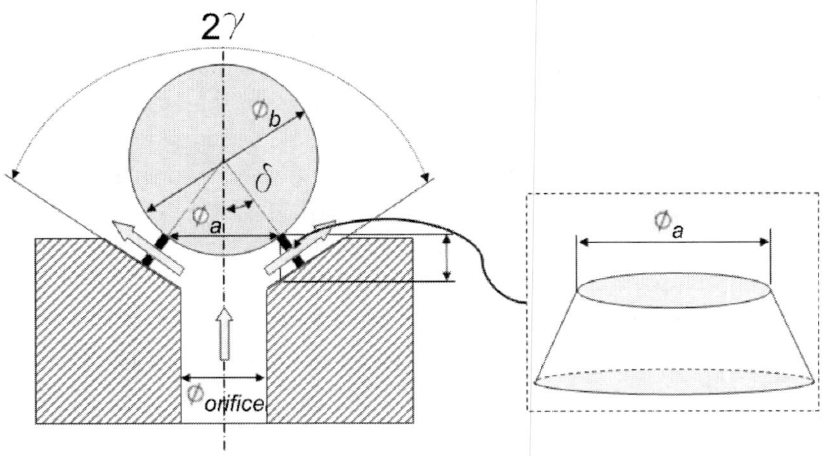

Figure 8. Ball seat geometry.

In this chapter, for illustrating purposes, only the ball valve seat model is explained, since the principles of the other variable section orifices are analogous.

Figure 8 depicts the geometry of a ball seat valve. It can be seen that, in this case, the area of the restriction introduced by the seat is determined by the curved surface of a truncated cone, as follows:

$$A_{eff} = \pi \cdot x \cdot (\phi_b + x \cdot \cos\delta) \cdot \sin\delta \cdot \cos\delta \quad (24)$$

where δ is defined as $(\pi/2-\gamma)$, and γ is the semi angle of the valve seat. Thus, the flow rate is computed applying Equation (8), where the area is replaced by the effective area of Equation (24). The discharge coefficient C_d is a function of the flow number λ following Equation (9), where the hydraulic diameter is given by:

$$\phi_{hydraulic} = 2 \cdot x \cdot \cos\delta \tag{25}$$

The pressure forces are also computed taking into account the diameter ϕ_a.

An example of a complete model of a common-rail injector, together with a representation of the real injector, is shown in the Appendix. It can be seen how the different explained components are linked together to conform the whole multi-domain system. It must be noted that the electromagnetic part of the system has been omitted in the present chapter, due to the different typologies of command systems present in common-rail injectors (solenoid, piezoelectric, etc.). For further reference on these topics, the reader may refer to the literature [14,15].

Numerical Resolution of the Equations

When developing one-dimensional models of physical systems, several kinds of equations need to be solved: ordinary differential equations (ODEs), differential algebraic equations (DAEs), discontinuities and partial differential equations (PDEs).

In order to solve ODEs, two main groups of methods are used. On one hand, Runge-Kutta methods, which use a fixed time step to solve the equations and are easier to implement, but are not appropriate for systems with numerical stiffness. On the other hand, Linear Multistep Methods (LMMs), which can use variable time steps and can solve stiff problems, although they present some issues when treating discontinuities. One of the most known LMMs methods is the Gear method. Some algorithms to implement these methods are Adams, Gear and LSODA. DAEs can be solved through the DASSL algorithm, which is an extension of the Gear algorithm. Finally, PDEs are usually reduced to ODEs and solved through a simple temporal discretization via finite-difference methods, finite-element methods or the method of characteristics.

AMESim does not give the user the choice of the integration algorithm. It automatically analyses the characteristics of the equations that describe the model in order to make the choice. If the model contains any implicit variables, the DAE integration algorithm DASSL is used. Otherwise, the ODE integration algorithm LSODA is used.

LSODA uses both non-stiff integration methods (Adams-Moulton multistep) and stiff integration methods (backward differentiation formulae multistep). The algorithm monitors the characteristics of the governing equations and switches from one method to the other accordingly. This capability allows LSODA to be an efficient solver independently of the characteristics of the equations.

DASSL solves the set of constraint equations and implicit equations by using specific Newton-base iterative methods. The user can choose between the LU algorithm (based on the Gaussian elimination method, which solves the system in a direct way) and a Krylov method (which minimises residuals on Krylov subspaces and then iterates over the subspaces towards the real solution of the initial system). In several occasions, the solution of DAEs may require the use of a preconditioner.

In order to deal with discontinuities, AMESim employs a specific treatment in each component. Finally, to solve PDEs, it makes it possible to choose among finite-difference methods or finite-element methods.

Validation

The model capabilities need to be proved by comparing its results against experimental data under a wide range of operating conditions. The standard validation is performed in terms of mass flow rate injected to the combustion chamber. Figure 9 shows a validation of the injector model with experimental mass flow rate curves obtained at different conditions. The total mass injected in an injection event, computed as the integration of the previous curve, is also displayed. This figure highlights the ability of the model to reproduce the dynamic behaviour of the injection system.

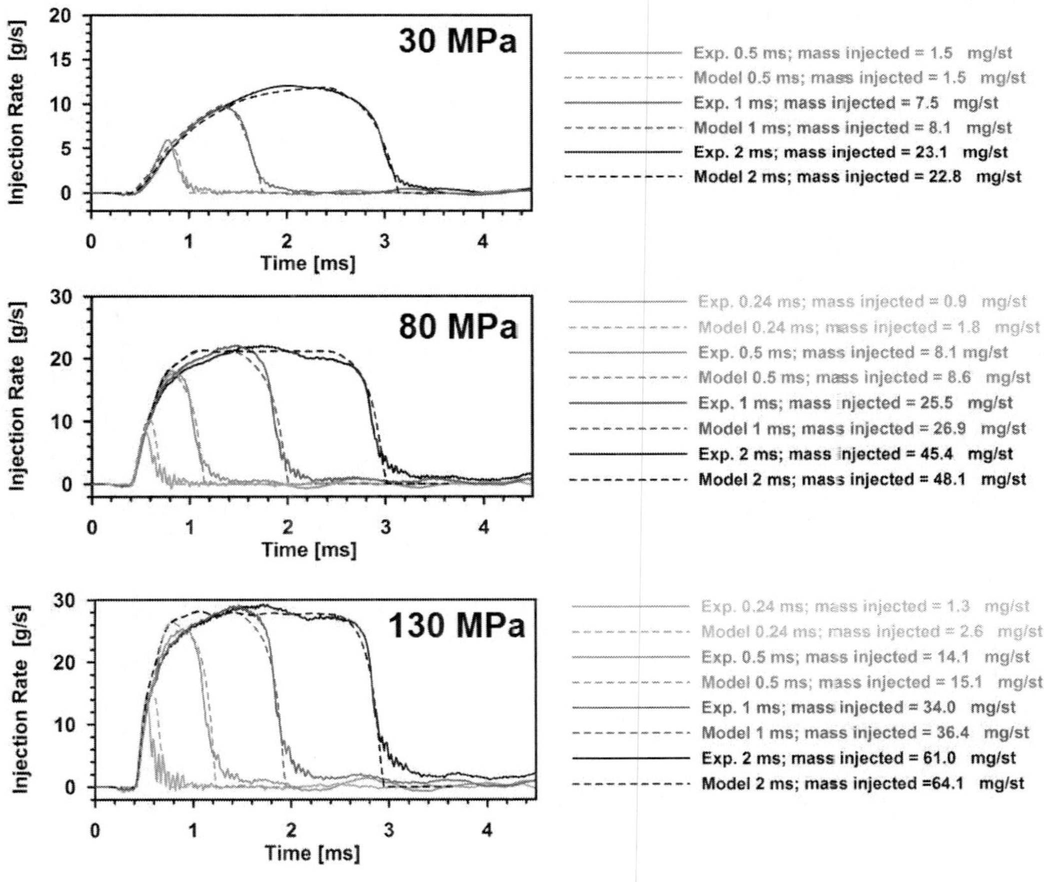

Figure 9. Model validation with experimental mass flow rate curves at different conditions of injection pressure and energizing time.

Appendix

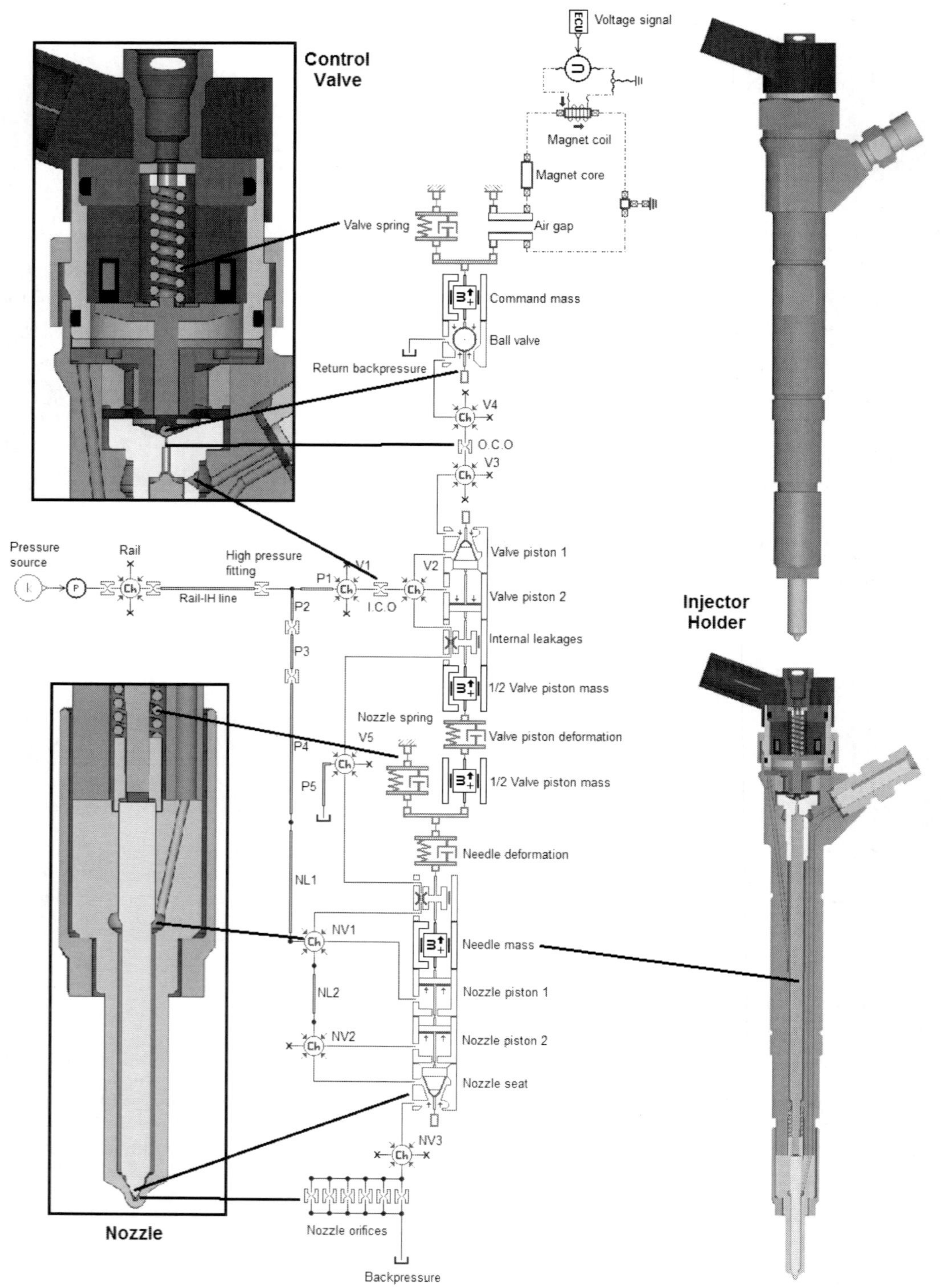

Figure A.1. Complete AMESim model of a common-rail injector, together with a representation of the injector.

Conclusion

A methodology to generate one-dimensional models of common-rail diesel injector systems based on the bond graph technique has been described and implemented in AMESim. Two main factors have been identified as the key in order for the model to be capable of reproducing the behaviour of the real system: on one hand, a correct abstraction of the system in different elements and the way they relate to each other must be performed. On the other hand, a thorough characterization of the physical components of the system must be carried out in order to feed the model with appropriate values of the parameters involved in the governing equations. Validation of the models against experimental data support this affirmation, demonstrating the potential of the model as a reliable tool to predict the behaviour of the system under different operating conditions.

References

[1] Heywood, J.B. *Internal combustion engine fundamentals*; McGraw-Hill: New York, NY, 1988; pp 491-566.

[2] Payri, F., Desantes, J.M. *Motores de combustión interna alternativos*; Reverté: Barcelona, 2011; pp 579-618.

[3] LMS Imagine.Lab AMESim v.10. User's manual, 2010.

[4] Payri, R., Climent, H., Salvador, F.J., Favennec, A.G. *P. I. Mech. Eng. D-J Aut.* 2004, *218*, 81-91.

[5] Payri, R., Tormos, B., Salvador, F.J., Plazas, A.H. *Int. J. Veh. Des.* 2005, 38(1), 58-78.

[6] Salvador, F.J., Gimeno, J., De la Morena, J., Carreres, M. *Energy Convers. Manage.* 2012, 54, 122-132.

[7] Karnopp, D.C., Margolis, D.L., Rosenberg, R.C. *System Dynamics – Modeling, Simulation and Control of Mechatronic Systems*; John Wiley&Sons, Inc.: New Jersey, NJ, 2012.

[8] Moody, L.F., Princeton, N.J. *T. ASME* 1944, 66(8), 671-684.

[9] Nurick, W.H. *J. Fluid Eng.-T. ASME* 1976, 98(4), 681-687.

[10] Macián, V., Bermúdez, V., Payri, R., Gimeno, *J. Exp. Techniques* 2003, 27(2), 39-43.

[11] Karnopp, D. *J. Dyn. Syst.-T. ASME* 1985, 107(1), 100-103.

[12] Desantes, J.M., Arrègle, J., Rodríguez, P.J. *SAE Tech. Paper* 1999-01-0915.

[13] Blackburn, J.F., Reethof, G., Shearer, J.L. *Fluid Power Control*; The MIT Press: Cambridge, MA, 1960.

[14] Hayt, W.H., Buck, J. *Engineering Electromagnetics*; McGraw-Hill: New York, NY, 2010; pp. 230-276.

[15] ANSI/IEEE, *IEEE Standard on Piezoelectricity*; IEEE, Std. 176, New York, 1987.

In: Mathematical Modeling in Social Sciences ...
Editors: J. C. Cortés López et al.
ISBN: 978-1-63117-335-6

Chapter 3

MATHEMATICAL MODELLING OF FILTRATION AND CATALYTIC OXIDATION OF DIESEL PARTICULATES IN FILTER POROUS MEDIA

N. V. Vernikovskaya[1,2,3,*]***, T. L. Pavlova***[1]***, N. A. Chumakova***[1,2]
and A. S. Noskov[1,3]
[1]Boreskov Institute of Catalysis SB RAS, Novosibirsk, Russia
[2]Novosibirsk State University, Novosibirsk, Russia
[3]Novosibirsk State Technical University, Novosibirsk, Russia

Abstract

The soot abatement in diesel exhaust is a very important task in particular because fine soot particles inhaled by humans do serious damage to their health. The mathematical modelling of soot filtration and catalytic oxidation in diesel particulate filters can help in finding more promising filter design, filter porous materials, running regime, and so on.

The used mathematical model consists of nonlinear partial and ordinary differential equations. The nonlinearities can cause rather steep non-stationary profiles of concentrations and temperature which propagate (rather slowly move) along the filter. So the standard numerical schemes are not appropriate here. The new modelling methodology is proposed which is based on the three existing methods and takes advantages of each of them: the method of lines, the running scheme, and the second-order Rosenbrock method with stepsize adjustment algorithm. Verification of the mathematical model and numerical method is done by means of comparison of the numerical results with the experimental data. The mathematical modelling is realized with taking into account the mass transport of soot particles of each diameter from a log-normal particle size distribution. For estimating the kinetic parameters the method of Errors-in-Variables Model with numerical integration is applied.

Keywords: Soot abatement, diesel filter, mathematical modelling

---

*E-mail address: vernik@catalysis.ru

1. Introduction

Diesel engine exhaust contains large amount of soot. Use of a catalytic filter seems to be most promising for soot abatement in the diesel exhaust. Filtration through porous medium occurs in all types of filters in a varying degree. In this case the porous medium retains the solid particle as it passes. Three basic mechanisms of soot capture exist: Brownian diffusion, particle interception, and inertial impaction. Accumulated in the pores of filter material, soot increases the pressure drop over a filter. Thereafter, the burning of soot inside the filter is needed. To force the soot burning at not high enough temperature of the diesel exhaust gases the catalytic oxidation is efficient.

Surface of the soot particles always contains adsorbed hydrocarbons [1, 2, 3], and the amount of hydrocarbons strongly depends on the engine-operation conditions. For instant, under low loads in idling, the fuel and lubricants are only partially oxidized, which leads to emission of some saturated hydrocarbons [4, 5]. Soot burning in presence of a catalyst starts at 350–400°C. Under these conditions, oxidation of the soot particles, evaporation, and oxidation of the adsorbed hydrocarbons occur simultaneously affecting the dynamics of soot oxidation in the filter and producing sharp profiles of temperature and concentrations.

Different size particles are present in exhaust diesel engine: from a few nm up to tens μm. Particles size depends on the engine type, the fuel kind, and the operating conditions. The particulate size distributions obtained with a modern Common-Rail Diesel (Peugeot 406 HDI) at different operating conditions are shown in [6].

The size of these particles ranges from 20 to 250 nm or even larger, the mean aerodynamic diameter is nearly 0.1 μm. It is well known that Brownian diffusion is efficient in capturing small particles, while the interception and the inertial impaction are efficient for large particles.

There are a few publications devoted to mathematical modelling of the soot filtration through a porous medium and the catalytic oxidation [7, 8, 9, 10]. The authors did not take into account the presence of high reactive hydrocarbons on the surface of soot particles. This means that the more weak conditions of the process performance is assumed. Moreover, the mean aerodynamic diameter of the soot particles is used in those models.

The goal of this chapter is to propose a robust numerical algorithm for solving the differential equations of the model and to carry out the mathematical modelling, at first, the filtration of the polydisperse diesel soot particles through a filter porous medium and, at second, the catalytic oxidation of the soot particles captured within the filter together with the hydrocarbons adsorbed on their surface.

For estimating the kinetic parameters (activation energy and Arrenius pre-exponential factor) of the reaction of soot oxidation over different catalysts the EVM (Errors-in-Variables Model) method with numerical integration is applied. As usual authors use some more simple and less accurate methods for estimation of the kinetic parameters.

2. Mathematical Model

A transient two phase mass transfer and one phase heat transfer model is used in studying the trapping of the polydisperse soot particles in the porous material of the filter and the catalytic oxidation of the captured soot via different catalysts. The model developed in

[3, 11] is supplemented by the mass-balance equations describing the mass transport of soot particles of each diameter from a log-normal particle size distribution (from a few nm up to tens μm).

The following assumptions and simplifications are made:

- The gas flow linear velocity, the gas density and heat capacity remain unchanged along the filter porous media;
- The density and heat capacity of the filter porous media remains unchanged during the process;
- The axial thermal conductivity of the filter is not taken into account;
- The change of nitrogen concentration in the gas mixture is ignored because nitrogen is inert to soot oxidation reactions.

The resulting equations are as follows:

- Equations for the gas phase concentration of the soot particles of each diameter from a log-normal particle size distribution

$$\epsilon(m)\frac{\partial c_i^s}{\partial t} + u\frac{\partial c_i^s}{\partial x} = -c_i^s u\eta(d_{p_i}, m)\varphi(m), \qquad i = 1, M, \tag{1}$$

- Equation for the concentration of the soot particles trapped inside the pores of the filter

$$\frac{\partial m_1}{\partial t} = (1-\gamma)\varphi(m)u \int_{d_{p_{\text{left}}}}^{d_{p_{\text{right}}}} c^s(d_p)\eta(d_{p_i}, m)dd_p - r_{p_1} m_{\text{cat}}, \tag{2}$$

- Equation for the concentration of octadecane adsorbed on the soot particles and trapped inside the pores of the filter

$$\frac{\partial m_2}{\partial t} = \gamma\,\varphi(m)u \int_{d_{p_{\text{left}}}}^{d_{p_{\text{right}}}} c^s(d_p)\eta(d_{p_i}, m)dd_p - r_{p_2} m_{\text{cat}}, \tag{3}$$

- Equations of mass balance for O_2 and CO_2

$$\epsilon(m)\frac{\partial c_i}{\partial t} + u\frac{\partial c_i}{\partial x} = m_{\text{cat}}\sum_{j=1}^{2}\mu_{ij}\frac{r_{p_j}}{M_j}, \qquad i = 1, 2, \tag{4}$$

- Equation of heat balance

$$\rho_f c_{p,f}\frac{\partial T}{\partial t} + \rho_g u c_{p,g}\frac{\partial T}{\partial x} = m_{\text{cat}}\sum_{j=1}^{2} r_{p_j}(-\Delta H_j). \tag{5}$$

Boundary conditions:

$$x = 0: \qquad c_i^s(0,t) = c_{i_{\text{in}}}^s; \qquad c_i(0,t) = c_{\text{in,i}}, \quad i = 1,2; \qquad T(0,t) = T_{\text{in}}. \tag{6}$$

Initial conditions:

$$t = 0: \qquad m_i(x,0) = m_{i0}(x), \quad i = 1,2; \qquad c_i^s(x,0) = c_{i0}^s(x), \quad i = 1,M,$$
$$c_i(x,0) = c_{i0}(x), \quad i = 1,2; \qquad T(x,0) = T_0(x). \tag{7}$$

The particle size distribution in the exhaust gases upstream of the filter are given in the next manner [12]:

$$c_{i_{\text{in}}}^s = \left(c_{\text{in}}^s/\sqrt{2\pi}\ln\sigma_p\right)\exp\left[-0.5\left(\left(\ln d_{p_i} - \ln\mu_p\right)/\ln\sigma_p\right)^2\right]. \tag{8}$$

The expression for the efficiency of soot capturing by the filter porous material $\eta(d_{p_i}, m)$ is given as a sum of the efficiencies due to the inertial impaction, the interception, the Brownian diffusion, together with the interactions between interception and Brownian diffusion and the efficiency reduction caused by the erosive phenomenon [7, 9]. The expression for the pressure drop calculation over the filter porous material is taken as [13]:

$$\Delta P = \Delta P^0 \cdot \left(\frac{\epsilon^0}{\epsilon}\right)^3 \cdot \left(\frac{(1-\epsilon)}{(1-\epsilon^0)}\right)^{1.5}. \tag{9}$$

The expression for the reaction rate of soot oxidation is as follows [14]:

$$r_{p_1} = k_0 e^{-E_A/RT}(Py_1)^\alpha m_{10}\left(\frac{m_1}{m_{10}}\right)^\beta. \tag{10}$$

The used mathematical model consists of the partial differential equations (PDEs) and the ordinary differential equations (ODEs). PDEs (1), (4), and (5) represent the transient nonlinear heat and mass transport equations. ODEs (2) and (3) represent the transient nonlinear equations of soot and hydrocarbons accumulation or their loss due to oxidation reactions. The nonlinearities in all equations are caused by the accumulation of soot or soot with hydrocarbons particles and/or very rapid catalytic reaction of the soot or soot with hydrocarbons oxidation. A distinctive feature of the equations (1), (4), and (5) consists in generation of quite steep non-stationary profiles of concentrations and temperature moving along the filter. Moreover, the polydispersity of soot particles results in solving equation (1) for the concentration of the soot particles of each diameter from a log-normal particle size distribution.

For estimation of the kinetic parameters k_0 and E_A of the reaction rate (10) over CeO_2/θ-Al_2O_3, Pt-CeO_2/θ-Al_2O_3, and Fe-Mn-K-O/γ-Al_2O_3 catalysts we use the dependencies of the outlet CO_2 concentration on temperature obtained in the home-made reactor during the temperature programmed soot oxidation experiments [15]. The catalyst-containing sample was loaded into a quartz reactor and heated from 50 to 700°C using a programmed linear temperature elevation at rate of 5°C/min in these experiments. The height of sample was 5.2 mm in the reactor of 7 mm diameter. The gas flow rate was 2 ml/s.

The mathematical model of the temperature programmed soot oxidation in the reactor is given below:

$$\frac{dy_i}{dt} = \left[(y_i^0 - y_i)\sum_{j=1}^{N_1} n_j^* - r_{p1}\cdot m_{\text{cat}}\right] \Big/ \sum_{j=1}^{N_1} n_i, \qquad i = 1, 2, \tag{11}$$

$$\frac{dm_1}{dt} = -r_{p1}\cdot m_{\text{cat}}, \tag{12}$$

$$\frac{dT}{dt} = 0.0833, \tag{13}$$

$$t = 0: \qquad m_1 = m_{10}, \qquad y_i = y_{i0}, \quad i = 1, 2, \tag{14}$$

$$r_{p1} = k_0 e^{-E_A/RT}\cdot (P\cdot y_1)^{\alpha}\cdot m_0\cdot\left(\frac{m}{m_0}\right)^{\beta}. \tag{15}$$

The expression (15) for the reaction rate of hydrocarbons oxidation and kinetic constants for this reaction rate are given in [3].

3. Numerical Methods

3.1. Discretization of the Differential Equations

We used the method of lines to semi-discretize the PDEs into the system of ODEs with time as independent variable. Consider for example the semi-discretization of Eq. (1) for concentration of the soot particles of fixed diameter from the log-normal particle size distribution and, for instance, let i = 1. Assume that the function is discretized in space but not time to N+1 points on a uniform grid. Then $c_1^s(x,t)$ is approximated in the next manner:

$$c_1^s(k\Delta x, t) \approx c_{1k}^s(t), \qquad k = 0, 1, \ldots, N, \tag{16}$$

and $c_{1k}^s(t)$ becomes a discrete variable with index k instead of a continuous dependence on x, though $c_{1k}^s(t)$ still is a continuous function on time. At the left boundary we have: $c_{1k=0}^s(t) = c_1^{s0}$, while initially $c_{1k}^s(0) = 0$ at $t = 0$. For discretization of the derivative $\frac{\partial c_1^s}{\partial x}$ we use the backward difference $\frac{c_{1k}^s - c_{1k-1}^s}{\Delta x}$. So for one PDE the system of ODEs was obtained:

$$\begin{cases} c_{10}^s = c_1^{s0}, \\ \frac{dc_{11}^s}{dt} = \frac{u_1}{\epsilon_1}\left(-\frac{c_{11}^s - c_{10}^s}{\Delta x} + c_{11}^s\cdot\eta_1\cdot\varphi_1\right), \\ \vdots \\ \frac{dc_{1k}^s}{dt} = \frac{u_k}{\epsilon_k}\left(-\frac{c_{1k}^s - c_{1k-1}^s}{\Delta x} + c_{1k}^s\cdot\eta_k\cdot\varphi_k\right), \\ \vdots \\ \frac{dc_{1N}^s}{dt} = \frac{u_N}{\epsilon_N}\left(-\frac{c_{1N}^s - c_{1N-1}^s}{\Delta x} + c_{1N}^s\cdot\eta_N\cdot\varphi_N\right) \end{cases} \tag{17}$$

with the initial conditions

$$c_{1k}^s(0) = 0, \qquad k = 0, 1, \ldots, N. \tag{18}$$

Equations (4) and (5) were discretized in a similar manner.

Equations (2) and (3) were discretized as follows:

$$\begin{cases} \frac{dm_{i0}}{dt} = c^{s0}(1-\gamma)u_0 \cdot \eta_0 \cdot \varphi_0 - r_{p_1 0} \cdot m_{cat} \\ \vdots \\ \frac{dm_{ik}}{dt} = c^s_k(1-\gamma)u_k \cdot \eta_k \cdot \varphi_k - r_{p_1 k} \cdot m_{cat}, \\ \vdots \\ \frac{dm_{iN}}{dt} = c^s_N(1-\gamma)u_N \cdot \eta_N \cdot \varphi_N - r_{p_1 N} \cdot m_{cat} \end{cases} \tag{19}$$

with the initial conditions

$$m_{ik}(0) = 0, \qquad i = 1, 2. \tag{20}$$

Thus, we obtained the system of the first order differential equations:

$$\frac{dy}{dt} = F(y), \tag{21}$$

for the vector with $N(M+5)+2$ components

$$y = (c^s_{11}, \ldots, c^s_{1N}, \ldots, c^s_{M1}, \ldots, c^s_{MN}, \ldots, m_{10}, \ldots, m_{1N}, m_{20}, \ldots, m_{2N}, \\ c_{11}, \ldots, c_{1N}, c_{21}, \ldots, c_{2N}, T_1, \ldots, T_N)^T$$

with the initial conditions:

$$y = y_0 \text{ at } t = t_0. \tag{22}$$

3.2. Solution of the System of Ordinary Differential Equations (21) with Initial Conditions (22)

The ODEs (21) with initial conditions (22) may in turn be solved using well-known methods, for example, the forward stepping Runge-Kutta method. But since the nonlinearities of right-parts of equations (1)–(5) can cause rather steep non-stationary profiles of concentrations and temperature which move along the filter, the standard numerical schemes are inappropriate here. The following algorithm is applied for solving the ordinary differential equations (21):

1. the vector y is reorganized in the next way

$$y = (\underbrace{m_{10}, m_{20}}_{k=0}, \underbrace{c^s_{11}, \ldots, c^s_{M1}, m_{11}, m_{21}, c_{11}, c_{21}, T_1}_{k=1}, \ldots, \\ \underbrace{c^s_{1N}, \ldots, c^s_{MN}, m_{1N}, m_{2N}, c_{1N}, c_{2N}, T_N}_{k=N})^T$$

2. the running scheme is used for solving this system: the corresponding subsystem of ODEs for each space grid point k is solved from t to $t+\Delta t$ in series in the streamwise direction (successively for $k = 0, 1, \ldots, N$), where the internal time-steps within Δt can vary from some small value to Δt independently for different k;

3. a special case of the second-order Rosenbrock method and an automatic adjustment of the internal integration step (from some small value to Δt) are employed for solving the ODE system at each space-grid point k from t to $t + \Delta t$. The needed current values from $(k-1)$-th point are obtained by using an interpolation procedure.

4. the item 2 is repeated till $t = t_{process}$.

3.3. Stepsize Adjustment Algorithm

Let us consider the item 3 of the algorithm in more detail for the example of solving the subsystem of ODEs in space-grid point $k = 1$ from t to $t + \Delta t$:

$$\frac{dy}{dt} = f(y), \qquad y = \Bigg(\underbrace{c_{11}^{s}, \ldots, c_{M1}^{s}, m_{11}, m_{21}, c_{11}, c_{21}, T_1}_{k=1}\Bigg)^{T}, \tag{23}$$

with the initial conditions

$$y = y_0 \text{ at } t = t_0. \tag{24}$$

We seek the solution with the stepsize h as follows [16]:

$$y_{n+1} = y_n + P_1 k_{n1} + P_2 k_{n2}, \tag{25}$$

where the vectors k_{n1} and k_{n2} satisfy

$$(I - \gamma \cdot h \cdot f_y(y_n))\, k_{n1} = h f(y_n), \tag{26}$$
$$(I - \gamma \cdot h \cdot f_y(y_n))\, k_{n2} = h f(y_n + \alpha \cdot k_{n1}), \tag{27}$$

with $P_1 = 0.5 + 0.25\sqrt{2}$, $P_2 = 0.5 - 0.25\sqrt{2}$, $\gamma = 1 - 0.5\sqrt{2}$, and $\alpha = \sqrt{2}$, in order to obtain the global truncation error ε such that

$$|k_{n2} - k_{n1}| < 15\varepsilon. \tag{28}$$

The used stepsize adjustment algorithm (Fig. 1) is based on the computed value of right-hand side of (25) and difference between k_{n1} and k_{n2} [16, 17]. We use the componentwise notation for $i = 1, 2, \ldots, M + 5$:

$$R_i = u_n + \big(0.5 + 0.25\sqrt{2}\big)k_{n1}^{i} + \big(0.5 - 0.25\sqrt{2}\big)k_{n2}^{i},$$
$$RE_i = |k_{n1}^{i} - k_{n2}^{i}|, \qquad RQ_i = |k_{n1}^{i} - k_{n2}^{i}|/|u_n^{i}|.$$

For solving the systems (26) and (27) we use the LU Decomposition.

3.4. Kinetic Parameters Estimation

Some partial order in respect to oxygen and order one in respect to soot in the expression of the reaction rate of soot oxidation (15) are taken from literature. Fixing α and β, the kinetic parameters are estimated using the method of Errors-in-Variables Model (EVM method) [18]. Equations (11)–(13) are integrated numerically, and the root-mean-square relative error does not exceed 3% for all variants.

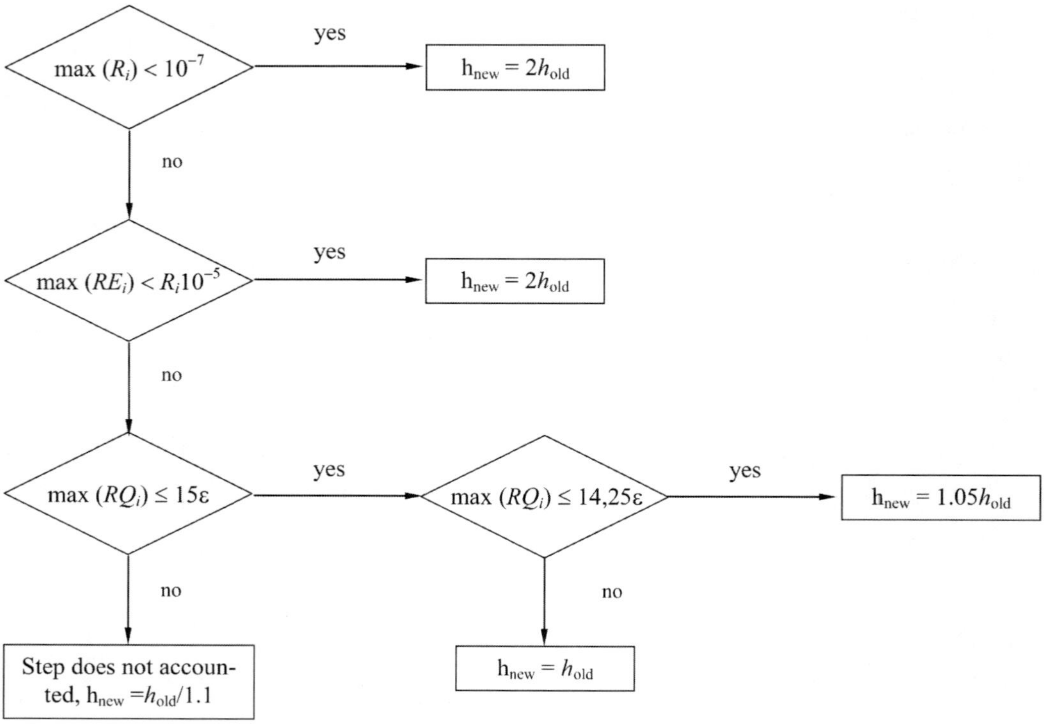

Figure 1. Stepsize adjustment algorithm.

4. Verification of the Mathematical Model and the Numerical Methods

For the purpose of verification of the mathematical model and numerical methods we compare the numerical results and the experimental data obtained with mullite foam as a filter substrate with and without catalyst [7]. The soot concentration in the flue gas is 0.13 g/m^3. The filter thickness and diameter are 17 mm and 50 mm respectively, the linear velocity is 0.82 m/s. The particle size distribution corresponds to the gases generated by the acetylene-burner with $\mu_p = 0.14$ μm and $\sigma_p = 1.24$. In Fig. 2 the dependencies of pressure drop on time for mullite foam filter at different temperatures with and without catalyst are shown. A rather good agreement between the numerical data and the experimental results is achieved. Both in the experiments and in the modelling with the use of the catalyst the slower pressure drop rises more slowly and the plateau values are achieved, while if a non-catalytic filter is used then the pressure drop rise is higher.

5. Results and Discussion

Some results of the mathematical modelling of a catalytic particulate filter of a fibrous material in the case of presenting the adsorbed hydrocarbons (octadecane) on the surface of soot particles were presented earlier [3]. Now we show the results of mathematical modelling of filtration of the polydisperse diesel soot particles through different filters and

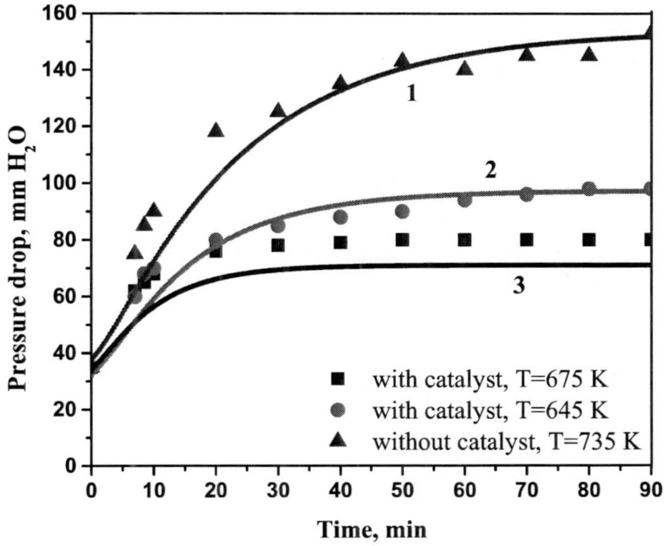

Figure 2. The dependencies of pressure drop on time for mullite foam filter at different temperatures with and without catalyst. Symbols represent the results of the experiments taken from the paper [7]. Lines correspond to the modelling results on the base of equations (1)–(5): 1 – without catalyst at T = 735 K, 2 – with catalyst at T = 645 K, 3 – with catalyst at T = 675 K.

catalytic oxidation of captured soot over CeO_2/θ-Al_2O_3, Pt-CeO_2/θ-Al_2O_3, and Fe-Mn-K-O/γ-Al_2O_3 catalysts using the estimated kinetic parameters.

The kinetic parameters obtained by using the mathematical model (11)–(15) of the temperature programmed soot oxidation in the reactor are listed in Table 1.

Table 1. Estimated kinetic parameters

Catalyst	k_0, 1/(g·s·atm)	E_A, kJ/mole	α	β
CeO_2/θ-Al_2O_3	$15.0 \cdot 10^7$	160.8	0.5	0.9
Pt-CeO_2/θ-Al_2O_3	$2.5 \cdot 10^5$	111.4	0.6	0.8
Fe-Mn-K-O/γ-Al_2O_3	$13.0 \cdot 10^7$	128.7	0.5	1.0

In our calculation we consider the two filter porous materials: mullite foam and fiber quartz, and the two different schemes of catalytic traps for diesel particles. First scheme is the cylinder of 23 mm thickness, and the gas linear velocity is 0.7 m/s. Second one is the monolith filter with alternate cells opening/plugged on one end of the unit in checkerboard fashion and plugged/opening on the other end in a similar manner, where the exhaust gas is forced through the porous wall. In the last case the flow through channel walls only is modelled. The thickness of the walls is 0.43 mm, and the gas linear velocity is 0.1 m/s. The particle size distribution under study corresponds to the gases generated by the acetylene-burner with $\mu_p = 0.14$ μm and $\sigma_p = 1.24$ in both cases. Our research shows that the mullite foam is more suitable material for the first scheme of catalytic traps. In this case the filtration efficiency (θ) is 100 % and pressure drop is as low as 1.3 kPa. On the other hand, the fiber quartz is the best material for the channel walls. The filtration efficiency inside the walls is 38 %, and this value is good because the most soot accumulates on the channel walls, pressure drop is 2.8 kPa. The differences in capturing behaviour and pressure drop for these two materials are explained by the lower fiber diameter 4.4 μm for quartz in comparison with the "equivalent fiber diameter" for mullite foam 6.3 μm.

The next evidence shows the influense of the particle size distribution on the filter performance. Fig. 3 illustrates the results of the mathematical modelling of soot collecting in the fiber quartz wall-flow monolith with the following operating conditions: exhaust flow rate is 0.1 m/s, exhaust temperature is 200°C, soot feed concentration is 0.13 g/m^3, particle size distribution corresponds to the diesel exhaust gases of modern passenger cars with median particle diameter 0.17 μm and standard deviation 1.74 [12], O_2 concentration is 8 % vol., and wall thickness is 0.43 mm. It is useful to plot the filtration efficiency as a function of time (Fig. 3, a). The solid curve shows the filtration efficiency obtained with taking into account the polydispersity of soot particles, while the dash, dot, and dash-dot curves are obtained with mean aerodynamic diameter of particles. We see the significant deviation of solid curve from three others. Fig. 3, b shows the particle size distribution of the exhaust aerosol upstream and of the filter and downstream of the filter as the log-normal function. The mean particle diameter decreases after 2400 s from the beginning of the filtration process.

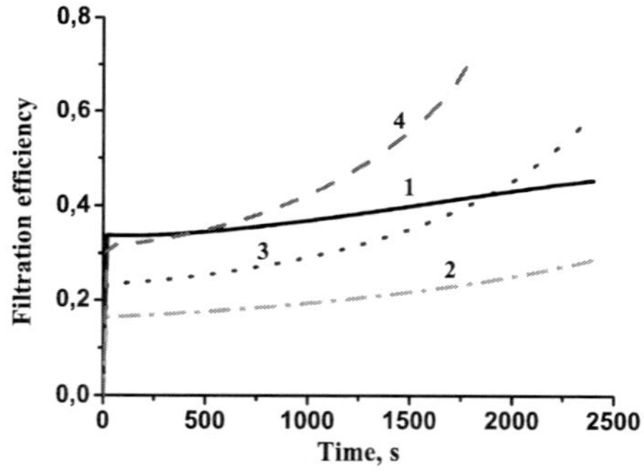

(a)

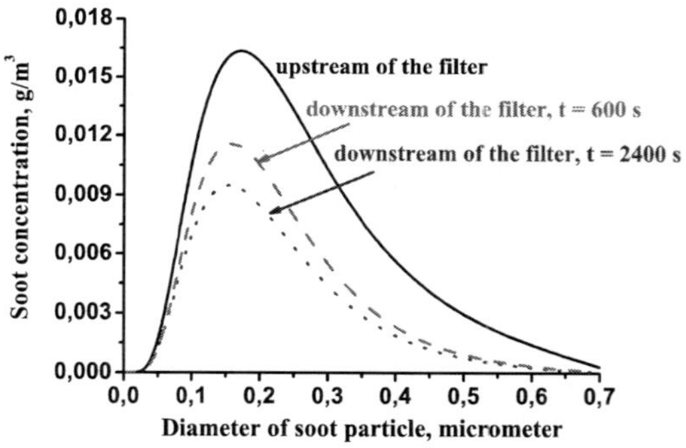

(b)

Figure 3. The influence of polydispersity of soot particles on the filtration efficiency: a) the dependence of the filtration efficiency on time: 1 – with taking into account polydispersity of the particles, the curves 2–4 with mean aerodynamic diameter, 2 – $d_p = 0.2$, 3 – $d_p = 0.3$, 4 – 0.4 μm; b) particle size distribution.

The results of the trapped soot oxidation over CeO_2/θ-Al_2O_3, Pt-CeO_2/θ-Al_2O_3, and Fe-Mn-K-O/γ-Al_2O_3 catalysts using the estimated kinetic parameters are listed in Table 2. The duration of the previous soot accumulation in the fiber quartz wall-flow monolith is 40 min with the parameters listed above.

Table 2. The results of trapped soot oxidation

Catalyst	CeO_2/θ-Al_2O_3	Pt-CeO_2/θ-Al_2O_3	Fe-Mn-K-O/γ-Al_2O_3
T_{in}, K	713	683	583
Regeneration degree, %	74	89	92
T_{max}, K	730	706	746

By using CeO_2/θ-Al_2O_3, Pt-CeO_2/θ-Al_2O_3, and Fe-Mn-K-O/γ-Al_2O_3 catalysts, the filter regeneration should be done at temperatures equal to 713, 683 and 583 K, respectively. The filter regeneration using Fe-Mn-K-O/γ-Al_2O_3 catalyst under the temperature of exhaust gases 583 K or even smaller with simultaneous soot accumulation and oxidation can be possible.

The results (pressure drop and filtration efficiency) of the alternating processes of collecting the soot particles and their oxidation are shown in Fig. 4. Mathematical modelling of soot collecting in fiber quartz wall-flow monolith is done at the above-listed parameters. Fe-Mn-K-O/γ-Al_2O_3 catalysts is used for soot oxidation at the input temperature 583 K. Already after the third cycle the process becomes practically periodic.

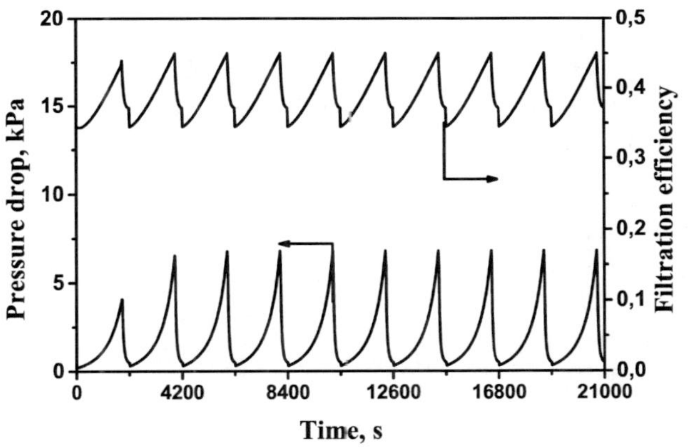

Figure 4. The dependencies of pressure drop and filtration efficiency on time at alternating the collecting of soot particles and their oxidation.

6. Conclusion

There is presented the mathematical model describing the filtration of the diesel soot particles with size from a few nm up to a few μm through a porous medium and the catalytic oxidation of the soot particles captured within the filter together with the hydrocarbons adsorbed on the particle surfaces. The system of nonlinear partial and ordinary differential equations to be solved is characterized by the rather steep moving profiles of the concentrations and temperature. The developed algorithm bases on the method of lines, the running scheme, and the second-order Rosenbrock method with an efficient stepsize adjustment procedure. Verification of the mathematical model and numerical algorithm is performed by comparison of the numerical results and the experimental data which are in a good agreement.

Note that a wide distribution of the diesel soot particle sizes also should be taken into account for design and optimization of the catalytic traps because the polydispersity of the particles strongly affects the filtration efficiency. Downstream of the filter, the mean particle diameter decreases with time. It is shown that the process with alternating the stages of collecting the soot particles and their oxidation over Fe-Mn-K-O/γ-Al_2O_3 catalyst is efficient and becomes practically periodic already after the third cycle.

6.1. Nomenclature

c_i^s – concentration of soot particles of i-th diameter in the gas phase, kg/m^3;
c^s – concentration of soot particles of all diameters in the gas phase, kg/m^3;
c_i – concentration of i-th compound in gas phase, mol/m^3;
$c_{p,g(f)}$ – specific heat capacity of the gas phase and the filter material, J kg^{-1} K^{-1};
d_p – soot particle diameter, μm;
d_{p_i} – the diameter of soot particle of i-th size, μm;
E_A – activation energy, J/mole;
$f_y(y_n)$ – Jacobian matrix;
h – integration step within Δt;
$-\Delta H_j$ – enthalpy of oxidation of carbon or octadecane, J kg^{-1};
I – identity matrix;
k_0 – pre-exponential factor, (g s atm$^{\alpha}$)$^{-1}$;
M_i – molecular weight of i-th gas compound, kg mol^{-1};
$m_{1(2)}$ – current concentration of carbon (hydrocarbons) in the filter pores, kg/m^3;
$m = m_1 + m_2$ – current concentration of carbon and octadecane in the filter pores, kg/m^3
or mole in eqs. (11)–(15);
m_{cat} – mass of the catalyst, kg/m^3 or g in eqs. (11)–(15);
M – number of partitions in a log-normal particle size distribution;
N_1 – number of compounds in the gas mixture, eq. (11);
$\sum n_i$ - total moles in the reactor, mole;
$\sum n_i^*$ - total moles flow in the reactor, mole/min;
P – pressure, atm;
ΔP – pressure drop across the filter, atm;
R – universal gas constant, J/mole/K;
r_{p_1} (r_{p_2}) – reaction rate of carbon (octadecane) oxidation kg/kg$_{\text{cat}}$/s or mole/g$_{\text{cat}}$/min in eqs. (11)–(15);
T – temperature, K;
t – time, s;
u – superficial velocity under normal conditions, m s^{-1};
x – the axial coordinate over the filter depth, m;

y_1 – concentration of oxygen in gas phase, mole fraction;

Greek letters:
$\alpha(\beta)$ – partial order of oxygen (soot);
γ – fraction of octadecane in soot particle;
ε – void fraction at a current time;
η – total efficiency of soot capturing;
θ – filtration efficiency, $\theta = (c^s_{\text{in}} - c^s)/c^s_{\text{in}}$;
μ_{ij} – stoichiometric coefficients;
$\rho_{f(g)}$ – density of the filter and gas phase;
φ – depth of particles penetration into the filter pores;

Subscripts:
^{0} – clean filter;
$_0$ – initial value;
$_{in}$ – inlet value.

References

[1] Van Gulijk, C. *Rational design of a robust diesel particulate filter*, Ph. D. Thesis; Technical University of Delft: Delft, 2002; pp 1–199.

[2] Johnson, J. E.; Kittelson, D. B. *Appl. Catal.* 1996, 10, 117–137.

[3] Pavlova, T. L.; Vernikovskaya, N. V.; Chumakova, N. A.; Noskov, A. S. *Combust. Expl. Shock Waves*. 2006, 42 (4), 396–402.

[4] Sharma, M.; Agarwal, A. K.; Bharathi, K. V. L. *Atmos. Environ.* 2005, 39 (17), 3023–3028.

[5] Kittelson, D. B. *J. Aerosol Sci.* 1998, 29 (5–6), 575–588.

[6] Ambrogio, M.; Saracco, G.; Specchia, V.; Van Gulijk, C.; Makkee, M.; Moulijn, J. A. *Separation and Purification Technology*. 2002, 27, 195–209.

[7] Ambrogio, M.; Saracco, G.; Specchia, V. *Chem. Eng. Sci.* 2001, 56, 1613–1621.

[8] Fino, D.; Saracco, G.; Specchia, V. *Chem. Eng. Sci.* 2002, 57, 4955–4966.

[9] Saracco, G.; Badini, C.; Specchia, V. *Chem. Eng. Sci.* 1999, 54, 3035–3041.

[10] Bensaid, S.; Marchisio, D. L.; Fino, D. *Chem. Eng. Sci.* 2010, 65, 357–363.

[11] Pavlova, T. L.; Vernikovskaya, N. V.; Chumakova, N. A.; Noskov, A. S. *Combust. Expl. Shock Waves*. 2004, 40 (3), 263–269.

[12] Oh, S. H.; MacDonald, J. S.; Vaneman, G. L.; Hegedus, L. L. *Mathematical modeling of fibrous filters for diesel particulates – theory and experiment*; SAE Technical paper series 810113; Society of automotive engineers: Detroit, MI, 1981; pp 1–12.

[13] Pontikakis, G. N.; Koltsakis, G. C.; Stamatelos, A. M. *Particulate Science and Technology*. 1999, 17, 79–200.

[14] Darcy, P.; Da Costa, P.; Mellottee, H.; Trichard, J.-M.; Djega-Mariadassou, G. *Catalyst today*. 2007, 119, 252–256.

[15] Ivanova, A. S.; Litvak, G. S.; Mokrinskii, V. V.; Plyasova, L. M.; Zaikovskii, V. I.; Kaichev, V. V.; Noskov, A. S. *Journal of Molecular Catalysis A: Chemical*. 2009, 310, 101–112.

[16] Novikov, E. A. *In Mathematical methods in chemical kinetics*; Bykov, V. I.; Ed.; Nauka: Novosibirsk, 1990; pp 53–68 (in Russian).

[17] Bibin, V.N. In *Proceedings XII International Conference on Chemical Reactors*; Boreskov Institute of Catalysis: Novosibirsk, 1996; Vol. 1, pp 127–132 (in Russian).

[18] Pavlova, T. L.; Vernikovskaya, N. V.; Ermakova, A.; Mokrinskii, V. V.; Kashkin, V. N.; Noskov, A. S. In *Proceedings International Conference Nanostructured catalysts and catalytic processes for the innovative energetics and sustainable development*, Boreskov Institute of Catalysis: Novosibirsk, 2011; Abstracts: Print-CD volume, p. 62.

In: Mathematical Modeling in Social Sciences ...
Editors: J. C. Cortés López et al. ISBN: 978-1-63117-335-6

Chapter 4

WATER DEMAND SIMPLIFICATIONS USED TO BUILD MATHEMATICAL MODELS FOR HYDRAULIC SIMULATIONS

J. Izquierdo,* E. Campbell, I. Montalvo, R. Pérez-García and D. Ayala-Cabrera
Instituto Universitario de Matemática Multidisciplinar, I.M.M. Fluing,
Universitat Politècnica de València, Valencia, Spain

Abstract

Mathematical models that describe the complex behavior of water distribution systems are bound to use various simplifications. Simplified models are used systematically in management, monitoring, and understanding of water utilities, sometimes without any awareness of the implications of the assumptions used. One of the reasons for that complexity stems from the vast spatial distribution characteristic of water supply. Thus, some simplifications are derived from the different levels of granularity at which a network can be considered. This chapter addresses the specific question: to what extent can consumptions associated with a line be allocated to the ends of the line? We present examples of situations where certain simplifications produce models that are very unrealistic. Moreover, we develop guidelines that enable to establish whether some simplifications are acceptable or, on the contrary, produce information that differs substantially from reality. We also develop formulae easy to implement that enable to distribute inner line demand between the line ends with minimal error.

Keywords: Mathematical models, error analisys, water demand, water distribution systems

1. Introduction

Mathematical modeling of water distribution networks (WDNs) uses simplifications aimed to optimize the development and use of the models. Such simplifications stem from the complexity of the modeled infrastructure and, at the same time, are related to the widespread spatial distribution typical of WDNs. Mathematical/numerical models are applied in all the

*E-mail address: jizquier@upv.es

areas of hydraulics – including urban hydraulics [2, 3]. Currently, with the generalized use of geographic information systems, models containing even hundreds of thousands of pipes and nodes are being built [4].

Among others, simplifications derived from the different levels of granularity at which a network is considered, make that real water distribution networks, especially those of large cities, cannot be efficiently modeled in their entireties.

One of these simplifications is the grouping of the consumptions associated with interior points of a line in one or both ends of the line. These points concentrate all the existing consuming points (users) within the line.

In a branched network with few consumers, this would not represent a major problem. However, problems arise in WDNs with up to 30 connections per line (e.g. a street pipeline). In a large WDN (e.g. a 500 km WDN) it would amount to considering about 150,000 nodes, which is impractical when it comes to the construction of the network, the performance of the calculations, as well as the display and understanding of the results. This simplification copes with the continuity principle or conservation of mass. However, the energy aspects are completely mislaid.

This chapter analyzes the error that may derive from the effect of using the widespread 50% rule, which allocates half of the in-line demand to each line end. We analyze to what extent this simplification is acceptable. Also, we obtain formulae that enable to distribute inner line demand between the line ends with minimal error.

Our proposal involves simple, direct methods that can be easily applied by any user of any WDN analysis package, since emphasis is not placed on programming ability. Users, having already developed models of their networks may revise the allocation rule used, and replace it, if necessary, with the values provided by the new formulae, what will enable them to obtain more reliable results. Also, users starting the model of a new network may make an a priori decision about how to simplify the network and suitably implement the associated simplifications.

We first present a simple case that enables us to shed light into the problem: a single line with variable distribution and granulation consumption is considered. Then an example of a real network is analyzed using the lessons learned. The conclusions section closes the chapter.

2. Line with Associated Demand

Let us firstly consider the case of a single line associated with some internal consumption under steady state condition. Such a line is representative of the simplest installation (a line between two nodes) with a given in-flow rate. The characteristics of the line are: length: L; diameter: D; upstream head (boundary condition at the upstream node): H_0; friction factor: f, with its associated line resistance: $K = 8f/(g\pi^2 D^5)$; and inflow: Q_{in}.

Various demand scenarios of consumption in the line may be tested. Such scenarios are associated with two characteristics: total demand in the line with regard to inflow, and specific distribution of the demand along the line.

Let us assume that the flow consumed within the line (total in-line demand) represents a percentage of the line inflow. If this fraction is represented by F_Q, $0 < F_Q \leq 1$, the actual demand in the line is given by the expression $Q_d = F_Q Q_{in}$.

2.1. Uniformly Distributed Demand along the Line

To start with the study, we will assume that the actual demand of the line is uniformly distributed into a number n of equally spaced interior points (nodes); n can take a value ranging from 1 (in the case of a line with a single demand node in the middle) to a large integer number (in the case of an equally distributed demand throughout the line).

Fig. 1 shows various distributions of piezometric head corresponding to values of F_Q equal to 1, 0.8, 0.6 and 0.4, for a set of values of L, D, H_0, f and Q_{in}. To build Fig. 1 we have used the specific values $L = 500$m, $D = 300$mm, $H_0 = 50$m, $f = 0.018$ and $Q_{in} = 0.25\text{m}^3/\text{s}$. As said, the demand has been equally distributed among n equally spaced interior nodes. Specifically, in Fig. 1, n takes the values 1, 3, 7, 11, 19 and 'infinity', this last case representing the uniform continuous demand ideal case. The various curves in Fig. 1 have straightforward interpretation.

The calculations for the polygonal hydraulic grade lines (HGLs) made out of segments between consumption points correspond to the usual hydraulic calculation of losses. The polygonals start at the boundary condition $(0, H_0)$, the other vertices being the $n+1$ points:

$$\left(j\frac{L}{n+1}, H_0 - K\frac{L}{n+1}Q_{in}^2\sum_{k=1}^{j}\left(1-\frac{k-1}{n}F_Q\right)^2\right), j = 1, \ldots, n+1. \tag{1}$$

Let us call $H_R(F_Q, n; x)$ these HGLs, the subindex '$_R$' standing for 'real' distribution of piezometric head along the line, as real demands are used to calculate (1).

The calculation for the ideal HGL corresponding to a uniform continuous demand, which is used here as the limit for $n \to \infty$ of the discrete uniform distribution of demands, is performed by integrating the differential loss $\Delta H = -K\left(Q_{in} - \frac{Q_d}{L}x\right)^2 \Delta x$, along the line. The value Q_d/L is the (constant) demand per unit length, and x is the distance to the upstream node. By integrating, the piezometric head corresponding to this continuous loss is given by

$$H_R(F_Q; x) = H_0 - K\frac{L}{3}\frac{Q_{in}^2}{F_Q}\left[1 - \left(1 - F_Q\frac{x}{L}\right)^3\right], \tag{2}$$

which corresponds to the upper curve – a cubic – in each of the graphs in Fig. 1.

It becomes clear that the greatest discrepancies occur for values of F_Q close to 1 (e.g. when a high percentage of the inflow is consumed along the line).

As said, these 'real' HGLs in Fig. 1 have been calculated according to the demand distribution at the various inner points in the line. However, models of large WDNs do not usually take intermediate demands into account; while, on the contrary, the demand of each line is distributed between the end nodes of the line, the 50% rule being generally used.

2.2. Allocation of In-Line Demand to the Line Ends. Is the 50% Rule Adequate?

Let F_{Qd} be the factor that allocates a part of the line distributed demand, Q_d, to its upstream end. Thus, the demand assigned to this upstream node is $Q_0 = F_{Qd}Q_d$. As a result, $Q_l = Q_{in} - Q_0$ is the flowrate through the line. Then, the expression of H_C, where

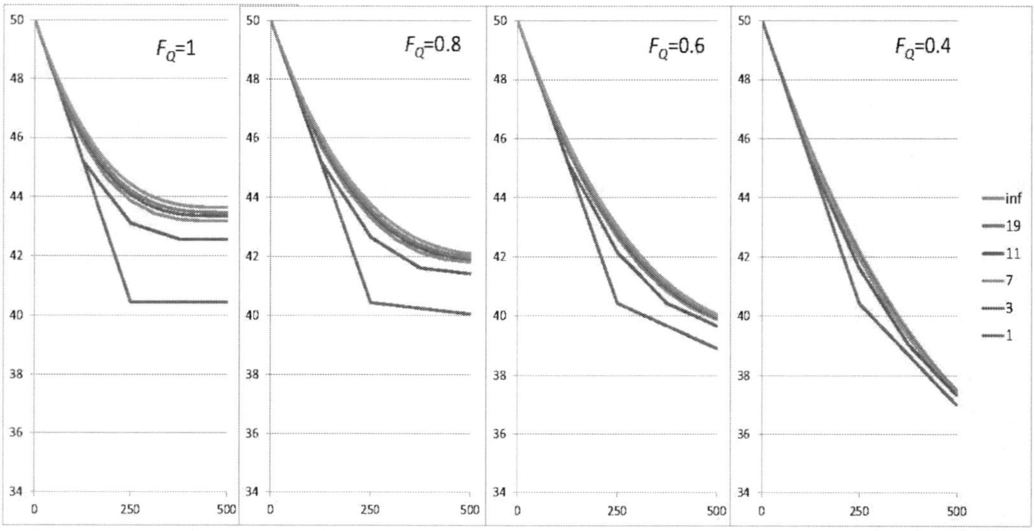

Figure 1. HGLs in one single line for various uniform distributions of demand.

subindex '$_C$' stands for 'calculated', for a given value of F_{Qd}, is

$$H_C\left(F_{Qd}; x\right) = H_0 - KQ_l^2 x. \tag{3}$$

The HGL obtained is, thus, a straight line that connects the point $(0, H_0)$ with the point $(L, H_C\left(F_{Qd}; L\right))$. This last value corresponds to the calculated head at L, the downstream node.

In Fig. 2, dashed lines have been added to the first two charts of Fig. 1. These new HGLs have been calculated to give the same piezometric head at the downstream node as the line corresponding to a demand concentrated in the middle point, $n = 1$, for the case $F_Q = 1$ (left chart); and as the line corresponding to a continuous demand for $F_Q = 0.8$ (right chart). These lines have been obtained by allocating a fraction of the interior line demand to the upstream node and the rest to the downstream node. Analogous dashed lines can be obtained for other combinations of F_Q and n. Should the allocated fractions to the line ends have been different, the corresponding (straight) lines (the lines given by the numerical model with lumped demands at the ends of the line) would have also been different.

The problem is, thus, to know the best allocation of the total in-line demand to the line ends; that is to say, to know the value F_{Qd} that solves the following problem:

$$\text{Solve } H_C\left(F_{Qd}; L\right) = H_R\left(F_Q,^*; L\right) \text{for } F_{Qd}. \tag{4}$$

This statement is constrained by the nature of the problem: we have to adhere to the fact that one or more lines (pipes) may be connected to the downstream end of the considered line. The connected lines need the right piezometric head at L - upstream end for them - to suitably perform their respective calculations. It means that the piezometric head at L, given by H_C and H_R must coincide. By solving (4) the following expressions for F_{Qd} are obtained:

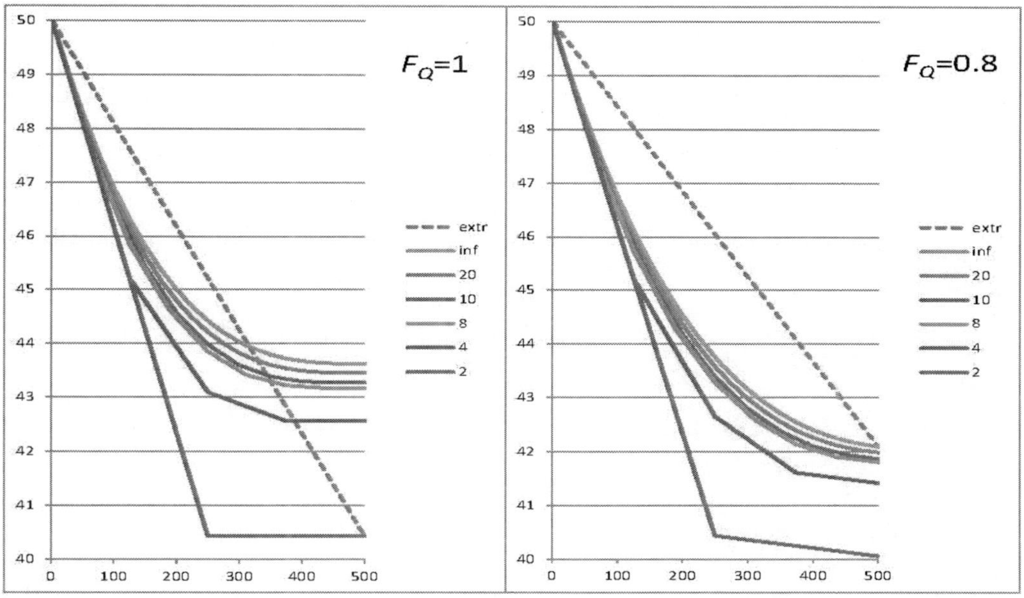

Figure 2. Examples of discrepancy between distributed and lumped demand models.

- Case of continuous demand:

$$F_{Qd}^{\infty}(F_Q) = \frac{1}{F_Q}\left(1 - \sqrt{\frac{1-(1-F_Q)^3}{3F_Q}}\right) \tag{5}$$

- Case of demand distributed among n equally distributed nodes:

$$F_{Qd}^{(n)}(F_Q) = \frac{1}{F_Q}\left(1 - \sqrt{\frac{1}{n+1}\sum_{k=1}^{n+1}\left(1-\frac{k-1}{n}F_Q\right)^2}\right). \tag{6}$$

In Table 1, values for (5) and (6) for F_Q values between 0.2 and 1, and for n varying along the previously used values, namely 1, 3, 7, 11, 19 and infinity, are presented.

The following facts are remarkable:

- These values are independent of the problem data, namely, H_0, Q_{in}, L, D and f, and depend only on F_Q – and n, in the case of (6) –, that is to say, they depend only on the magnitude and the pattern of the distributed demand. This is a very remarkable result, since the present study thus becomes non-dimensional and, as a result, completely general.

- The values range from approximately 0.3 to 0.5. This means that about 30%-50% of the total demand must be allocated to the upstream node, and the remainder to the downstream node.

Table 1. Values of F_{Qd} as a function of F_Q and n

n (order) \ F_Q	0.2	0.4	0.6	0.8	1.0
1	0.472	0.438	0.397	0.349	0.293
3	0.485	0.466	0.442	0.413	0.376
7	0.488	0.473	0.455	0.432	0.402
11	0.489	0.475	0.457	0.435	0.407
19	0.490	0.477	0.461	0.441	0.415
inf	0.491	0.479	0.465	0.446	0.423

- The lowest values correspond to the most awkward cases: the rate of demand is close to or equals the total inflow in the line, and the demand is highly concentrated at a few points (upper right corner of the table).

- The highest values, closer to 50%, correspond to the less problematic cases, meaning little total distributed demand in relation to the total inflow in the line, and widely distributed demand (lower left corner of the table). This value approaches 50% as the rate of inflow consumed in the line approaches zero. (Observe that for both (5) and (6) $\lim_{F_Q \to 0} F_{Qd}^{(\cdot)}(F_Q) = 0.5$).

As said, it is common practice in mathematical modeling of WDN engineering to distribute the line flow into two parts: 50% for the upstream node and 50% for the downstream node - which approximately coincides with what is observed in Table 1, except for cases where the demand line is highly concentrated and represents a large percentage of the total flow through the line.

As a result of what has been presented so far, it can be said that, for uniformly distributed in-line demand, the usual 50% rule seems, in principle, an appropriate solution provided that both the inner demand of the line is small compared with the pipe inflow and that such a demand is widely distributed. However, equal demand allocation to the end nodes of the line may produce important discrepancies, since the study highlights the need for other allocations in certain cases.

2.3. Arbitrary Demand along the Line

To state the problem in its more general form, let us now consider a demand distribution on the line whose accumulated demand is given by a function $Q(x) = Q_d q(x)$, where $q(x)$ is the accumulated demand ratio, a function increasing monotonically from 0 to 1. While H_C is calculated as in (3), H_R is calculated by integrating the loss $\Delta H = -K(Q_{in} - Q(x))^2 \Delta x$ through the line [0, L]:

$$H_R(F_Q; q(x); x) = H_0 - \int_0^x K(Q_{in} - Q(u))^2 \, du. \tag{7}$$

Observe that this function is monotonically decreasing and concave upwards.

Example 1. $q(x) = x/L$, for the case of continuous uniform demand, which gives (2). ▲

Using (7) and the expression (3) for H_C in L, the equation $H_C = H_R$ in L gives

$$F_{Qd}(F_Q) = \frac{1}{F_Q}\left(1 - \sqrt{\frac{1}{L}\int_0^L (1 - F_Q q(x))^2 \, dx}\right).$$

The general solution for the problem in hand when an arbitrary demand through the line is considered may only be solved after having a specific expression for $q(x)$. As a consequence, to pinpoint the new aspects that may arise when considering arbitrary demands, we will restrict ourselves to considering the case of discrete demands on a finite number of points of the pipe, as in real life happens.

Let us consider $q(x) = \sum_{k=1}^{n} d_k \delta(x - x_k)$, where d_k are the demands at points x_k, with $0 < x_1 < x_2 < \ldots < x_{n-1} < x_n < L$, such that $\sum_{k=1}^{n} d_k = Q_d$. $\delta(\cdot)$ is the Dirac delta.

Example 2. $q(x) = \frac{1}{n}\sum_{k=1}^{n} \delta\left(x - \frac{k}{n+1}L\right)$ in the case of uniform demand at n equally distributed points in the pipe. This demand produces the expression in (1). ▲

In the general case, H_R is calculated by

$$H_R(F_Q, L) = H_0 - K x_1 Q_{in}^2 - K(x_2 - x_1)(Q_{in} - d_1)^2 - K(x_3 - x_2)(Q_{in} - (d_1 + d_2))^2 - \cdots \tag{8}$$

$$-k(x_n - x_{n-1})\left(Q_{in} - \sum_{k=1}^{n-1} d_k\right)^2 - K(L - x_n)\left(Q_{in} - \sum_{k=1}^{n} d_k\right)^2$$

By denoting

$$\mu_i = \frac{d_i}{Q_d}, \mu_0 = 0, \lambda_i = \frac{x_i}{L}, \lambda_0 = 0, \lambda_{n+1} = 1, \text{ for } i = 1, \ldots, n, \tag{9}$$

this expression can be written

$$H_R(F_Q, L) = H_0 - KLQ_{in}^2 \sum_{k=0}^{n} (\lambda_{k+1} - \lambda_k)\left(1 - F_Q \sum_{j=0}^{k} \mu_j\right)^2.$$

Then, by equating again $H_C = H_R$ at L gives

$$F_{Qd}(F_Q, \lambda, \mu) = \frac{1}{F_Q}\left(1 - \sqrt{\sum_{k=0}^{n} (\lambda_{k+1} - \lambda_k)\left(1 - F_Q \sum_{j=0}^{k} \mu_j\right)^2}\right). \tag{10}$$

This expression can be easily calculated using, for example a worksheet.

3. Illustrative Example on a Real-world Network

This section is aimed to present some significant results obtained when applying the lessons learned in the previous section to a practical situation. The assessment is carried out in a real water supply network from Tegucigalpa City (the capital of Honduras). The network, which is intended to be a district metered area (DMA), is supplied by one main pipe connected to one of the water tanks administered by the local water authority.

The network has 161 nodes and 154 pipes. The total length of the network is 14,669 meters. Fig. 3 shows the network model in EPANET [1]. To illustrate the aspects previously studied, two intermediate nodes are placed in the mentioned main pipe, which is 500m long. Both points have a high level of demand that could be representative of a major consumer, such as a factory. The first point is located in the middle of the line (250m from the upstream node); its consumption amounts to 62.5l/s representing 25% of the total line demand of 250l/s. The second intermediate node is located 150 meters downstream (at a point located at 80% of the total length). At this point 75% of the total consumption of the line is extracted (187.5 l/s). The objective is to evaluate how the re-allocation of this intermediate consumption using one criterion or another affects the operation of the global model - under one specific load condition.

Three different calculations have been performed:

Case A. The performance of the network is evaluated with the actual demands placed at the intermediate nodes. This scenario represents the actual case and, as a consequence is the target for the other two calculations, which use different lumped allocations to the line ends.

Case B. The performance was evaluated by allocating 22.45% of the consumption of the line to the upstream node and 77.55% to the downstream node, obtained using formula (10).

Case C considers the general practice of distributing the line demand using the 50% rule.

To perform the evaluation of the network, eight nodes, uniformly distributed along the WSN, were selected (light circles in Fig. 3. The downstream node of the initial main line belongs to this set, Node A.

Table 2 shows clearly the differences that may represent the use of a demand distribution rule properly devised over the use of a 50-50% demand distribution rule. While the target is perfectly met in Case B, in Case C, the difference in pressure across the network can approach 1 bar, which would represent an uncalibrated network. Clearly, this example shows how critical the use of an unadjusted demand distribution rule may be when a major consumer is located in a main section of a WDN, since the resulting head at its downstream node will strongly affect the rest of the network.

Conclusion

This chapter focuses on the study of the influence that the concentration of a distributed demand in a line on the line ends represents in modeling steady state conditions in WDNs. By using a practical approach, we show the importance of the relation between the total inflow and the flow that is extracted due to the load of demands in the line. The distribution

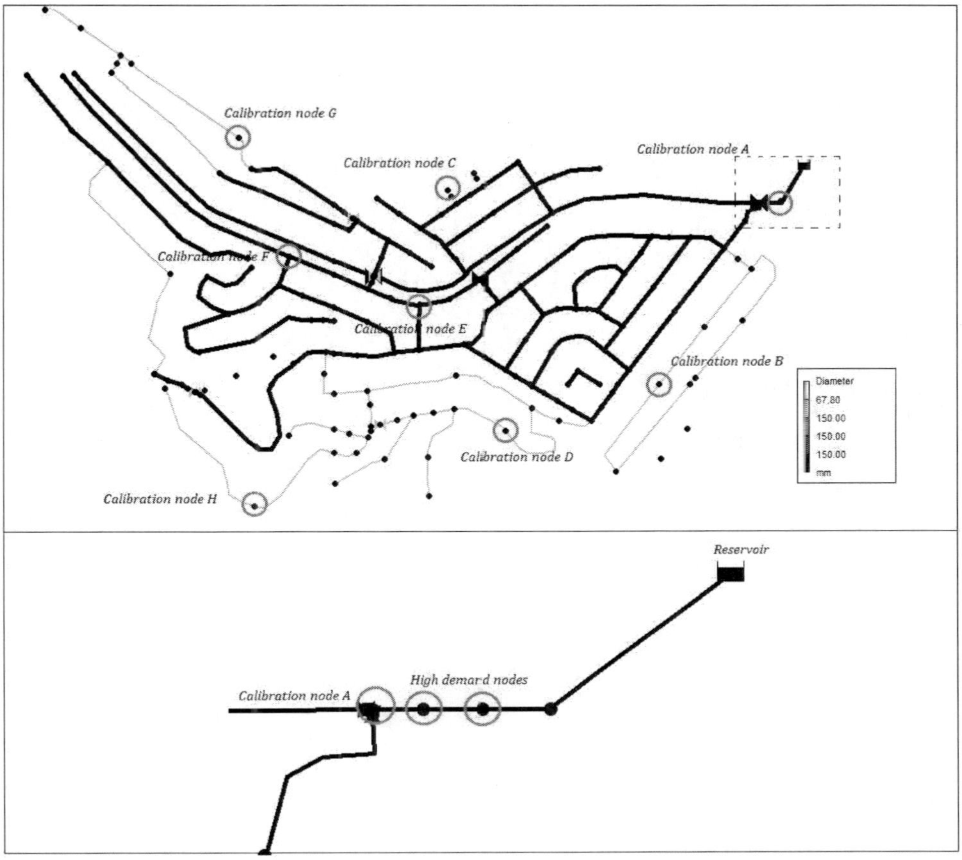

Figure 3. Water supply system used for the practical example.

of the consumptions along the line also greatly influences the validity of the model. After assessing the error that may derive from using the generalized 50% rule, we obtain the general formula (10) that enables to distribute arbitrary inner line demand between the line ends with zero error at the downstream end of the line.

We have also shown that our proposal is straightforward and involves direct methods that can be easily applied by any user of any WDN analysis package, for example by a using a very simple worksheet. Network models already developed may be revised to replace, if necessary, the used allocation values with he values provided by (10), which produce more reliable results. In the case that a new network model is started, users may make an a priori decision on implementing the right simplifications so that a suitable tradeoff between model simplification and accuracy is obtained.

The study performed applies only to branched networks, in which the flow direction is predefined. The results do not strictly apply to looped networks in which the definition of upstream and downstream nodes cannot be known a priori and even may vary depending on the hourly demand in the network and on the demand pattern used. Nevertheless, a little of iteration will enable to get the final design. If the expert is not happy with the obtained results for one of the lines, he or she has the opportunity to include in the model one additional point of the line. This will divide the line into two new lines to which the

Table 2. Comparison among scenarios

NODES ID	CASE A Pressure (m) Demand in the intermediates (real) nodes	CASE B Pressure (m) Using the calculated demand distribution rule	CASE C Pressure (m) Using a 50%-50% demand distribution rule	Differences (CASE A vs CASE B)	Differences (CASE B vs CASE C)
NODE A	56.01	56.01	69.25	0.00	13.24
NODE B	44.05	44.05	57.28	0.00	13.23
NODE C	11.13	11.13	24.37	0.00	13.24
NODE D	10.02	10.02	23.26	0.00	13.24
NODE E	23.96	23.96	37.20	0.00	13.24
NODE F	26.92	26.92	40.15	0.00	13.23
NODE G	15.04	15.04	28.29	0.00	13.25
NODE H	23.68	23.68	36.90	0.00	13.22

same criteria may be applied. This solution of including an interior point of one line into the model will be also useful in the case that a line is fed by both ends.

Some further development would be of interest. For example, calculation of the maximum head point discrepancy associated with the use of the allocation rule would be of interest to help in the abovementioned iterative process, to provide the user with suitable candidate interior points when they are deemed necessary. One can also consider the inclusion of variable values for the factor F_{Qd} in the demand curves assigned to the ends of the challenging lines, so that variations of the flow direction in certain lines during the day may be considered when running extended period simulations.

References

[1] EPA (Environmental Protection Agency) 2012. *EPANET*, http://www.epa.gov/nrmrl/wswrd/dw/epanet.html (Dec. 17, 2012).

[2] Izquierdo, J.; Pérez, R.; Iglesias, P.L. Mathematical Models and Methods in the Water Industry, *Math Comp Mod.* 2004, 39, 1353-1374.

[3] Izquierdo, J.; Montalvo, I.; Pérez-García, R.; Matías, A. On the Complexities of the Design of Water Distribution Networks, *Math Probl Eng.* 2012, Article ID 947961.

[4] Savić, D.A.; Banyard, J.K. *Water Distribution Systems*, ICE Publishing, London, UK, 2011.

In: Mathematical Modeling in Social Sciences ...
Editors: J. C. Cortés López et al.
ISBN: 978-1-63117-335-6

Chapter 5

DYNAMIC PREDICTION OF FAILURES: A COMPARISON OF METHODOLOGIES FOR A WIND TURBINE

S. Carlos[1,*]***, A. Sanchez***[2]***, I. Marton***[1] ***and S. Martorell***[1]
[1]Departament d'Enginyeria Química i Nuclear,
Universitat Politècnica de València, Valencia, Spain
[2]Departament d'Estadística i Investigació Operativa Aplicades i Qualitat,
Universitat Politècnica de València, Valencia, Spain

Abstract

Nowadays, there is an increasing demand for Condition Based Maintenance (CBM) activities as time-directed maintenance is observed to be inefficient in many situations. CBM is a maintenance strategy based on collecting information concerning the working condition of equipment, such as vibration intensity, temperature, pressure, etc., related to the system degradation or status in order to prevent its failure and to determine the optimal maintenance. Prognosis is an important part of CBM. Different methodologies can be used to perform prognosis and can be classified as: model-based or data-driven. Data-driven methods make use of the available monitoring data to train a learning algorithm. Two kinds of data driven methods are Principal component analysis-Partial least Squares and Neural Networks. Mathematical models based on these two methodologies are constructed, and from data monitored for a wind turbine the failure of this kind of systems can be detected and predicted. A discussion of the strength and weakness of each one of the methodologies is presented.

Keywords: Failure prediction, wind turbine, dynamic neural networks, partial least squares

1. Introduction

Energy production systems are becoming increasingly complex and require new methods to diagnose and anticipate unexpected failures in order to avoid revenue losses. Traditionally, time directed maintenance activities have been performed to avoid equipment degradation,

*E-mail address: scarlos@iqn.upv.es

but in many situations they are inefficient, for example, they are performed even if the equipment does not present degradation. To avoid such situation, Condition Based Maintenance (CBM) activities are becoming more demanded to plan the maintenance activities[1]. CBM maintenance strategy is based on collecting information concerning the working condition of equipment, such as vibration intensity, temperature, pressure, etc., related to the system degradation or status in order to prevent its failure and to determine the optimal maintenance. So, in order to plan an efficient CBM strategy it is necessary to perform a prognosis on the degradation level. The prognosis methods can be classified as: model-based and data-driven methods. Model-based methods use the process physical model, or a statistical estimation, based on state observers, such as Kalman filters, particle filters, etc. [2] data-driven methods make use of available monitored data to train learning algorithms to forecast the equipment failures. In this work, two data-driven approaches to detect the failure of a Spanish commercial wind turbine farm are compared. The first one combines principal component analysis (PCA) and partial least squares (PLS), and the second method is based on dynamic neural networks.

2. Principal Component-Partial Least Squares

Principal component analysis (PCA) is one of the most popular multivariate statistical techniques, used in fault detection. It is based on a linear transformation that produces new uncorrelated variables, named components, from the original correlated measured variables. This transformation implies a dimensionality reduction of the original data so, a few of these components are sufficient to adequately represent the hidden sources of variability in the process [2]. PCA is based on an orthogonal decomposition of the covariance matrix of the process variables along directions that explain the maximum variation of the data matrix, X, that contains the measured data. The projection of X down on a subspace by means of the projection matrix, P', gives the object coordinates in the plane, T. The columns in T are called score vectors and the rows in P' are called loading vectors. The deviations between projections and the original values coordinates are the residuals, E. In matrix form, PCA is expressed as:

$$X = TP' + E. \tag{1}$$

PCA can be extended when it is desirable to include all data available in the monitoring procedure, X, to predict and to detect changes in the output variables Y. This extension is known as Partial Least Squared (PLS) [3]. The PLS model is used to obtain the relationship between the matrix data, X, and the variables, Y, which is also decomposed as

$$Y = UC' + F, \tag{2}$$

where U is the scores matrix of Y, C is the loading matrix of Y, and F is the residual matrix.

PLS model consists of the regression between the scores for the X and Y. The scores matrix of Y, U, has an internal relationship with the score matrix of X, T, which can be represented as:

$$U = TB + R, \tag{3}$$

where B is the matrix of regression coefficients and R is the residual matrix. The PLS technique computes the regression coefficient in such a way that covariance between T and U is maximized. Here the PLS model is computed using the Non-linear Iterative Partial Least Squares (NIPALS) algorithm [3].

3. Dynamic Neural Networks

A linear model in a stochastic environment, that is in presence of noise, can be expressed as [4]:

$$y(t) = G\left(q^{-1}, \theta\right) u(t) + H\left(q^{-1}, \theta\right) e(t),$$

where, θ is a vector of p adjustable parameters, $u(t)$ are the input values, $e(t)$ is a white noise process and G and H are matrix polynomials in the delay operator q^{-1}.

A common linear dynamic model is the ARX model given by:

$$A\left(q^{-1}, \theta\right) y(t) = q^{-d} B\left(q^{-1}, \theta\right) u(t) + e(t),$$

where

$$\begin{aligned} A\left(q^{-1}, \theta\right) &= 1 + a_1 q^{-1} + \cdots + a_n q^{-n}, \\ B\left(q^{-1}, \theta\right) &= b_0 + b_1 q^{-1} + \cdots + b_m q^{-m}. \end{aligned}$$

So the output of the system can be expressed as

$$\begin{aligned} y(t) &= q^{-d}\left(b_0 u(t) + b_1 u(t-1) + \cdots + b_m u(t-m)\right) \\ &- a_1 y(t-1) - a_2 y(t-2) - \cdots - a_n y(t-n) + e(t), \end{aligned}$$

and a predictor of this model is given by

$$\begin{aligned} \hat{y}(t) &= q^{-d}\left(b_0 u(t) + b_1 u(t-1) + \cdots + b_m u(t-m)\right) \\ &- a_1 \hat{y}(t-1) - a_2 \hat{y}(t-2) - \cdots - a_n \hat{y}(t-n). \end{aligned}$$

The coefficients of the ARX model usually are estimated using a least square error criterion from time series data. When the time series is noisy and the underlying dynamical system is nonlinear, models based on the neural networks frequently outperform standard linear techniques. Neural networks constitute a large research field that has been successfully applied to many engineering problems. Most of the existing neural networks can be covered by the definition: *A system of simple processing elements, called neurons, that are connected into a network by a set of weights.* There exist different kind of neural networks, which are determined by the network architecture, the magnitude of the weights and the processing elements mode of operation [4][5]. The neuron is the simplest processing unit that takes a number of input variables, weights them and sum them up and uses the result as an argument for a singular valued function, called the activation function. Essentially, the activation function, f_i, can take any form but often it is monotonic and the most usual activation functions are the linear function, the hyperbolic tangent, the sigmoid function or the step function. The neurons can be combined in different fashion, but the most common network is the multilayer perceptron (MLP). The MPL is constructed by ordering the

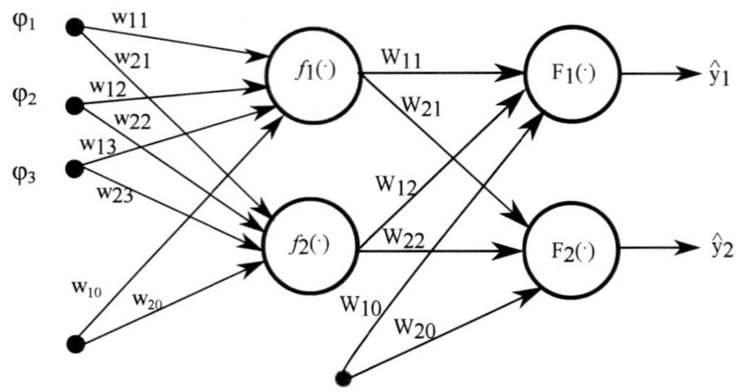

Figure 1. Multilayer Percentron Neural Network.

neurons in different layers. The inputs of a neuron in a certain layer are the outputs of the neurons in the previous layers, or external inputs. Figure 1, shows a two layer MPL with three inputs and two outputs.

For this network the output produced is given by the expression.

$$y_i = g_i\left(\varphi, \theta\right) = F_i\left(\sum_{j=1}^{n_h} W_{i,j} f_i\left(\sum_{l=1}^{n_\varphi} w_{j,l}\varphi_l + w_{j,0}\right) + W_{i,0}\right),$$

where θ is the parameter vector, $\theta = (w_{j,l}, W_{i,j})$ of the weights and bias, and φ is the vector of inputs. The task of determining the parameters from a series of examples that correlate inputs and outputs is called training. In the training process the number of layers as well as the number of neurons per layer are parameters that have a great influence on the neural network performance, and have to be optimized for each particular application.

Non-linear Autoregressive models models with exogenous inputs (NARX) neural architecture can be represented as:

$$y(t) = f\left(u(t), u(t-1), \ldots, u(t-m), y(t-1), y(t-2), \ldots, y(t-m)\right), \tag{4}$$

where $f()$ is an unknown non-linear function that can be approximated, for example, by means of a multilayer perceptron neural network.

4. Case of Application

A wind turbine is an equipment in which all the components are in a series configuration, so the failure in one of them causes the failure of the complete system. One of the most important and most expensive component of a wind turbine is the gearbox, so the objective of this work is focused on detecting the failures in this component making use of the information provided by the SCADA system implemented in a commercial wind farm [6],[7]. A wind farm is a set of wind turbines, generally identical, working independently. The power produced by the wind turbine depends principally on the wind velocity. Thus, for two identical and neighbour wind turbines with the same wind velocity recorded by SCADA system the power produced should be the same but, Figure 2 shows a real SCADA data of two neighbour wind turbines of a Spanish commercial wind farm. In this Figure, we observe that although the wind velocity is almost identical for both wind turbines, the power of one of them goes to zero, what means that it is not producing energy with high wind velocity values, that is, the wind turbine is failed.

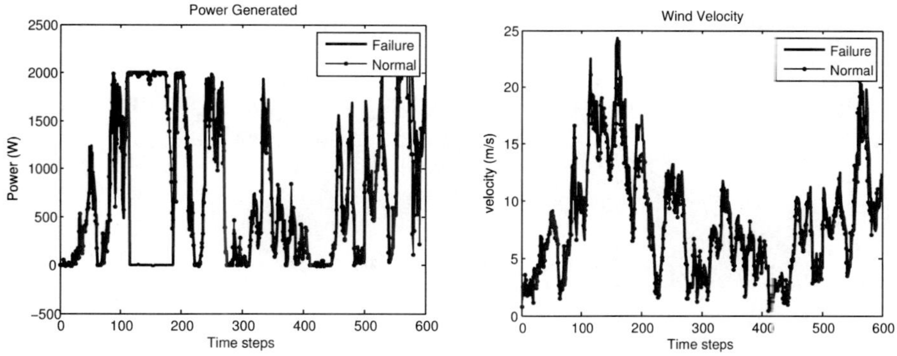

Figure 2. Power generated and wind velocity.

In particular, the failed component that caused the wind turbine failure is the gearbox, and the failures in this component can be diagnosed and predicted by the evolution of the gearbox oil temperature monitored by the SCADA system. So the analysis using the PCA-PLS and Neural networks proposed is focused on reconstructing the gearbox oil temperature to detect the gearbox failure [6], [7].

4.1. PCA-PLS Analysis

In this application the output variable predicted by the PCA-PLS technique is the gearbox oil temperature, T_{oil}. To reconstruct the value of T_{oil} in an time t the input variables used are the wind velocity in t and $(t-1)$ and the values of the gearbox oil temperature at times $(t-1)$ and $(t-2)$. Using this data the PCA obtains three principal components which are linear combinations of the input variables and are expressed as:

$$\begin{aligned} T1\,(t) &= b_1 v\,(t) + b_2 v\,(t-1) + b_3 Toil\,(t-1) + b_4 Toil\,(t-2)\,, \\ T2\,(t) &= c_1 v\,(t) + c_2 v\,(t-1) + c_3 Toil\,(t-1) + c_4 Toil\,(t-2)\,, \\ T3\,(t) &= d_1 v\,(t) + d_2 v\,(t-1) + d_3 Toil\,(t-1) + d_4 Toil\,(t-2)\,, \end{aligned}$$

where b_1, b_2, b_3, b_4, c_1, c_2, c_3, c_4, d_1, d_2, d_3, d_4 are constants. And the predicted value of the gearbox oil temperature in time t is given by a linear combination of the three principal components as:

$$\hat{Toil}\,(t) = K_1 T1\,(t) + K_2 T2\,(t) + K_3 T3\,(t)$$

where K_1, K_2, K_3 are constants determined by the PLS method. The model has been obtained with the data of the non failed wind turbine and has been used to reconstruct the failed behaviour. In order to analyse the model performance the mean square error and the maximum relative error have been calculated and are shown in Table1. In this Table, it can be observed that the errors obtained using the PCA-PLS technique for the normal behaviour reconstruction are quite low so a good prediction is obtained as it is shown in Figure 3. For the failed wind turbine the errors calculated with PCA-PLS method are larger than the ones obtained for normal behaviour. For failure detection large errors between real and predicted values of the time series are needed to set an alarm. Thus, in Figure 3 a difference between predicted and real values for the failed wind turbine are evidenced, but only at the final time steps what makes difficult to establish an alarm based on these data.

Table 1. Error prediction

	Normal behavior		Failed behavior	
	Mean square error	Maximum relative error	Mean square error	Maximum relative error
PCA-PLS	0.009102	0.033717	0.075052	0.154856
NARX	0.019991	0.086893	0.328079	0.736849

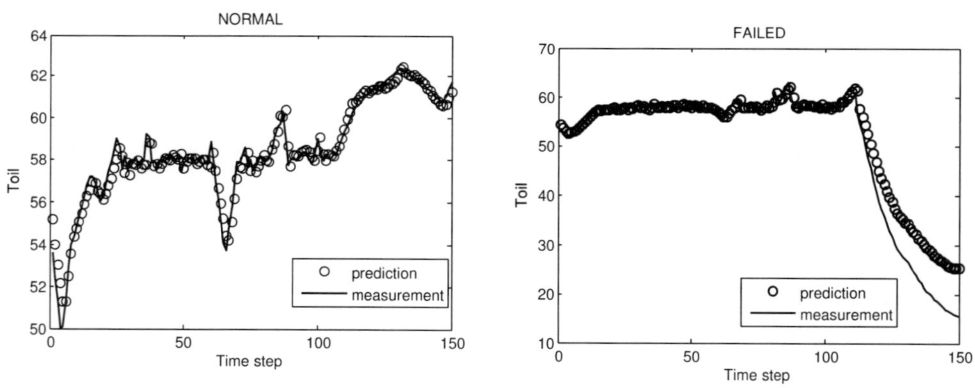

Figure 3. PLS reconstruction.

4.2. Neural Network Analysis

A NARX model has been also constructed to reconstruct the gearbox oil temperature. The model is based on an MLP of three layers and uses as input data the wind velocity measurements at time t and $(t-1)$ and the gearbox oil temperature at $(t-1)$ and $(t-2)$. Thus, the equation that follows the NARX model to obtain the output value is given by

$$\hat{Toil}(t) = f(v(t), v(t-1), Toil(t-1), Toil(t-2)),$$

where the $f()$, is approximated by the multilayer perceptron.

The NARX model predictions for normal and failed wind turbine behaviour have been performed and the errors obtained are exposed also in Table 1. In this Table, it is observed that both error measurements calculated, the mean square error and the maximum relative error, are quite low for normal behaviour reconstruction, what means the model prediction is quite good. This is evidenced in Figure 4 in which it can be observed that the measurements and the NARX predicted values present a good agreement. The errors obtained with the NARX technique for the failed wind turbine are larger that the ones obtained for normal behaviour, what, as said before, is a necessary feature to detect the failure and set the alarm. In Figure 4 it can be observed that, just after the failure is produced the error between the measured signal and the prediction becomes very large.

When comparing the error values calculated by PCA-PLS and NARX models presented in Table1 it can be observed that the NARX model provides larger errors, specially for the failed wind turbine predictions. In fact, comparing the failed behaviour reconstruction provides by PCA-PLS and NARX models in Figure 3 and Figure 4, it is observed that the failure is earlier detected by the NARX model what is an advantage from the practical use as the repairing task can be launched earlier and the time until the wind turbine is available to produce electricity is reduced.

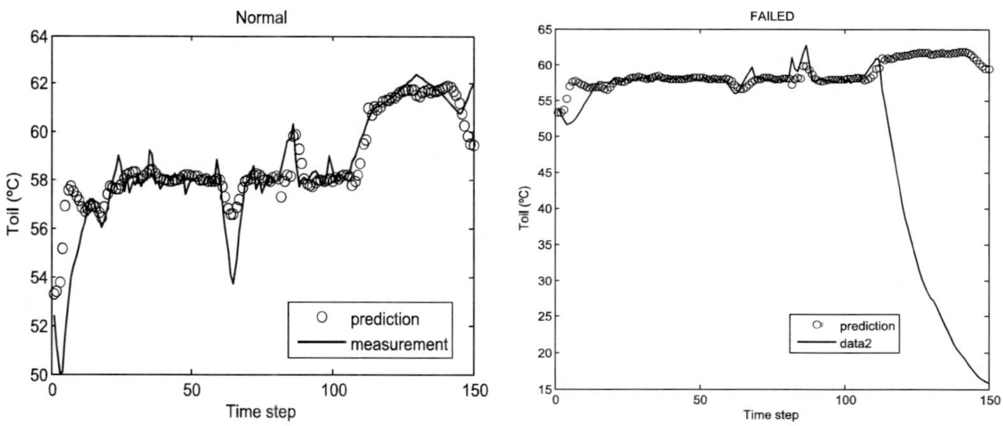

Figure 4. NARX model reconstruction.

Conclusion

SCADA data are available in wind farms and they can be used to perform condition based maintenance what will improve the maintenance planning reducing costs. In this work, two mathematical techniques used in fault detection have been used to analyse the normal and failed behaviour of real wind turbines. Both techniques are capable of detecting the failure. PCA-PLS detects the failure but, if its capability of reducing the number of variables is used, the physical meaning of the dominant principal components may be lost. The NARX model constructed provides good results to set the alarms as the failure is detected just when it is produced while PCA-PLS technique presents a delay in detecting the failure. The models presented in this work perform a one-step prediction and, for practical application, there is not enough time to perform the maintenance task and restore the wind turbine to normal operation. So, for real plant maintenance planning the models presented here should be extended to perform k-steps predictions, but for constructing the k-steps model a good wind velocity model prediction is needed.

References

[1] Dragomir O. E.; Gouriveau R.; Dragomir F.; Minca E.; Zerhouni N. Review of Prognostic Problem in Condition Based Maintenance, *European Control Conference*. Budapest, Hungary, 2009.

[2] Wold S.; Esbensen K.; Geladi P. *Chemometrics Intell. Lab. Syst.* 1987, 2, 37-52.

[3] Geladi P.; Kowalsky R.B.; *Anal. Chim. Acta*. 1986, 185, 1-17.

[4] Ljung L.; T. Glad. *Modelling of Dynamic Systems*; Information and Systems Science series; Prentice Hall: Englewwod Cliffs, NJ, 1994.

[5] Norgaard M.; Ravn O.; Poulsen N.K.; Handsen L.K. *Neural Networks for modelling and control of dynamic systems: A practitioner's handbook*; Springer: London, UK, 2000.

[6] Garcia M.C.; Sanz-Bobi M.A.; Del Pico J. *Comput. Ind.* 2006, 57, 552-568.

[7] Yang W.; Court R.; Jiang J.; *Renew. Energy* 2013, 53, 365-376.

In: Mathematical Modeling in Social Sciences ...
Editors: J. C. Cortés López et al.
ISBN: 978-1-63117-335-6

Chapter 6

ADVANCES IN MATHEMATICAL MODELING OF SUPERCRITICAL EXTRACTION PROCESSES

Florian Meyer[1,*], Marko Stamenic[2,†], Irena Zizovic[2,‡] and Rudolf Eggers[1,§]

[1]Institute of Thermal Separation Processes, Heat and Mass Transfer, Hamburg University of Technology, Hamburg, Germany
[2]Faculty of Technology and Metallurgy, University of Belgrade, Belgrade, Serbia

Abstract

A new trend in mathematical modeling of supercritical extraction process from plant material is presented.The most of the previously published models considered properties of the fixed bed of plant material during the extraction to be constant. This assumption might be quite true in the case of plant materials with relatively low quantity of extractables. However, the fixed bed property change may occur in the case of extraction from material with high quantities of extractables. Recently, mathematical model which took into account particle density, bed porosity and particle diameter as variables was derived. In this chapter, the model is further improved by introducing variable solubility of the extract in supercritical fluid. On the basis of new experimental results on the binary equilibrium of the system supercritical fluid – extract present in solid, a relationship between solubility in supercritical fluid and oil content in solid was established. The new model was derived which took into account particle density, bed porosity, particle diameter and solubility in supercritical fluid as variables. The model was verified on results from experiments in which the kintecis of supercritical extraction from rapseed was analyzed. Parameters of the model showed expected behaviour with respect to the change of particle size. Further analysis, shown in this chapter, showed that, if the dependency of solubility from the concentration in the solid phase is not accounted for, the model tends to overestimate the yield of the extraction.The new model is so far, the most realistic model and one of the most demanding models regarding experimental work and mathematical tools for the description of mass transfer in fixed beds.

* E-mail address: f.meyer@tuhh.de.
† E-mail address: stamena@tmf.bg.ac.rs.
‡ E-mail address: zizovic@tmf.bg.ac.rs.
§ E-mail address: r.eggers@tuhh.de.

Keywords: Supercritical extraction; Mathematical modeling; Solubility; Fixed bed

List of Symbols Nomenclature

Symbol	Explanation	Unit
a	specific surface area of the solids	[1/m]
b	parameter in eq. 6	
d_p	mean particle diameter	[μm], [m]
ε	bed porosity	
G	fraction of cells opened by pre-treatment	
J(x,y)	mass transfer term	[kg/m³ s]
k_s	effective mass transfer coefficient during the slow extraction period	[m/s]
k_f	effective mass transfer coefficient during the fast extraction period	[m/s]
$m_{solids,0}$	initial mass of solids	[kg]
m_{CO2}	mass of CO2	[kg]
p	Parameter in eq. 6	
r	Parameter in eq. 6	
ρ_s	particle density	[kg/m³]
ρ	density of the fluid phase	[kg/m³]
t	time	[s], [min]
x	concentration of the solute in the solid phase	[kg/kg]
x_0	initial concentration of extract in the solid phase	[kg/kg]
x_{res}	residual concentration of extract in the solid phase	[kg/kg]
y	concentration of the solute in the fluid phase	[g/kg]
y_{exp}	CO_2 loading with extract (experimentally determined)	[g / kg]
y^*	binary solubility of oil in CO_2	[g/kg]
y_r	equilibrium solubility of oil in CO_2 in presence of solids	[g/kg]
z	axial coordinate of the extractor	[m]

Introduction

Nowadays extractions with supercritical fluids (SFE) play an essential role in isolation of natural substances for application in food and pharmaceutical industry. Main advantages of these processes are in obtaining high quality extracts with no traces of organic solvents at low processing temperatures. At the same time, these processes have no hazardous byproducts.

Along with exploring the possibilities which SFE from natural matter provides, mathematical models for their simulation were derived. They can be classified as empirical, analogue to heat transfer, based on a solution on differential mass balances, etc. The most numerous are models based on the solution of differential mass balances for fluid and solid phase for the fixed bed of extractor vessel. While the description of extraction mechanisms varied broadly, all of these models took into account the same simplification – properties of the fixed bed of plant material during the extraction were considered to be constant. This assumption might be quite true in the case of plant materials with relatively low quantity of extractables. However, the fixed bed property change may occur in the case of extraction from material with high quantities of extractables, like in the case of oilseeds where high yields are expected. The first model describing mass transfer in fixed beds with properties that change along the extraction time e.g. by swelling of particles (bed porosity, particle density, mean particle diameter) was recently introduced in the literature by our research groups [1]. Thorough experimental work was performed in order to determine the influence of extraction on fixed bed properties. On the basis of experimental data linear relationships between the particle density/bed porosity/mean particle diameter and oil content in solid were established. Mathematical model which took into account particle density, bed porosity and particle diameter as variables was derived. The proposed model described experimental data with high accuracy.

In this chapter, further improvement of the previously published model [1] will be presented. On the basis of new experimental results on the equilibrium of the system supercritical fluid – extract present in solid, a relationship between solubility in supercritical fluid and oil content in solid was established. The new model was derived which took into account particle density, bed porosity, particle diameter and solubility in supercritical fluid as variables. The new model is so far, the most realistic model and one of the most demanding models regarding experimental work and mathematical tools for the description of mass transfer in fixed beds.

1. Experiments

1.1. Pre-treatment and $scCO_2$ Extractions

The experimental methods for pre-treating the solids and their subsequent extraction with supercritical CO_2 were described in our recent studies [1, 2]. For the experiments discussed in this chapter, the pre-treatment method of impact-milling was applied with rapeseed as natural material. After milling the seeds fractions of mean particle sizes of 305, 425 and 810 μm were produced by sieving. The extraction of these fractions was carried out using $scCO_2$ at 30 MPa and 50 °C at a mass flow rate between 9 and 9.5 kg/h.

1.2. Experimental Set-up for Solubility Determination

The experimental set-up shown in Figure 1 is based on the extraction plant with closed CO_2 cycle that was described in detail in our former studies [1, 2].

In order to investigate the partition of the extract between the solid and the supercritical CO_2 phase it is necessary to prolong the residence time of CO_2 in the extractor in order to achieve an equilibrium state. For this purpose the extractor was equipped with a recirculation loop as depicted in the figure. The recirculation loop allows pulse-free pumping of the CO_2. The pipes in this part of the plant are equipped with a controlled electrical heating and thermal insulation towards the environment to ensure isothermal conditions in the entire system. With knowledge of the initial oil content of the natural material under investigation it is possible to calculate its residual oil content at steady state by a mass balance. For this purpose, the extract concentration in the compressed fluid has to be determined.

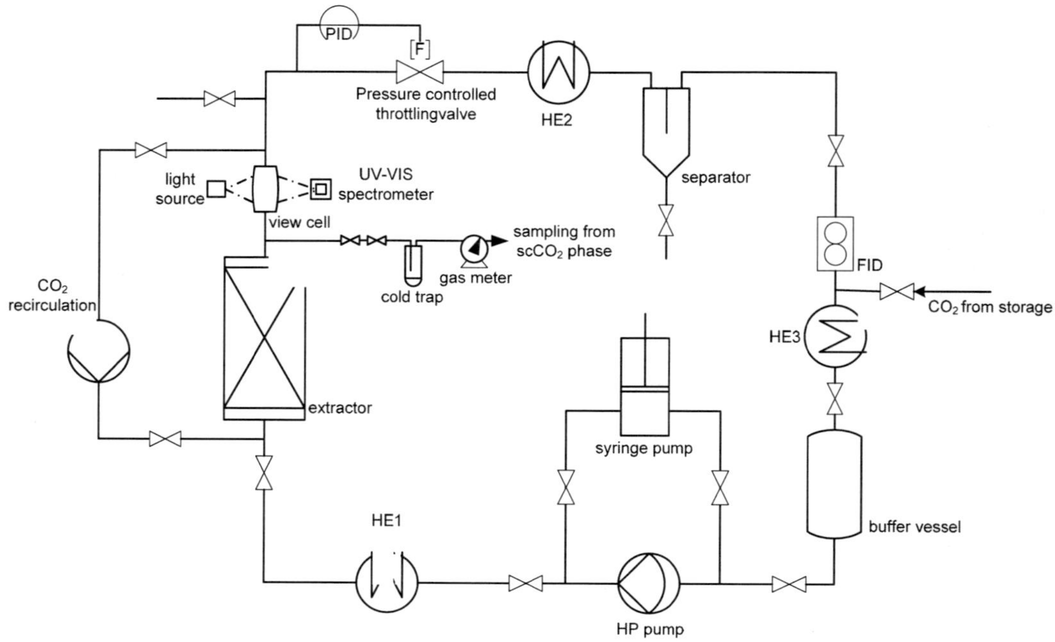

Figure 1. Experimental set-up for the $scCO_2$ extractions and equilibrium measurements.

In this study two sets of experiments were performed. In the first set of experiments, the extract concentration was measured by taking samples from the fluid phase at certain time intervals. Multiple samples have to be taken to ensure that an equilibrium state was reached. This was realized by a system of heated needle valves and a controlled syringe pump that continuously replaced the amount of removed CO_2 and thus kept constant the pressure during sampling. A pre-heater (HE1) avoids temperature decrease inside the system due to injection of cold CO_2 during sampling. The oil is precipitated in a cold trap and determined gravimetrically while the mass of CO_2 is measured with a drum-type gas meter. This method is more accurate when higher extract concentrations are measured. For this reason, conditions of 40 MPa and 60 °C with relatively high extract solubility (approx. 12 g/kg) were chosen for this series of experiments. However, a risk of this method is to influence the state of equilibrium in the system due to the repeated sampling procedure. For this reason, a non-invasive measurement technique was developed that determines the extract concentration in the compressed gas online using a UV-VIS spectrometer (*USB 4000, OceanOptics, Inc.*) in combination with a high pressure view cell (V = 20 mL). In addition to its invasiveness it provides the particular advantage of monitoring the achievement of the state of equilibrium. It is however essential to perform extensive calibrations. For this set of experiments, a pressure of 30 MPa and a temperature of 50 °C were applied.

1.3. Equilibrium Measurements

For the determination of the partition of extract between the solid phase and the supercritical phase a designated amount of pre-treated rapeseed is placed into the extractor and pressurized with CO_2. Then, the extractor is isolated from the rest of the extraction plant with valves upstream and downstream and the recirculation pump is put into operation. The minimum amount of oil that is necessary to achieve the binary equilibrium is determined in reference experiments in the absence of plant matrix using a fixed bed of glass beads as an inert carrier for rapeseed oil. The corresponding minimum amount of rapeseed was calculated from the initial oil content of the material (45 wt. %). At least this 'equilibrium amount' of oil present in the plant matrix of rapeseed was provided to the extractor for the subsequent equilibrium measurements.

The results are shown in figure 2. Although a five-fold excess amount of impact-milled rapeseed is fed to the extractor the binary equilibrium loading of the CO_2 is not achieved. When the equilibrium amount of impact-milled rapeseed is provided to the extractor the discrepancy to the binary equilibrium is even more pronounced. It has been shown in our former study that this kind of pre-treatment results in minor destruction of the plant matrix and a large percentage of oil cells remain intact [2]. For this reason the difference to the binary equilibrium is likely due to non-accessibility of oil in the solids. In addition, a part of the oil is expectably retained by adsorptive solute-matrix interactions.

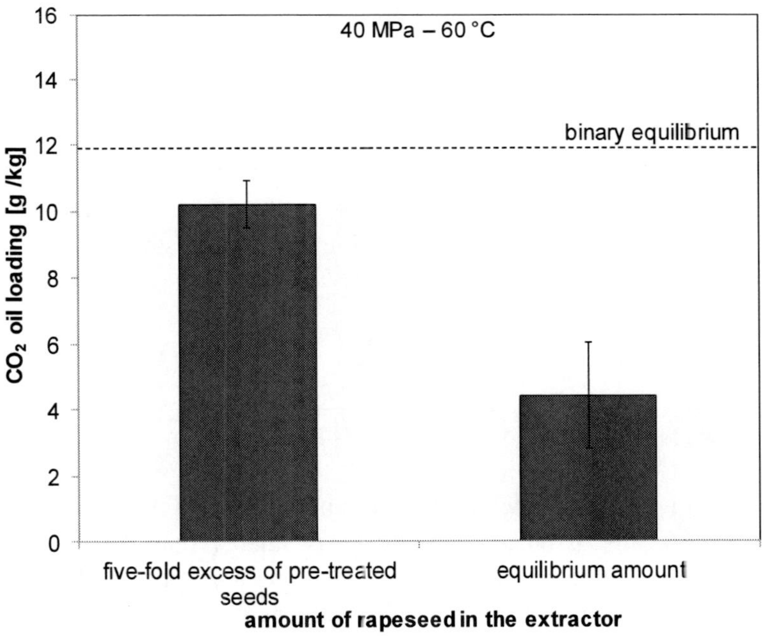

Figure 2. Equilibrium behavior of impact-milled rapeseed at 40 MPa – 60 °C.

Further experiments were carried out using UV-VIS spectroscopy to determine the concentration of oil in the compressed CO_2. These equilibrium measurements were carried out at 30 MPa and 50 °C.

The results can be expressed as oil loading of the CO_2 as a function of residual solid oil content (Figure 3). In order to realize different residual oil amounts in the natural material phase the initial amount of solids fed to the system was varied.

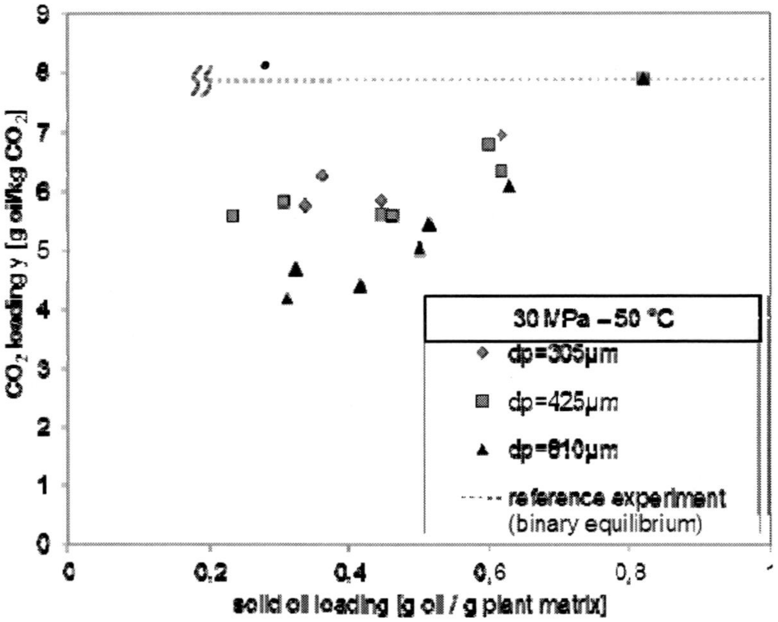

Figure 3. Equilibrium behavior of impact-milled rapeseed in contact with compressed CO_2 at p=30 MPa and T = 50 °C.

Due to unfeasible sampling from the solid phase under pressure, the residual concentration in the solid phase after reaching the steady state is calculated by a global mass balance around the solid and fluid phase (Equation 1). The mass of CO_2 inside the system is known from the volume of the system and the CO_2 density at the respective conditions.

$$m_{solids,0} \cdot x_0 = m_{solids,0} \cdot x_{res} + y_{exp} \cdot m_{CO_2}$$
$$\Leftrightarrow x_{res} = \frac{m_{solids,0} \cdot x_0 - y_{exp} \cdot m_{CO_2}}{m_{solids,0}} \quad (1)$$

The results in Figure 3 indicate that equilibrium depends on the particle size. For smaller particles, the deviation from binary equilibrium is smaller than for larger particles. This indicates that non-accessibility of oil is predominant over adsorption effects on the surface. If adsorption was the predominant mechanism the influence on the equilibrium would be more pronounced with increasing surface area and respectively smaller particle sizes. The results are lacking experimental points for low residual oil contents due to missing possibility of comparing with the binary equilibrium and high experimental error for determining low concentrations in the CO_2. However, exhausting Soxhlet extractions showed different extraction yield dependent on the pre-treatment method [2]. This indicates that for small residual oil loadings of the solid the equilibrium loading of the solvent tends to zero. In contrast, for high residual solid oil loadings no significant effect of the plant matrix on the equilibrium loading of CO_2 is expected. It was found it our recent study that impact-milling leads to a certain cell breakage and thus produces ‘free oil’ on the surface [2]. Furthermore, it is discussed in various studies in literature that the resulting CO2 loading with extract at the beginning of the extraction process often equals the binary solubility.

2. Modeling

The model developed in this chapter is based on the model that was recently published [1]. For this chapter, not only the bed porosity and particle density but also the equilibrium was considered as variable with the course of extraction. The model equations for the solid (Equation 2) and the fluid phase (Equation 3) are as follows.

$$\rho_s(1-\varepsilon)\frac{dx}{dt}-\rho_s x\frac{d\varepsilon}{dt}+x(1-\varepsilon)\frac{d\rho_s}{dt}=-J(x,y) \tag{2}$$

$$\rho\left(y\frac{d\varepsilon}{dt}+\varepsilon\frac{dy}{dt}\right)+\rho u\frac{dy}{dz}=J(x,y) \tag{3}$$

The variable equilibrium concentration is introduced into the mass transfer term that is based on Sovová’s model [3, 4].

$$J(x,y)=k_f a\ \rho(y_r-y) \quad for \quad x>(1-G)x_0 \tag{4}$$

$$J(x,y)=k_s a\rho x\left(1-\frac{y}{y_r}\right) \quad for \quad x<(1-G)x_0 \tag{5}$$

The equations were discretized and solved numerically. The standard deviation between experimental and modeling results was minimized using G, k_f and k_s as adjustable parameters.

2.1. Mathematical Description of the Equilibrium

The equilibrium is approximated with a sigmoid-shaped curve (Figure 4). The corresponding equation (Equation 6) involves two adjustable parameters r and p. The binary solubility is represented by parameter b in g/kg. At low solid oil loadings the corresponding CO_2 loading tends to zero and no extraction is possible because the extract is not accessible to the solvent CO_2. For high residual oil contents, close to the initial oil content of the solid, the corresponding oil loading of the CO_2 is close to the binary solubility. The amount of non-accessible oil decreases with decreasing particle sizes because, in general, a higher degree of exhaustion can be achieved when smaller particle sizes are applied [5, 6].

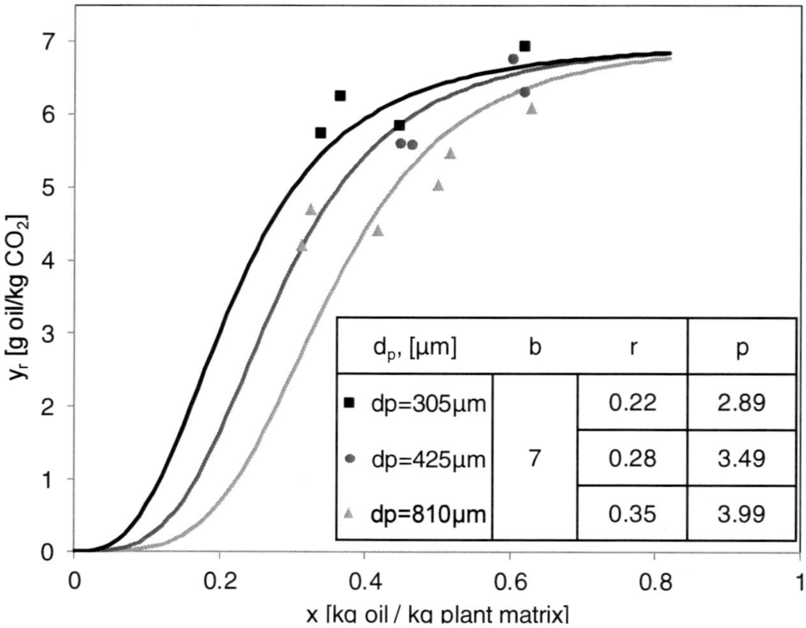

Figure 4. Mathematical description of the equilibrium between rapeseed oil in impact-milled seeds and CO_2 at 30 MPa and 50 °C.

Furthermore, the fraction of free-oil present predominantly on the surface of the particles is larger because the surface-to-volume ratio is larger. The interactions of free-oil and plant matrix are presumably negligible. For this reason, y_r remains on the level of the binary equilibrium down to lower solid oil loadings when the particle size decreases.

$$y_r = \frac{-b}{1+\left(\frac{x}{r}\right)^p} + b \tag{6}$$

2.2. Modeling Results

Experimental extraction kinetics are shown in Figure 5together with modeling results. The experimental data is represented well by the model. The results are physically consistent because the resulting effective mass transfer coefficient $k_s$is lower by three orders of magnitude compared to the solid-fluid mass transfer coefficient $k_f$of the fast extraction period at the beginning of the process.

Further, the resulting parameter G, which corresponds to the fraction of ruptured oil cells during pre-treatment, is decreasing with increasing particle sizes. This can be explained by the fact that for larger particles the surface-to-volume ratio decreases. It can be seen in Figure 6 that both show a similar trend. This indicates that impact-milling ruptures predominantly the oil cells at the surface of the particles and is consistent with the findings from the former study [2].

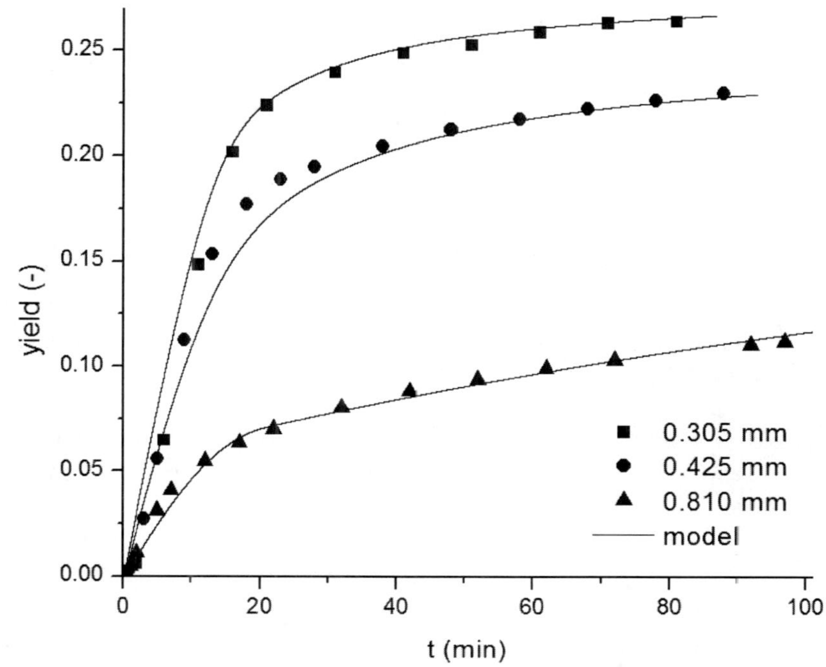

d_p, mm	x_0	G	$k_f \cdot 10^5$, m/s	$k_s \cdot 10^8$, m/s
0.305	0.70	0.75	1.95	7.3
0.425	0.62	0.67	1.90	6.8
0.810	0.25	0.30	1.70	1.2

Figure 5. Modeling of extraction of impact-milled rapeseed with CO_2 at 30 MPa and 50 °C.

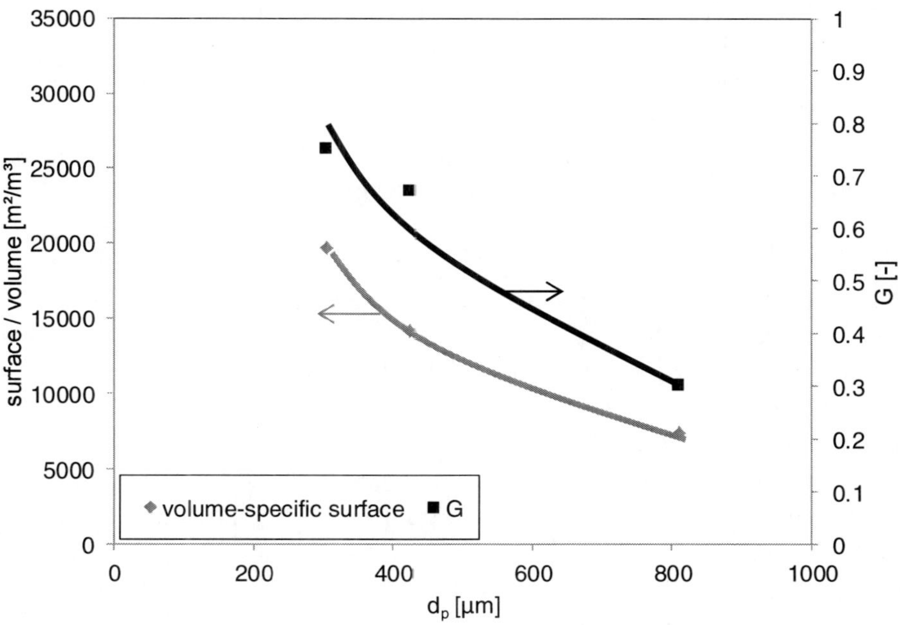

Figure 6. Dependence of specific surface and grinding efficiency from particle size.

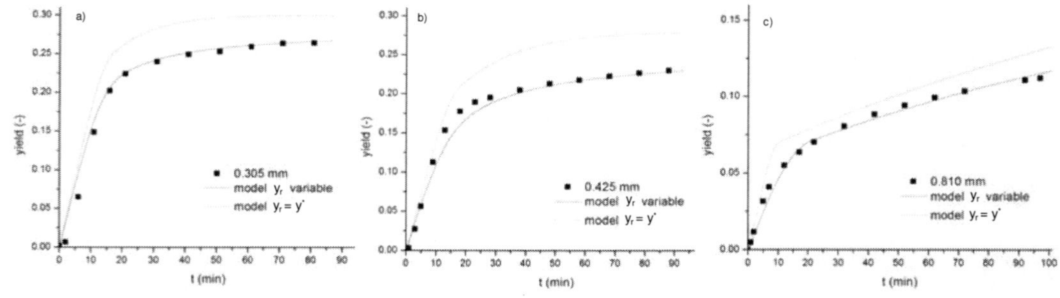

Figure 7. Influence of variable equilibrium on extraction of impact-milled rapeseed with CO_2 at 30 MPa and 50 °C.a) d_p=305µm b) d_p=425 µm c)d_p=810µm.

The influence of a non-constant y_r is shown in Figure 7. The curves of constant y_{sat} were produced by calculating extraction kinetics with the resulting parameters from the complete model with variable y_r.

It is evident that the model with constant y_{sat} overestimates the extraction yield. The effect is most pronounced in the slow extraction period and can be explained by non-accessible oil that is described with the variable equilibrium. It is also possible to model the extraction kinetics with constant $y_r = y^*$. In this case, the adjustable parameters have to be adapted appropriately. More exactly, in order to neutralize the yield overestimation the initial oil content x_0 has to be reduced, thus loses significance.

Conclusion

It was shown with equilibrium measurements performed with two independent experimental procedures, that the plant matrix of the natural material influences the binary equilibrium between extract and solvent. Moreover, a partition of extract between the solid and solvent occurs that depends on the de-oiling degree and particle size. This effect is related to adsorption effects and non-accessibility of oil.

Intensive experimental work is necessary for the complete investigation of the equilibrium between extract present in natural matter and the solvent. The accuracy of the experimental equilibrium data is limited by measuring the extract concentration in the solvent. Further, this quantity is subsequently used to calculate the residual solid oil content due to the impracticality of taking samples from the pressurized solids.

The knowledge of the equilibrium behaviour is important for modeling solid fluid extraction processes since it defines the driving force for mass transfer during extraction.A model was developed that is feasible to describe extraction kinetics of impact-milled rapeseed taking into consideration the equilibrium dependent on the de-oiling degree of the material.

The results indicated that for low de-oiling degrees the plant matrix does not influence the equilibrium and the obtainable oil loading of the CO_2 tends to its binary value. Below certain residual oil content, the corresponding resulting oil loading of the solvent expectably tends to zero due to non-accessible oil in the solids. The decrease from binary solubility down to zero for intermediate solid oil contents can be described with a sigmoid-shaped equilibrium curve or with a basic linear relationship.

The experimental extraction kinetics can be described by the model with good accuracy. The optimized parameters as modeling results gave reasonable values especially regarding the description of efficiency of the pre-treatment method.

References

[1] Meyer, F.; Jaeger, P.; Eggers, R.; Stamenic, M.; Milovanovic, S.; Zizovic, I. *Chem. Eng. Proc.* 2012, 56, 37–45.

[2] Meyer, F.; Stamenic, M.; Zizovic, I.; Eggers, R. *J. Supercrit. Fluids* 2012, 72, 140–149.

[3] Sovová, H. *Chem. Eng. Sci.* 1994, 49, 409–414.

[4] Sovová, H. *J. Supercrit. Fluids* 2005, 33, 35–52.

[5] Snyder, J.M.; Friedrich, J.P.; Christianson, D.D. *J. Am. Oil Chem. Soc.* 1984, 61, 1851–1856.

[6] del Valle, J.M.; Uquiche, E.L. *J. Am. Oil Chem. Soc.* 2002, 79, 1261–1266.

In: Mathematical Modeling in Social Sciences ...
Editors: J. C. Cortés López et al.
ISBN: 978-1-63117-335-6

Chapter 7

PIPE DATABASE ANALYSIS TRANSDUCTION TO ASSESS THE SPATIAL VULNERABILITY TO BIOFILM DEVELOPMENT IN DRINKING WATER DISTRIBUTION SYSTEMS

***E. Ramos-Martínez*[1,*], *J. A. Gutiérrez-Pérez*[1], *M. Herrera*[2], *J. Izquierdo*[1] and *R. Pérez-García*[1]**
[1]Instituto Universitario de Matemática Multidisciplinar
I.M.M. Fluing, Universitat Politècnica de València, Valencia, Spain
[2]BATir Department, Université libre de Bruxelles, Brussels, Belgium

Abstract

Biofilm develops in drinking water distribution systems (DWDSs) as complex communities of microorganisms bound by a matrix of organic polymers and attached to pipe walls. Biofilm can lead to various undesirable problems such as deterioration of bacterial water quality, generation of bad tastes and odors, and biocorrosion, among others. Biofilm formation in DWDSs is dependent on a complex interaction of water quality, infrastructure, and operational factors. However, all these factors have not been studied to the same extent, those associated with the design and operation of DWDS being the most forgotten. Several studies have shown the influence that various physical and hydraulic characteristics of DWDSs have on biofilm. However, due to the complexity of the community and the environment under study, their joint influence, apart from few exceptions, has been scarcely studied. To help bridge this gap an agent-based label propagation technique is applied to a previously obtained database to detect in a given DWDS significant pre-relationships among the characteristics of the pipes related to design and operation, their location, and biofilm development. Besides, the edge betweenness algorithm, which is one of the standard measures of centrality and is able to quantify the importance of a pipe in a network, is implemented. This algorithm enable us to know, according to the design and operation of the DWDS (physical and hydraulic aspects), which are the hot spots of the network, where management efforts must be focused to reduce biofilm formation.

*E-mail address: evarama@upv.es

Keywords: biofilm, drinking water distribution systems, agent-based label propagation, edge betweenness centrality

1. Introduction

Biofilm is a complex community of microorganisms bound by a matrix of organic polymers that develops in drinking water distribution systems (DWDSs) attached to the inner pipes surface. These communities form spontaneously due to the presence of moisture; bind strongly against initial repulsion, and modify the pipe as they capture nutrients and new bacteria. Besides the health risk that biofilm implies due to its role as a pathogen shelter, a number of additional problems associated with biofilm development in DWDSs are universally recognised. These problems include: aesthetic deterioration of water; proliferation of higher organisms; biocorrosion; and disinfectant decay, among others.

Thus, biofilm represents a real paradigm in DWDSs management. Although residual disinfectants are used to avoid the growth of bacteria in these systems, these substances are harmful for human health and sometimes regulated doses are not enough to avoid biofilm formation. So, the only way left to evade DWDSs' water and service quality degradation caused by these communities is to maintain biofilm at the lowest levels. In this context, this chapter focuses on how the physical and hydraulic characteristics of DWDSs affect the development of these communities of microorganisms, seeking the problematic points associated with biofilm development within a distribution system.

Biofilm formation in DWDSs is dependent on a complex interaction of water quality, infrastructure, and operational factors. However, all these factors have not been studied to the same extent, those associated with the design and operation of the DWDS being the most forgotten. Several studies have shown the influence that various physical and hydraulic characteristics of DWDSs have on biofilm. However, due to the complexity of the community and the environment under study, their joint influence, apart from few exceptions, has been scarcely studied.

This chapter presents a proposal to help bridge this gap. Firstly, by compiling data from different studies and using machine learning techniques we have generated a comprehensive and extensive enough database to do inference by posterior analysis [2]. Thus, we are able to study the effect that the interaction of the relevant hydraulic and physical characteristics of DWDSs has on biofilm development. In this chapter we resort to label negotiation. This is accomplished via discriminant analysis and label propagation in a given DWDS, being able to identify groups of pipes, in space, according to its vulnerability to suffer high biofilm development. To perform this label negotiation we propose multi-agent systems (MASs) as an appropiate tool. MASs are composed of multiple interacting computing elements, known as agents. Each agent is a computer system situated in some environment that is capable of autonomous action in this environment to meet its design objectives.

With these techniques, according to the design and operation of the DWDSs (physical and hydraulic aspects), groups of pipes with similar vulnerability to biofilm development would be defined. In this way, these groups of pipes can be turn into management areas where distinct levels of effort must be carried out to reduce biofilm development. Finally, in addition, the edge betweenness algorithm is implemented. This algorithm is one of the standard measures of graph centrality and is able to quantify the importance of a pipe in a network. Thus, we would know which are the most important elements of the network to be specially considered in DWDS management.

The road map of this chapter is as follows. Section 2 presents the materials and methods, discriminant analysis via label propagation, and the *edge betweenness* centrality measure. A case study is presented and the results of the application of this label propagation on a DWDS are discussed in Section 3. Section 4 closes the chapter with conclusions and recommendations.

2. Assesing DWDSs Vulnerability to Biofilm Develoment

To help get this chapter's goal, biofilm data from different sources were compiled and preprocessed using machine learning approaches and preparing a case-study database. This compilation was not at all straightforward. Data provided by different studies and information sources was ambiguous, difficult to compare, and incomplete. Thus, data were preprocessed to generate a complete database. This preprocessing involved a detection stage, where outliers were found and removed using clustering techniques, and a transformation stage, where lost data were reconstructed by suitable imputation using mainly artificial neural networks [2]. Finally, the continuous variables were discretized to normalize the database. This resulted in a database of 210 complete cases. The variables of the database were found relevant to biofilm development in DWDSs when individually studied by various researchers. The used variables are described below.

(i) Flow velocity. The nutrient mass transfer increases with flow velocity as this favours biofilm development. Nevertheless, specific velocities of between 3-4 m/s may favour biofilm release.

(ii) Hydraulic regime. This may be turbulent or laminar (see Table 1). Biofilm is likely to be more active in turbulent flow, having more mass per cm^2, increased cell density, and distinct morphology, than biofilm in laminar flow.

(iii) Pipe material. Pipe material may be metal, plastic, or cement (see Table 1). In general, metal pipes tend to develop more biofilm than cement pipes, and these more than plastic pipes. This is because pipes with a rough surface have greater potential for biofilm growth. Rough surfaces provide more area for biofilm growth and protect biofilm from water shear forces.

(iv) Pipe age. The accumulation of corrosion and dissolved substances in older pipes can increase their roughness, thus favouring biofilm development. In addition, older deposits may have greater biomass and bacteria content. We divide the pipes into *young*, *medium*, and *old* (see Table 1).

(v) Biofilm. We chose the heterotrophic plate count (HPC/cm^2) as the biofilm quantification method. Although there are other methods, this is the most commonly used, and so more data are available. Based on the observed biofilm data distribution and expert criteria, these data were divided into *low*, *medium* and *high* biofilm development (see Table 1).

Table 1. Variables and categories of the database

P.MATERIAL	P.AGE (years)
metallic cement plastic	high [$\geq$ 31] medium [11-30] low [0-10]
BIOFILM (HPC/cm^2)	FLOW VELOCITY (m/s)
high [$\geq 10^7$] medium [10^4-10^6] low[0-10^3]	high [1.8-3.5] medium [0.8-1.7] low[0-0.7]
HYDRAULIC REGIME	
laminar turbulent	

This database represents the starting point that allows the implementation of the methodology described below in this chapter. Now we are able to apply the label negotiation through discriminant analysis via label propagation. Machine learning and multi-agent systems are the chosen tools to implement efficiently and accurately the proposed methodology. Finally, the edge betweenness centrality index based on the graph theory is also put into practice. In this way, we could identify the critical elements of the network.

2.1. Discriminant Analysis via Label Propagation

The label propagation associated with discriminant analysis clustering is used to approach a discriminant analysis in a practical case-study. Thus, pipes of a given DWDS can be classified depending on the similarities of the constructed database. Once the DWDS pipes have been classified by the aforementioned discriminant analysis, an agent-based method is launched. An agent is any entity in a system that generates events that affect itself and other agents. Once agents have been defined and their relationships established, a schedule of combined actions on these objects defines the process to occur, in our case, the assessment of the vulnerability level to biofilm development [5]. So, in this case, pipes properties are inherited by the nodes and node membership to the clusters are renegotiated [3, 5]. Thus, this process can be understood as a *label propagation method* methodology. Table 2 summarizes the process.

The agent-based model performs a mixture of individual and collective actions. It can explore good network sectorization layouts by trying to meet the equation

$$\sum_{i=1}^{n}\sum_{c=1}^{C}[\alpha_v(v_i-\overline{v}_c)+\alpha_w(w_i-\overline{w}_c)+\alpha_p(p_i-\overline{p}_c)+\alpha_m(m_i-\overline{m}_c)] \quad (1)$$

where n is the number of pipes of the DWDS, C the total number of clusters and the α's are the associated weights to velocity (v), water age (w), pipe age (p), and pipe material (m); and $\overline{v}_c$, $\overline{w}_c$, $\overline{p}_c$ are the respective averages by cluster, and $\overline{m}_c$ the median for the discrete variable material. The model is validated by the corresponding stabilization of this value that we attempt to minimize.

Table 2. Method for label propagation in practice

MAS method for label propagation
1. Discriminant analysis based on theoretical database clustering
2. Membership negotiation
2.1. Facilitate sharing the same label by neighboring pipes such that:
- have more similar *velocity* than the average of their current cluster.
- have more similar *water age* than the average of their current cluster.
- have more similar *pipe age* than the average of their current cluster.
2.2. Facilitate sharing the same label by neighboring pipes such that:
- have more similar *pipe material* than their neighboring pipes.
3. *If* there are not changes in last iteration *then* stop. *Otherwise* go to 2.

It is worth to notice that, when we extend our point of view from a pipe to a network, the spatial distribution of the pipes can be influential. This is specially important in our case where we focus on the effect that the physical and hydraulic characteristics of DWDS have on biofilm development. That is why the variable "water age" has been introduced as a correction factor in the calculation of membership in the negotiation explained above (Table 2). It is known that the residual disinfectant is consumed along the distribution system, therefore the pressure on biofilm decreases. Thus, the older the water, the greater the residual disinfectant decay and, moreover, sediment deposition and temperature increase [4]. All of them are factors that favour biofilm development. So, apart from the variables studied before in this spatial negotiation the "water age" variable is introduced. We suppose that in this network the water entries (reservoirs, tanks, etc.) correspond to the chlorination points and there are no more chlorination points in the system, so we calculate the shortest paths (km) between each node and each water inlet. Knowing the shortest path between a node and each of the water entries we calculate the weighted average for each node. Finally, we scale it, consequently obtaining a "water age" index between 0 and 1, which increases with the age of the water. This variable is discretized and introduced to the label propagation.

By this new complementary viewpoint of the more classical discriminant analysis, it is possible to achieve homogeneous groups where various characteristics in relation to biofilm development can be described. In addition, this new division offers an interesting starting point for further attempts to divide a given DWDS into hydraulic sectors.

2.2. Graph Theory Measurements to Assess the Edges Importance

Graph theory is a useful approach for the treatment of complex networks of real systems, whose techniques facilitate their representation and analysis. The framework is based on a set of measurements that enable to capture the global properties of such networks and model them as graphs. Formally, a graph $G = (V, E)$ is a pair that consist of two sets V and E, where $V \neq \varnothing$ is the set of vertices (nodes or points) $V = \{v_1, v_2, ..., v_n\}$ and E is a set of unordered (or ordered) pairs of vertices $E = \{(v_1, v_2), (v_2, v_3), ..., (v_j, v_k), (v_{n-1}, v_n)\}$ named edges $E = \{e_1, e_2, ..., e_n\}$ (links or lines). In this regard, DWDSs are complex networks, which can be abstracted and analysed as graphs; the nodes would represent junctions, reservoirs, tanks and pumps, while links would be the pipes and valves. In the context of DWDSs, we are interested in knowing the structurally important edges, which might have implications on where the impact of biofilm development is higher. Below, we introduce the concept of graph theory tipically used to measure the edge importance, the *edge betweenness centrality*.

2.2.1. Edge Betweenness Centrality

The *betweenness* is one of the standard measures of node centrality, originally introduced to quantify the importance of an individual in a social network. For such a reason, the concept *betweenness centrality* focus on the centrality of a node in terms of the degree to which the node falls on the shortest path between other pairs of nodes. If a node has a high betweenness centrality, then it lies on the path of many pairs of nodes. The communication of two non-adjacent nodes, j and k, depends on the nodes belonging to the connecting paths going through it, and defining the *node betweenness*. In this regard, the Girvan-Newman algorithm (by generalizing Freeman's proposal [6]) extends this definition to the case of edges and define the *edge betweenness centrality* as the number of the *shortest paths* that go through an edge in a graph or network [7]. If there is more than one shortest path between a pair of nodes, each path is assigned equal weight such that the total weight of all of the paths is equal to unity. Besides, each edge in the network can be associated with an *edge betweenness centrality* value. An edge with a high *edge betweenness* score represents a bridge-like connector between two parts of a network, and their removal may affect the communication between many pairs of nodes through the shortest paths between them.

The *edge betweenness* of edge i is defined by

$$b(e_i) = \sum_{i \neq j} \frac{n_{ij}(e_i)}{n_{ij}} \tag{2}$$

where $n_{ij}(e_i)$ is the number of paths from node i to node j through edge e_i, and n_{ij} is the total number of *shortest paths* of the network.

In this regard, in a DWDS a pipe with high *edge betweenness* would be between many potencial upstream contamination events and downstream receptor populations [8]. Also, pipes with high *edge betweenness* could be potential locations for chlorination points or sensors.

3. Case Study

Firstly, we performed a clustering analysis to the obtained database [2]. The medoids shown in Table 3 were obtained.

The desired number of clusters was specified before the analysis. We chose three, because it is a suitable number for the network size, and also because the variable of interest (biofilm) is also divided into three categories: low, medium, and high. We observe that the medoid of each cluster corresponds coincidentally with each of the states defined for biofilm development (Table 3). Cluster 0 is defined by having low biofilm development,

Table 3. Medoids of the clusters obtained in the biofilm development database

MEDOIDS	H. REGIME	F. VELOCITY	P. MATERIAL	P. AGE	BIOFILM
Cluster 0	Turbulent	High	Plastic	Young	Low
Cluster 1	Turbulent	Medium	Cement	Old	Medium
Cluster 2	Turbulent	Medium	Metal	Medium	High

Cluster 1 by medium, and Cluster 2 by high development. Each medoid is also defined by a type of pipe material and pipe age. The fact that each medoid corresponds to each of the categories of biofilm development studied, confirms that the variables selected are valid, since they are capable of explaining the differences in the degree of biofilm development. This result, due to its consistency, also validates the process of obtaining, processing, and transforming the data applied before.

In this chapter, the Example 3 of Epanet [9] (Figure 1a) was chosen as a given DWDS where apply this methodology. With the aim to make the network as real as possible, the material and the age of the pipes were randomly assigned - within the ranges indicated in Table 4 - depending on the average age of the area (see Figure 1b).

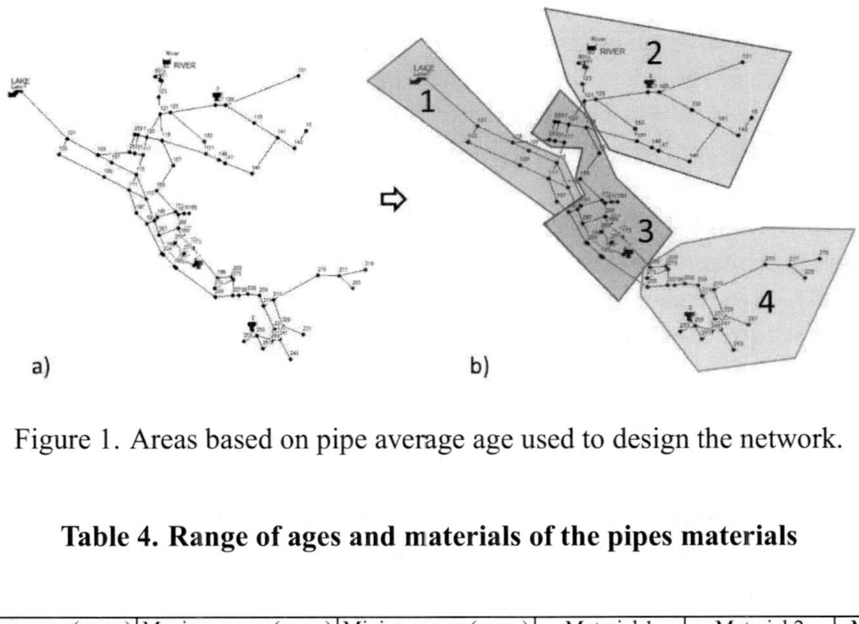

Figure 1. Areas based on pipe average age used to design the network.

Table 4. Range of ages and materials of the pipes materials

Area	Average age (years)	Maximum age (years)	Minimum age (years)	Material 1	Material 2	Material 3
1	60	86	54	concrete	asbestos cement	iron cast
2	45	58	33	asbestos cement	iron cast	-
3	30	38	24	asbestos cement	iron cast	polyethylene
4	15	25	5	iron cast	polyethylene	-

Once the network and the medoids were ready, discriminant analysis and label propagation were applied (Figure 2).

According to our theoretical database, after the discriminant analysis (Figure 2a) in the given DWDS, most of the pipes are prone to suffer high biofilm development. No pipes with a trend to have low biofilm development are found. As a result of the label propagation

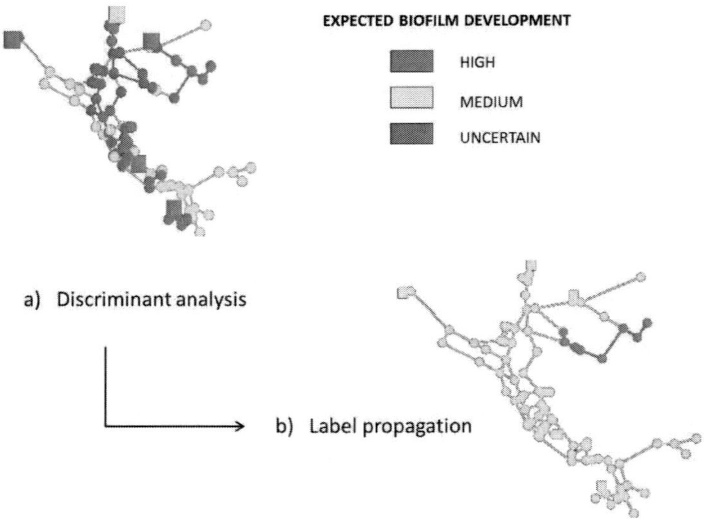

Figure 2. Results of the discriminant analysis via label propagation.

process (Figure 2b) two homogeneous areas associated with different degree of biofilm development appear. As previously, again there are no cases with low biofilm development. The biggest cluster corresponds with medium biofilm development and the small one with an area where high biofilm development is expected. This cluster prone to high biofilm development is formed by seven pipes, all of them are old pipes with high water age, low flow velocity and, except for two that are metalic pipes, the rest are all cement pipes. This is in strong agreement with what was theoretically expected, in the case of the pipe age, pipe material and water age variables. Instead, opposite to what was expected, all the pipes in this area have low flow velocity. This could be explained by the fact that although the water velocity increases the nutrient mass transfer, it also increases the shear forces and thus produces biofilm detachment.

When applying the *edge betweenness* algorithm to each sub-graph, the obtained values in each pipe were scaled to facilitate the observation of the results (Figure 3). It is worth to highlight that the proportion of main pipes is bigger in the area prone to high biofilm development than in the other area. This raises the importance of focusing management efforts in this zone. Because of the importance of these pipes in the network operation, avoiding, as much as possible, biofilm development within them must be crucial to guarantee a service of quality in DWDSs.

These highlighted pipes (Figure 3) are also important because they are strategic points where carrying out targeted monitoring to control the quality of the water that goes trough them, developing cleaning processes to remove the biofilm adhered to its walls, as well as, locating chlorination points to reduce the development of these communities. They represent the biofilm hot spots of the network, where the management efforts must be focused.

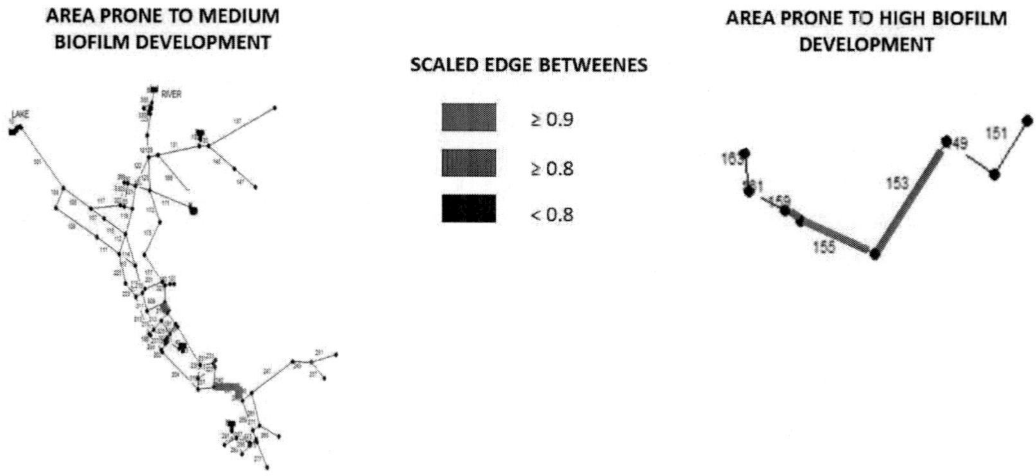

Figure 3. Results of the *edge betweenness* score.

Conclusion

This chapter provides an overview of an innovative perspective in the study of biofilm development in DWDSs. On one hand, the effect that the interaction among the hydraulic and physical characteristics of the DWDSs, relevant to biofilm development, has been introduced in this proposal. On the other, label negotiation has been carried out in this field, via discriminant analysis and label propagation. It has been resorted to MASs to apply this methodology, a highly multidisciplinary field with no previously known applications in this area.

Until now, the effect that the different physical and hydraulic characteristics of the DWDSs have on biofilm development was studied individually in a majority of cases. This is due to the complexity of the community and the environment under study, together with the scarcity of data. This chapter highlights the need to study the importance of studying the effect that the interaction of all the characteristics of DWDSs has on biofilm development. Thus, we can identify the joint conditions (physical and hydraulic) that determine the varying development of biofilm inside pipes.

It has been shown label negotiation via discriminant analysis and label propagation as interesting tools that enable the use of knowledge gained in the development of biofilm in DWDSs in a practical and efficient manner. This methodology enables an advanced visualization of the case-study database. According to the results obtained in this work, there are some areas within a DWDS more vulnerable to support high biofilm development, thus, biofilm is not uniform in space.

In the same way the introduction of the *edge betweenness* score has demonstrated to be a great help to improve the efficiency of DWDSs management. Thanks to it the most problematic pipes can be easily detected. These pipes represent the critical elements of the network. Thus, special attention must be focused on these elements to prevent its deterioration and mitigate as much as possible the negative effects derived of the biofilm development in DWDSs.

References

[1] Yu, J.; Kin, D.; Lee, T., Microbial diversity in biofilms on water distribution pipes of different material. *Water Sci Technol.* 2010, 61, 163–171.

[2] Ramos-Martínez, E.; Herrera, M.; Izquierdo, J.; Pérez-García, R., Pre-processing meta-data on biofilm development in drinking water distribution systems. *Water Res.* 2013, under review.

[3] Wooldridge, M., *An Introduction to MultiAgent Systems*, John Wiley & Sons. 2009.

[4] EPA, Effects of water age on distribution system water quality. *International Water Association* . 2002.

[5] Herrera, M.; Izquierdo,J.; Pérez-García, R.; Montalvo, I., Multi-agent adaptive boosting on semi-supervised water supply clusters. *Adv Eng Softw.* 2012, 50, 131–136.

[6] Freeman, L. C., A set of measures of centrality based upon betweenness. *Sociometry.* 1977, 40(1), 35–41.

[7] Girvan, M.; Newman, M. E. J., Community structure in social and biological networks. *Proceedings of the National Academy of Sciences of the United States of America PNAS.* 2002, 99(12), 7821–7826.

[8] Xu, J.; Fischbeck, P. S.; Small, M. J.; VanBriesen, J. M.; Casman, E., Identifying Sets of Key Nodes for Placing Sensors in Dynamic Water Distribution Networks. *J Water Res Pl-ASCE.* 2008, 134(4), 378–385

[9] Rossman, L., *EPANET–User's Manual.* United States Environmental Protection Agency EPA. 2002.

In: Mathematical Modeling in Social Sciences ...
Editors: J. C. Cortés López et al.
ISBN: 978-1-63117-335-6

Chapter 8

ON KERNEL SPECTRAL CLUSTERING FOR IDENTIFYING AREAS OF BIOFILM DEVELOPMENT IN WATER DISTRIBUTION SYSTEMS

***M. Herrera*[1,*]*, E. Ramos-Martínez*[2]*, J. A. Gutiérrez-Pérez*[2]
J. Izquierdo[2] *and R. Pérez-García*[2]
[1]Brussels School of Engineering,
Université libre de Bruxelles, Belgium
[2]Instituto Universitario de Matemática Multidisciplinar,
Fluing-IMM, Universitat Politècnica de València, Spain

Abstract

Nowadays biofilm develops in all drinking water distribution systems (DWDSs) and leads to various undesirable problems, representing a paradigm in management of these systems. Biofilm formation is dependent on a complex interaction of various factors, those associated with the DWDSs infrastructure being the most flexible and adaptable to minimize biofilm development in the inner pipe walls. One of the main objectives in the quality control of DWDSs is the analysis of biofilm development in the network, discovering areas more prone to it. This chapter proposes spectral clustering to achieve these issues after a kernel abstraction of the available data. The theoretical approach is enriched by both: an entropy based ranking to select the eigenvectors to work with, and a full tuning process that automatizes the proposal. Finally, a real case-study is introduced to check the performance of this division of the network. The results are promising, showing the relative trend to biofilm development for each sector. Using this information, managers can carry out, if neccesary, changes in the DWDS infrastructure to minimize biofilm development and perform biofilm control policies more efficiently.

Keywords: Spectral clustering, kernel methods, drinking water distribution systems, biofilm

---

*E-mail address: mherrera@ulb.ac.be

1. Introduction

In drinking water distribution systems (DWDSs) a complex mixture of microbes, organic and inorganic material accumulate amidst a microbially-produced organic polymer matrix attached to the pipes inner surface forming what is known as biofilm. Once developed, biofilm is very resistant and can lead to various undesirable problems such as deterioration of bacterial water quality, generation of water bad tastes and odors, proliferation of higher organisms, biocorrosion, and disinfectant decay, among others.

Biofilm is not necessarily uniform in time and space, indeed, biofilm formation is dependent on a complex interaction of water quality, operational factors and infrastructure associated with distribution systems. Since water quality depends on the water source and the operational factors are fixed, to satisfy the users demand, the distribution system's infrastructure is the most adaptable element. Thus, introducing the infrastructure characteristics that are known that affect biofilm development in the distribution system sectorization would allow to identify areas with different biofilm development trends. According to the obtained results, modifications in the infrastructure and different control policies could be carried out to reduce biofilm development as much as possible and thus, mitigate the problems associated with it. To achieve this aim we propose a division of a DWDS by kernel spectral clustering. Firstly, we define a kernel matrix [1] that captures the semantics inherent to the graph structure but, at the same time, is reasonably efficient for evaluation. In the first instance, the affinity graph matrix is transformed into a kernel matrix, carrying out the correspondence kernel abstraction of the essential characteristics of a DWDS. Another interesting features associated with biofilm development are included by their corresponding similarity matrices and also embedded into the kernel space. All the information is combined in just one kernel matrix. Next, spectral clustering techniques are applied to divide the DWDS depending on variables related to different levels of biofilm development. An eigenvector selection based on entropy measures attempts to adapt the spectral clustering to the underlying data information better than the usual Ng-Jordan-Weiss (NJW) proposal [2]. The chapter also includes a guide for tuning the related parameters to this process. Both the scale parameter of the Laplacian matrix and the weights of the kernel combination, are tuned by a new approach based on multistart trust-regions. This alternative searches the best combination of parameters for achieving spectral clustering under some optimality condition.

The road map of the chapter is as follows. Kernel spectral clustering, describing the self tuning of the Laplacian matrix and the eigenvector selection based on entropy is presented in Section 2 Next, an experimental study is introduced and the results of the kernel spectral clustering application are discussed in Section 3 A section of conclusions and future work closes the paper.

2. Tuning Spectral Clustering

To cluster the pipe attributes while maintaining graphs characteristics we should take into account both graph structure and pairwise similarities. The process starts by embedding the Laplacian of the graph together with the different similarity matrices into a kernel matrix. The process steps are as Table 1 shows.

Table 1. Kernel embedding process

1. Build the affinity matrix $A \in \mathbb{R}^{n \times n}$
defined by $A_{ij} = exp\left(-\frac{||x_i - x_j||^2}{2\sigma^2}\right)$ if $i \neq j$ and $A_{ii} = 0$.
2. Define D to be the degree diagonal matrix whose (i, i)-element is the sum of the entries in A's ith row.
3. Build the matrix $L = I - D^{-1/2} A D^{-1/2}$.
4. Embed into a kernel space the Laplacian and dissimilarity matrices associated with the problem.
   4.1 Scale data between 0 and 1.
   4.2 Plug a diagonal of 1's into the diagonal of each matrix.
   4.3 Mirror matrices through their diagonals to make them symmetric.
5. $K = w_{Lap} K_{Lap} + \sum_{i \in I} \omega_i K_i$

The scaling parameter of step 1, σ^2, controls how rapidly the affinity A_{ij} falls within the distance between x_i and x_j. $D = diag(d_1, \ldots d_{n_U})$ is the degree matrix and A is the adjacency matrix. Each ω_i (step 5) allows to give different importance to each dissimilarity matrix, K_i $(i \in I)$, involved in the performance of K. In step 5, ω_{Lap} and K_{Lap} are the weight and the kernel matrix, respectively, associated with the Laplacian. Finally, a spectral clustering process takes place in this kernel matrix, K, by the calculus of its related eigenvectors, the selection of the c more important ones, and the application of a division algorithm, such as *k-means* (see Figure 1).

In summary, we have to choose the following key parts of the kernel spectral clustering: the σ parameter, the weights ω_i, $(i \in I)$ and ω_{Lap}, and the eigenvector selection.

2.1. Local Scaling for the Laplacian Matrix

The scaling parameter is a measure of similarity between two points. Similarity provides an intuitive way for selecting possible values for σ. Ng et al. [2] suggested selecting σ automatically by running their clustering algorithm repeatedly for a number of values of σ and selecting the one which provides least distorted clusters. This increases significantly the computation time. Additionally, the range of values to be tested still has to be set manually. Moreover, when the input data includes clusters with different local statistics there may be not a single value of σ that works well for all the data.

The proposal is to calculate a local scaling parameter for each data point x_i [3]. The distance from x_i to x_j has two point of views: from x_i it is $d(x_i, x_j)/\sigma_i$, and from x_j it is $d(x_j, x_i)/\sigma_j$. Thus, the product is $d(x_i, x_j)d(x_j, x_i)/\sigma_i\sigma_j = d^2(x_i, x_j)/\sigma_i\sigma_j$; and the affinity matrix of step 1 of Table 1 can be re-written as Equation 1 indicates:

$$\hat{A}_{ij} = exp\left(-\frac{||x_i - x_j||^2}{2\sigma_i\sigma_j}\right) \tag{1}$$

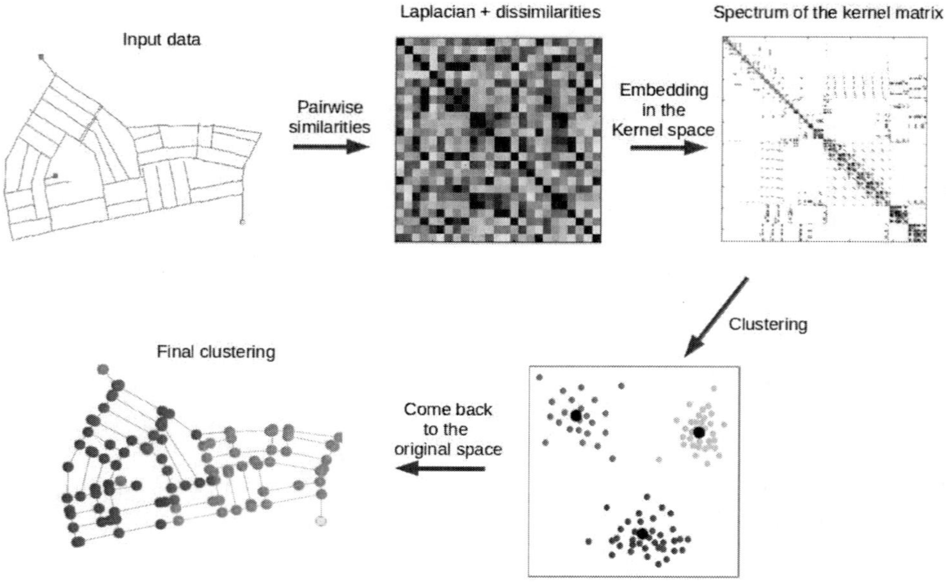

Figure 1. The overall process of kernel spectral clustering.

The local scale, σ_i, can be estimated by its closest environment using the k-nearest-neighbor points to x_i and then calculating

$$\sigma_i = d(x_i, x_k). \tag{2}$$

The selection of the number of neighbors is independent of scale and is a function of the data dimension of the embedding space.

2.2. Eigenvector Selection Based on Entropy

For a c-clustering problem, the NJW method [2] always splits data using c eigenvectors related to the lowest eigenvalues of the normalized affinity matrix of a dataset. Although the spectral relaxation approach of normalized cut criterion lies in the subspace spanned by these eigenvectors, it is not guaranteed that these method can detect well the structure of the data. Xiang and Gong [4] were the first to use eigenvector selection to improve the performance of spectral clustering based on how well their combination can separate data into clusters. The proposal to find the eigenvector with we work is by similar ranking ideas, but based on entropy measures. Thus, we directly select the first eigenvectors in the ranking list of entropy.

According to the entropy theory, entropy of a system measures the disorder in the system. In this part of the work, we utilize entropy to rank the eigenvectors for kernel spectral clustering. We use $V \in \mathbb{R}^{n \times n}$ to denote the matrix consisting of all the eigenvectors of the kernel matrix obtained following the process of Table 1. The entropy of V is defined as follows:

$$E = -\sum_{v_i \in V} [p(v_i)log(p(v_i)) + (1 - p(v_i))log(1 - p(v_i))] \tag{3}$$

where $p(v_i)$ denotes the probability at the point v_i. In practice, we substitute those probabilities by similarities

$$E = -\sum_{v_i \in V} \sum_{v_j \in V} E_{ij} = -\sum_{v_i \in V} \sum_{v_j \in V} [s_{ij} log(s_{ij}) + (1 - s_{ij}) log(1 - s_{ij})] \quad (4)$$

where s_{ij} is the similarity between two points v_i and v_j. In addition, s_{ij} is given by the expression:

$$s_{ij} = e^{-Dist_{ij}} \quad (5)$$

taking into account that the distance between v_i and v_j, $Dist_{ij}$, is calculated as

$$Dist_{ij} = \left[\sum_{l=1}^{n} \left(\frac{v_{il} - v_{jl}}{\max(v_{.l}) - \min(v_{.l})} \right) \right] \quad (6)$$

When the removal of an eigenvector causes more disorder than removing another eigenvector, then the first one will be more important than the rest and should be ranked higher in the list of eigenvectors. The selecting idea is based on this principle. Thus, in order to obtain the ranking list of the eigenvectors, each eigenvector is removed in turn and the corresponding entropy is calculated.

2.3. Automatic Search of Best Partition by Multistart Trust-Regions

Multistart algorithms are an option if global extrema are searched, since these algorithms can explore more than a single basin of attraction of the objective function. The starting points should be enough and well distributed onto the design space [5].

Since we are interested in tuning weights and parameters within other optimization process (clustering in this case), and we are not interested in the resulting model, we propose to apply derivative-free optimization algorithms [5]. Thus, only the objective function value is required. Our proposal is to adopt a linear second order model to interpolate (zones of) the parametric space by a simple surface where we can easily search its extreme. The use of a surrogate model, instead of the computation of the real objective function, reduces the computational time and extends the possible solutions from a set of points (Grid Search) to an entire surface. Thus, a so-called Design Of Experiments (DOE) is required at the begin of the algorithm by a double use: firstly, it searches different zones where locate our trust regions; next, DOE samples a number of starting locations into each of these regions.

The basic algorithm is outlined in Table 2, where the process starts selecting m regions of the parametric space (trust-regions). These trust-regions are specified with a center point and a radius, r. Once initialized, the trust region radius is dynamically adjusted by checking the quality of the clustering configuration for parameters at a certain distance from the search point.

The rule for redefining these trust-regions depends on the average silhouette width (ASW) [6]. Since the ASW is in the [-1,1] interval, it is possible to calculate its increase, for each combination of weights and parameters during the iterations. The trust-region size is updated by the following rule[1]:

[1]The value of h will directly affect to each trust-region radius.

- If the value of r is $<$ - 0.1, we set the value of the growth parameter, $h = h_{shrink}$.
- If the value of r is $>$ 0.1, we set the value of the growth parameter, $h = h_{grow}$.
- If the value of r is between -0.1 and 0.1, then the region size is not changed: $h = 1$.

Table 2. Multistart trust-region-based surrogate surfaces

1 - Compute a number, m, of initial areas from the whole parametric space (DOE).
2 - Sample points (parameters) in the selected regions (DOE).
3 - Compute the clusters for the sampled weights and parameters.
4 - Create m surrogates based on the sampled points: weights, parameters, and clusters.
5 - Search the maximum values of ASW on the m surrogate surfaces.
6 - Compute the spectral clustering with the new selected parameters.
7 - Redefine each trust-region and resize it by the factor h.
8 - Repeat steps 4-7 until reach a stop condition.

The trust-region process is stopped if one of the following criteria is met: the new increment is lower than 0.1 times the initial one; the new solution improves the last one by a number lower than 1.e-06; the number of iterations is greater than 30.

3. Experimental Study

This chapter focusses on kernel spectral clustering and its application, approaching drinking water distribution systems sectorization identifying areas with different trends of biofilm development. To test the developed methodology we apply it in a real case, the DWDS of Celaya (Guanajuato, Mexico) central district. The structure of the network is presented in Figure 2. This network is made out of 479 lines and 339 nodes; its total pipe length is 42.5 km.

Due to the aim of the application, namely identifying areas of biofilm development in water distribution systems, apart from the common parameters used in hydraulic sectorization (pipe length, pipe diameter, roughness, flow velocity, losses and friction factor), we have also introduced in the analysis the pipe age and pipe material. These parameters have been chosen because they are known as distribution system infrastructure variables that affect biofilm development. Pipe material can be classified into three main groups: plastic, cement and metal. In general, metal pipes tend to develop more biofilm than cement pipes, and these more than plastic ones [7]. This is mainly explained because pipes with a rough inner surface have greater potential for biofilm growth [8]. Rough surfaces provide more area for biofilm growth and protect biofilm from water shear forces. In the same way, the older the pipes the greater the accumulation of corrosion products and dissolved substances

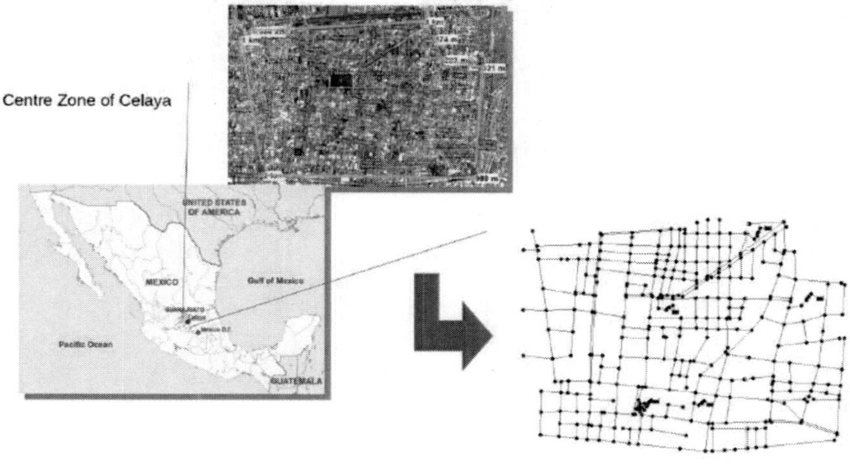

Figure 2. Location and structure of the experimental study drinking water distribution system.

that increases their roughness [9], thus favoring biofilm development. In addition, older deposits may have greater biomass and bacteria content [8]. So, the older the pipe the greater the biofilm development potencial.

Four parameters (three weights and the parameter, k, related to the Laplacian's scale parameter) will be tuned in order to illustrate the performance of the proposed process. We have different kernels depending on the nature of the variables: continuous (will be weighted by w_1), discrete, such as the case of pipe material, (weighted by w_2), and derived from the Laplacian associated with the affinity matrix (w_{Lap}). These three weights are positive numbers and we work without constraints respect to its sum (conic combination). In addition to these three weights, it is necessary tuning the local scaling parameter for the affinity matrix. This work has been reduced to search the more suitable number of k-nearest-neighbors to estimate the distances. This number, k, may vary between 2 and 6, but this number usually depends on the size of the database [3]. All these four parameters start from three different points in their parametric space running the trust-region process above proposed (see Subsection 2.3). The best clustering configuration is selected by the criterion of maximum ASW, which reaches to 0.32 in the final approach vs. 0.26 achieved by the same process changing the entropy for eigenvector selection by the more usual NJW method.

Once applied the kernel spectral clustering in the central area of the DWDS of Celaya, we obtained 3 well differentiated clusters (Figure 3). As expected, each cluster is characterized by the age and material of the pipes. The two clusters associated with metal pipes, also correspond to the clusters with the oldest pipes. One cluster corresponds mostly to 50 years old cast iron pipes while the other to 40 years old galvanized iron pipes. Thus the first cluster would be the most prone to biofilm development. In contrast, the third cluster, integrated mostly by 30 years old asbestos cement pipes, would be the one with less risk of biofilm development when comparing it with the other clusters.

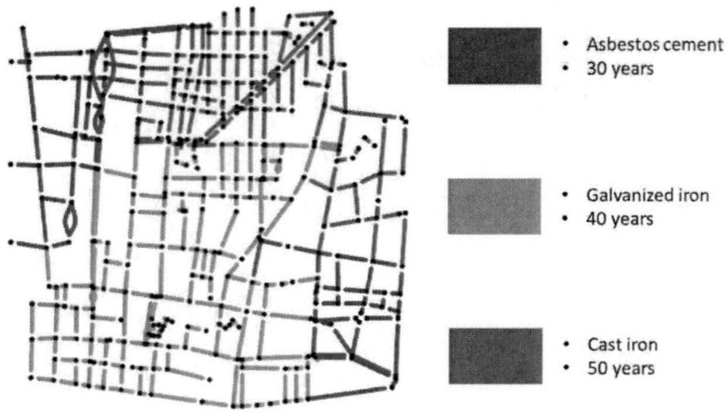

Figure 3. Obtained sectorization after kernel spectral clustering analysis.

Conclusion and Future Work

Knowing the relative trend to biofilm development in the different areas within a DWDS would serve as a decision support tool for the distribution system managers. The implementation of this instrument could facilitate and increase the effectiveness in biofilm control and mitigation policies by replacing and/or renewing the most vulnerable and problematic areas or pipes of the distribution system.

The kernel spectral clustering process has been tuned by a new methodology based on multistart trust-regions, which offers efficient results with a good computational performance. This tuning method has been proposed along with an interesting option for estimating the scale parameter of the Laplacian matrix by a k-nearest-neighbors approach and an eigenvector selection based on entropy measures with better detection ability than the usual NJW method for detecting the structure of the data.

Further research will be focused on a deeper understanding of the final kernel matrix to work with and the inclusion of semi-supervised techniques. From a more hydraulic point of view, we also plan to observe if cluster homogeneity remains when the distribution system is repaired and new pipes of different materials are introduced within DWDS areas constructed long time ago.

References

[1] Schölkopf, B.; Smola A. J. *Learning with kernels: Support vector machines, regularization, optimization, and beyond.* MIT Press, 2002.

[2] Ng. A.; Jordan, M.; Weiss, Y. On spectral clustering: Analysis and an algorithm, *NIPS.* 2001, 14, 849-856.

[3] Zelnik-Manor, L.; Perona, P. Self-Tuning Spectral Clustering, *NIPS.* 2004, 17, 1601-1608.

[4] Xiang, T.; Gong, S. G. Spectral clustering with eigenvector selection, *Pattern Recognition.* 2008, 41(3), 1012-1029.

[5] Peri, D.; Tinti, F. A multistart gradient-based algorithm with surrogate model for global optimization, *Communications in Applied and Industrial Mathematics*. 2012, 3(1).

[6] Rousseeuw, P. J. Silhouettes: a graphical aid to the interpretation and validation of cluster analysis, *Computational and Applied Mathematics*. 1987, 20, 53-65.

[7] Niquette, P. M.; Servais, P.; Savoir, R. Impacts of pipe materials on densities of fixed bacterial biomass in a drinking water distribution system, *Water Resources*. 2000 34(6), 1952-1956

[8] Chowdhury, S. Heterotrophic bacteria in drinking water distribution system: a review, *Environmental Monitoring and Assesment*. 2011, 2407-2415

[9] Christensen, R. T. *Age effects on iron-based pipes in water distribution system*. Utah State University, 2009.

In: Mathematical Modeling in Social Sciences ... ISBN: 978-1-63117-335-6
Editors: J. C. Cortés López et al.

Chapter 9

UNSUPERVISED METHODOLOGY FOR SECTORIZATION OF TRUNK DEPENDING WATER SUPPLY NETWORKS

E. Campbell,* R. Pérez-García, J. Izquierdo and D. Ayala-Cabrera
Instituto Universitario de Matemática Multidisciplinar,
I.M.M. Fluing, Universitat Politècnica de València, Valencia, Spain

Abstract

The increasing pressure over water resources has raised the need to establish procedures to efficiently control water losses in water supply networks (WSN). An example of such procedures is the segmentation of the networks (also known as sectorization), either by installing valves (closed) or by sectioning pipes. WSN division into sectors implies an improvement of active leakage control by comparing the metered demand (domestic and industrial) vs. the flow inlet in each sector. It also allows water companies to have better control over pressure, consumption and water quality, due to an implicit areal reduction. In most cases, sectorization projects are carried out using empirical criteria, following trial and error techniques. This work presents a novel combination of hydraulic, social, and economic aspects of WSNs, in order to establish sectorization layouts in WSNs with a reduced number of water sources. The methodology uses machine learning-based techniques. Through hierarchical clustering, an initial exploration of the natural clusters in the WSN is conducted. Then, by means of spectral clustering, the WSN is subdivided considering two aspects: the importance of the different partitioning criteria and the minimization of the number of pipes that should be closed. The connection points of the clusters in the distribution network to the trunk are established through an energy analysis aimed to minimize the pressure drop after sectorization. The methodology entails several advantages over other sectorization methodologies. First, it is one of the few that may be applied to WSNs depending on a trunk. Also, the procedure only considers the distribution networks, which offer two important advantages. One advantage is that it translates to a reduction of implementation costs as the closing valves are allocated in low diameter lines; additionally, the resulting active leakages control can be performed more effectively as the sectors are centered in the zones with highest incidence of leakages.

Keywords: sectorization, power, supply, leaks, networks, clustering

*E-mail address: encamgo1@upv.es

1. Introduction

1.1. Sectorization of WSNs: Concept, Advantages and Drawbacks

Water supply networks (WSNs) are infrastructures that transport drinking water from the water sources to consumers. Through them, water losses represent one of the main problems for water utilities. In some countries (specially developing countries), water losses may be of order of 50% of the water injected in WSNs [17, 18, 12, 6]. Given the concern of the increasing pressure over this natural resource and constant demand of a good quality water supply (including pressure, constancy and chemical quality), some techniques to improve water management have been put forward in recent years. One of them is sectorization, which entails the subdivision of WSNs into subsectors by closing some pipelines and installing flow meters in a single line of each sub-sector. The goal is permanently to control the inflow of each sector.

This technique was implemented for the first time in two cities in UK in the early 1980s. Since then, several reports and guidance have been published in this regard. The *Guidance Notes* document proposed by [13], is one of the most important and broadly known. Despite a good level of acceptance of the technique in many countries, most of the implementation cases follow a trial-error approach. The technical recommendations to select the size of the sectors are very general. For the size of sectors, the best-recognized guidance recommends the use of a wide range of number of connections (500-3000 connections)[1] [13]. When first introduced, the idea of WSN sectorization was focused on control of Non Revenue Water (NRW) by comparing the water that enters each sector vs. the registered consumed water quantity; However, since the 1980s, others goals have been proposed, including: water audits; chemical quality control; planning repairing activities, and others. In order to address those goals, and depending on the topological characteristics of the WSN, two types of sectorization schemes might be implemented. If the WSN features many water sources, each sector might be defined around one or a few sources (normally a maximum of three is recommended). In others networks, the establishment of this type of layout might be unfeasible if sources are located far from the consumers. In that case, water supply depends on a supply trunk. Therefore, the sectors do not have their own water source (or sources) and the water entrances should be connected to the supply trunk. The technique has remarkable advantages although also some drawbacks. The first advantage is the increased ease with which an anomaly is detected (pipe burst, for example) in the network. In small sectors, any change in flow is more easily detected. However, implementation of small sectors also increases the cost of each projects implementation as it represents the need of more boundary valves. Moreover, closing pipes increases the friction in the pipe walls, which conducts to a decrease in the overall network pressure. In a positive sense, this is translated in a decreasing in the level of actual losses (leaks in pipes, also known as physical losses) [18]; Nevertheless, an excessive pressure drop can cause supply problems, being this the most critical drawback of the technique. When the WSN are designed, a principle of hydraulic redundancy is followed in order to ensure their reliability. Closing pipes conflicts this principle. Therefore, sectorization reduces the capacity of WSN to overcome problem-

[1] A connection is each point that links the WSN with the consumers. A connection could be a house, a building, or a industrial facility.

atic situations and thus, its implementation must be supported by an exhaustive technical analysis. A trial-error approach based on hydraulic simulation does not necessarily represent a meaningful problem for sectorization of small WSN, especially in the case of fairly branched WSN. Nevertheless, for large extension looped networks (WSNs of more than one thousand km of pipelines), the level of complexity in the design of sectors is expected to be higher. In both cases, networks of large extension and small networks, it is important to ensure that sectorization will not generate supply problems.

Considering the advantages and disadvantages above, it is expected that a good layout of sectorization would balance level of pressure and leakage control aspects. It also has to minimize the number of required boundary valves and flow meters, in order to make the project economically implementable.

1.2. State of the Art

During last decade, several computer based methods have been put forward in order to create sectorization schemes for large WSNs. Essentially, these studies have used graph techniques along with mathematical optimization tools to treat the sectorization as a graph partitioning problem. Most of them have focused efforts on identification of the influence zone of each water source in WSN, so that each sector may rely on at least one exclusive source of water. [20] used graph theory-based algorithms to verify sectorization layouts topologically. [21] proposed the use of Dijkstra algorithm to identify the influence zone of each source. [14] use the same theory to find strongly and weakly connected clusters in WSN. It also have been combined with energy-based criteria and with heuristic techniques [2, 4, 5, 3, 1, 11]. [9] proposed the use of Kernel methods along with the spectral clustering technique. This method features an important advantage over others sectorization methods, since it allows the consideration of several operational characteristics of the WSNs in order to obtain a partitioning that balances the operational context of the WSNs, including the priorities of the water utilities. In the same line [10] proposed a boosting algorithm to obtain a more stable partitioning using spectral clustering. As mentioned above, most of the proposed sectorization methodologies start from sources of water, defining sectors around them. However, [7] presented a methodology in which sectors are not gathered from water sources. In this case the information storage in the nodes is used to cluster the nodes by means of k-means and multiagents techniques.

2. The Proposed Method

2.1. Transformation of WSN into Graphs

WSN in the software of hydraulic simulation EPANET 2.0 can be conceived as graphs. The consumption and supply nodes (tanks) are the vertices, and pipes and valves, edges (also named links or arcs). Depending of whether the direction of the pipes is taken into consideration or not, graphs of WSN can be categorized either as directed graph (or digraph if directions of the edges are taken into account), or undirected graphs (when pipes flow or head losses direction is avoided). In looped WSN, flow direction may vary along the day, thus digraph of looped WSN must be represented in a hourly basis. In WSN the period of maximal demand is the most critic in terms of supply. If supply is guaranteed in this scenario, then it is also guaranteed for the remaining time.

Some features may be added to graph nodes (*geographical coordinates, elevation, demand, emitter coefficient, pressure, head*) and to edges (*diameter, length, roughness, head losses*). When added, they become their respective weights. Also, for both, edges and vertices, new properties can be created. e.g. for vertices, hydraulic power (pressure or head times actual demand) of the nodes can be added.

2.2. Clustering in a WSN

Clusters (or sectors) in WSN, represent a collection of nodes with maximum intra-cluster similarity of profiles (characteristics) and maximum dissimilarities to nodes in others clusters. In the field of machine learning, kernel methods are a series of algorithms used for pattern analysis purposes [16]. [9] proposed the use of this technique in combination with spectral clustering to cluster WSN in subsectors with the maximum degree of uniformity and minimizing pipes cuts at the same time. Despite advantages of spectral clustering, this technique does not provides the number of cluster for the partitioning. In this sense, an initial exploration of the data through hierarchical clustering can be conducted before spectral clustering. The idea behind this technique is to build binary trees that merge into groups based on their similarity [8]. From the study of the tree that is generated, useful information that helps to understand data structure can be extracted, and from there, the number of clusters to conduct spectral partitioning may be estimated.

2.3. The Proposed Method: Step by Step

The proposed unsupervised method follows the machine learning-based method previously proposed by [9]. However, in this proposal, the trunk of the WSN is not considered. Thus, the partitioning problem is solved only over the nodes of the distribution network. Consequently, only pipes of the distribution network are candidates for being closed (boundary valve allocation). First, the method conducts a hierarchical clustering (unsupervised method) over the nodes of the distribution network. Through hierarchical clustering implementation, an initial exploration of the characteristics of the network is carried out. Then, by means of clustering validation measures (*Silhouette Index, Dun Index, Connectivity Index*) the number of clusters that better support the data is determined. This number is used in the following spectral clustering stage. It is worth noting two aspects. Firstly, hierarchical clustering algorithms use a dissimilarity matrix that contains the characteristics of the nodes (*geographic coordinates, demand, elevation and emitter coefficient*). Secondly, among all the existing algorithms to carry out the process, one is selected based on a ranking of all the methods according a resulting cophenetic correlation coefficient. This coefficient measures the correlation between the initial dissimilarity matrix and the resulting ultrametric matrix. Once the number of clusters for the partitioning is estimated, the spectral clustering process is carried out over a global kernel matrix (equation 1) that represents the sum of the kernelized laplacian matrix of the network graph and the kernelized dissimilarity matrix of each characteristic in the nodes (the sames used is the hierarchical clustering stage). In this global kernel matrix, each factor is weighted so the importance that decision makers give them and/or the operational priorities of the WSN can be reflected and thus, influence the resulting partitioning. These weights are estimated through Analytic Hierarchy Process (AHP) which is a mathematical methodology for decision making. In this, actors uses comparison to evaluate different alternatives based on different criteria [15]. Here, only partitioning criteria are compared, giving as result a vector containing the weights of such criteria.

The weight λ next to the kernelized laplacian matrix of the graph (k^*), is estimated by analysis of the costs for different partitions.

It is important to mention that including emitter coefficients as criteria allows the isolation of clusters based on the level of leakage in the different zones of the WSN. Figure 1 displays the flowchart of the process.

$$K_g = \lambda * k^* + (1 - \lambda)[\sum(\omega_i * C_i)] \tag{1}$$

Where K_g corresponds to the global kernel matrix; k^*, the kernelized laplacian graph matrix; ω_i, the weights of the characteristics (*geographical coordinates, demand, elevation, emitter coefficient*) and C_i their respective matrices of characteristics.

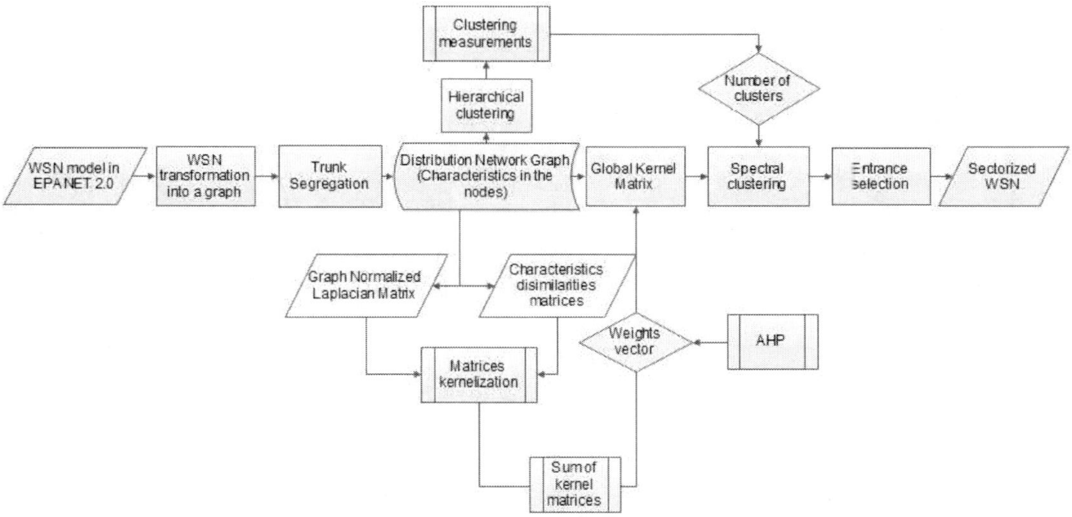

Figure 1. Flowchart of clustering process.

Once clusters in the distribution network are clearly defined, the entrances for each sector (connection of each cluster to the trunk) are determined through an iterative process (see Figure 2). In this process, the lines that are independently able to supply each sector along a 24 hours period are labeled as candidates lines (while the rest of edges lines are closed). Then, for one sector, all the boundary-lines are closed except one of the candidates. In this scenario, an energy evaluation is performed. This energy evaluation is based on two energy performance indices. The first index corresponds to a deviation of the resilience index before and after sectorization. The resilience index (I_r) proposed by [19] compares the total head loss required to reach an given pressure requirement and the actual head loss (see equation 2). Higher values of the index are indicative of more robust networks. The second indicator, named Loss Energy Coefficient (LEC), evaluates the percentage of energy loss in nodes due to sector implementation. This indicator compares the energy in the nodes of each sector (expressed as actual demand times head) before and after sectorization (see equation 3).

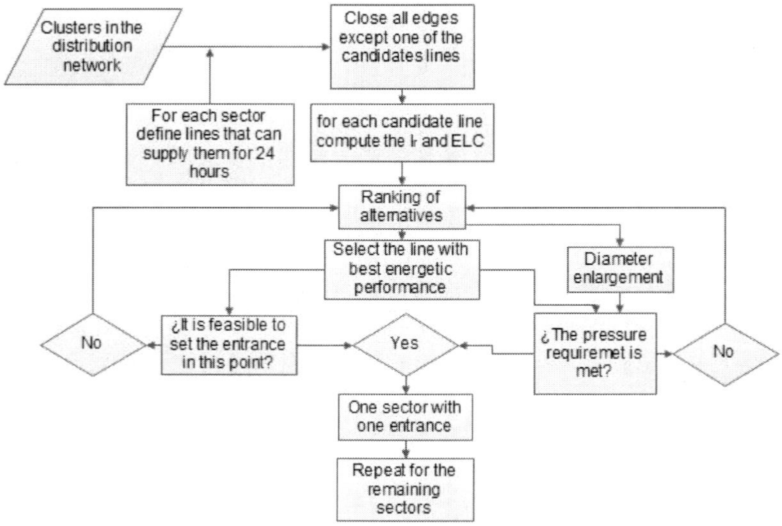

Figure 2. Iterative process for entrances selection.

By means of these two indices, a ranking of solutions may be proposed. In some contexts (cities) urbanistics or administratives features may prevent the allocation of sectors entrances in a given point (e.g. when is in a highly transited avenue). In that case, the next solution of the ranking may be considered, as long as it meet the established pressure requirement.

$$I_r = 1 - \frac{{H_L}^A}{{H_L}^R} \tag{2}$$

Where ${H_L}^A$ denotes the actual head loss in the pipes and ${H_L}^R$ denotes the required head loss in the pipes to reach the minimum pressure threshold.

$$ELC = \frac{\sum_{j=1}^{n} D^R * H^R}{\sum_{j=1}^{n} D^I * H^I} \tag{3}$$

Where D^R and H^R denote the demand (modulated based demand plus the fraction corresponding to leakages) and head after sectorization and D^I and H^I the same magnitude but before sectorization.

Finally, the solution is modelized in EPANET 2.0 to ensure that the system still performs acceptably, specially in term of pressure. Intuitively this verification should be done in the highest demand period, as it represents the most critical period.

3. Example of Implementation

To exemplify the described methodology a WSN network formed by 43 nodes and supplied only by one source is subdivided. The network has a trunk formed with the pipes of largest diameter (over 500 mm). Figure 3. shows the original network and the segregation of the trunk from the distribution network.

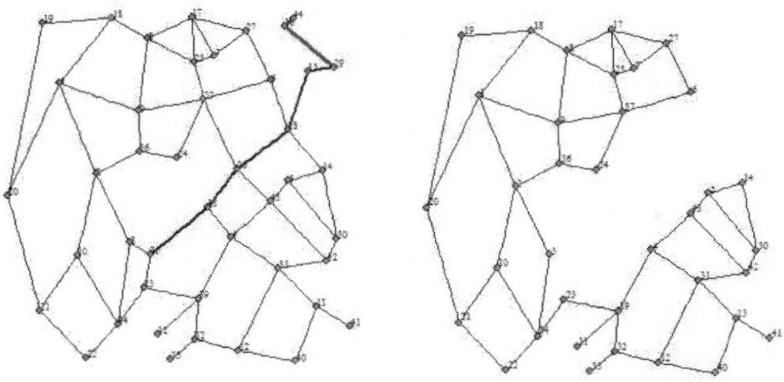

Figure 3. Example network with trunk (left) and distribution network (right).

When hierarchical clustering is implemented over the nodes of the distribution network, a dendrogram that represents nodes connectivity at different levels is obtained. From this dendrogram it was determined that the network would be subdivided into three clusters. By means of the energy assessment, the entrance to each sector is iteratively established and the remaining connections to the trunk closed. Finally, the resulting network layout is hydraulically compared with the layout before sectorization. As shown in Figure 4. the pressure values in the WSN after sectorization still are in an acceptable range.

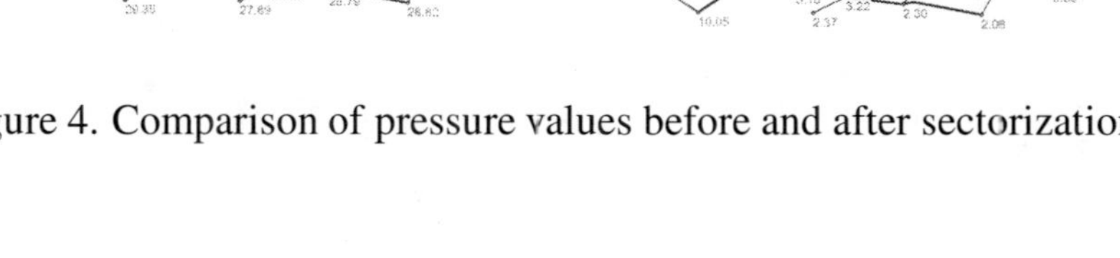

Figure 4. Comparison of pressure values before and after sectorization.

4. Conclusion

This paper presents a novel methodology to address the problem of sectorization in WNS depending of a trunk by means of non-supervised methods belonging to the field of machine learning techniques. Carrying out an initial exploration of the information of the WSN through hierarchical clustering allows the estimation of the number of sectors with highest possible degree of characteristics homogeneity. To have homogenous sectors in a WSNs translates to greater network efficiency. The process of spectral clustering improves the results of hierarchical clustering, as it takes into consideration the level of importance of the partitioning criteria and minimizes the number of edge-cut more efficiently. The use of indicators of energy dissipation through the network make it possible to find the entrance for each sector, in order to minimize the extra energy dissipated by the network (which lowers the pressure droop) as consequence of sector implementation.

This approach has several advantages over other sectorization methodologies previously presented. For one, is one of the few that is applicable to WSNs with a reduced number of water sources. Also, since it avoids considering the trunk, active leakage control after sectorization may be focused in mid-low diameter pipes, which improves efficacy in the range of pipes where leakages are more frequent. Also, the inlet flow measurements in each sector may be more precise and any change in flow (caused by a burst for example) can be instantly detected. In economics terms, as it can be expected, by avoiding the trunk, the required boundary valves and flow meters are of lower diameter. Therefore, the investment for the purchase of flow meters may be significantly reduced.

For the future, it would be interesting to set the number (amount) of connections as restriction when the size of the sector is determined. It would be also interesting to complement the methodology with a multiobjetive function to compare different solutions in monetary terms in long run.

References

[1] Alvisi, S.; Franchini, M. A heuristic procedure for the automatic creation of district metered areas in water distribution systems, *Urban Water j.* 2013, DOI: 10.1080/1573062X.2013.768681.

[2] Di Nardo, A.; Di Natale, M. A heuristic design support methodology based on graph theory for district metering of water supply networks. *Eng Optimiz.* 2011, 43(2), 193-211.

[3] Di Nardo, A.; Di Natale, M.; Santonastaso, G. F.; Salvatore, V. graph partitioning for automatic sectorization of a water distribution system, *in Proc. of the eleventh international conference on computing and control for water industry (CCWI).* 2011, 841-846.

[4] Di Nardo, A.; Di Natale, M.; Santonastaso, G. F.; Tzatchkov, V.; Alcocer-Yamanaka, V. Water network sectorization based on genetic algorithms and minimum dissipated power paths, *Water Sci Technol.* 2013. 13(4), 951-957.

[5] Di Nardo, A.; Di Natale, M.; Santonastaso, G. F.; Tzatchkov, V.; Alcocer-Yamanaka, V. Water network sectorization based on graph theory and energy performance indices. *J Water Res Pl-ASCE.*2013, DOI: 10.1061/(ASCE)WR.1943-5452.0000364.

[6] GIZ (Deutsche Gesellschaft fr Internationale Zusammenarbeit GmbH).; VAG (Armaturen GmbH).; FHNW (Fachhochschule Nordwestschweiz).; KIT (Karlsruhe Institute of Technology). *Guidelines for water loss reduction: a focus on pressure management*, Eschborn, 2011.

[7] Hajebi, S.; Barrett, S.; Clarke, A.; Clarke, S. Multi-agent simulation to automate water distribution network partitioning, *in Proc. of the 27th European simulation and modelling conference - ESM'2013*, 2013.

[8] Han, J.; Kamber, M.; Pei, J. *Data mining: concepts and techniques*, Morgan Kaufmann Publishers Inc, 2006.

[9] Herrera, M. *Improving water networks management by efficient division into supply clusters*, PhD Thesis-Universitat Politècnica de València, 2011.

[10] Herrera, M.; Izquierdo, J.; Pérez-García, R.; Montalvo, I. Multi-agent adaptive boosting on semi-supervised water supply clusters, *Adv Eng Softw*. 2012, 50(August 2012), 131-136.

[11] Izquierdo, J.; Herrera, M.; Montalvo, I.; Pérez-García, R. Agent-Based Division of Water Distribution Systems into District Metered Areas, *in Proc. Software and data technologies: 4th international conference ICSOFT*. 2009, 167-180.

[12] Kingdom, B.; Liemberger, R.; Marin, P. *The challenge of reducing non-revenue water (NRW) in developing countries. how the private sector can help: a look at perfomance-based service contracting*, The World Bank Group, 2006.

[13] Morrison, J.; Stephen, T.; Rogers, D. *District metered areas: guidance notes. Technical Report 1*, Water Loss Task Force- International Water Association (IWA), 2007.

[14] Parelman, L.; Ostfeld, A. Short communication: topological clustering for Water distribution systems analysis, *Environ Modell Softw*. 2011, 26(7), 969-972.

[15] Saaty, T.L. Decision making with the analytic hierarchy process, *Int. J. Services Science*. 2008, 1(1), 83-98.

[16] Shawe-Taylor, J.; Cristiani, N. *Kernel Methods for Pattern Analysis*, Cambridge University Press, 2004.

[17] (TSAWU-ADB) The Southeast Asian Water Utilities Network and Asian Development Bank. *Data book of southeast Asian water utilities 2005* , Asian Development Bank, 2007.

[18] Thornton, J.; Sturm, R.; Kunkel, G. *Water Loss Control*, McGraw-Hill, 2008.

[19] Todini, E. Looped water distribution networks design using a resilience index, *Urban Water*. 2000. 2(2), 115-122.

[20] Tzatchkov, V.; Alcocer Yamanaka, V.; Bourguett Ortiz, V. Sectorización de redes de distribución de agua potable a través de algoritmos basados en la teoría de grafos, *Tláloc AMH*. 2008, 40(Enero-Febrero 2008), 14-22.

[21] Tzatchkov, V.; Alcocer-Yamanaka V. Graph partitioning for automatic sectorization of a water distribution system, *in Proc. of the 10th international conference on hydroinformatics HIC*. 2012.

In: Mathematical Modeling in Social Sciences ...
Editors: J. C. Cortés López et al.
ISBN: 978-1-63117-335-6

Chapter 10

QUANTIFYING THE BEHAVIOR OF THE ACTORS IN THE SPREAD OF ANDROID MALWARE INFECTION

J. Alegre[1], *J. C. Cortés*[1], *F. J. Santonja*[2] *and R. J. Villanueva*[1,*]
[1]Instituto Universitario de Matemática Multidisciplinar,
Universitat Politècnica de València, Valencia, Spain
[2]Departamento de Estadística e Investigación Operativa,
Universitat de València, Burjassot, Valencia, Spain

Abstract

The ubiquity of smart-phones for personal and business use has increased the spread of mobile malware caused by malware applications that the user downloads to the smart-phone. In this chapter, we analyze the actors involved in the the Android malware infection, markets, users and Apps, quantifying the number and type of Apps in each market where a user may download Apps and the behavior of the users when downloading. This quantification will be focused in the Community of Valencia, Spain.

Keywords: Android malware infection, Quantification, App market evolution, User behavior

1. Introduction

The problem of security in computers and networks is well known and has long been considered by governments, companies and users all over the world. However, this awareness of the importance of security is not perceived by mobile users who often act without control and without knowledge of the potential risks, despite the amount of sensitive data stored in these devices. These risks are increasing more even now that it is increasing the policy of permitting employees to bring personally owned mobile devices to their workplace (Bring

*E-mail address: rjvillan@imm.upv.es

Your Own Device BYOD), and use those devices to access privileged company information and applications.

Nowadays, there are already documented many malware types for smart-phones that produce financial-charges, root control, etc. [1, 2]. It is a real problem that must be studied to determine the potential threat to the users. In this chapter, we focus on Android platform because is the most popular mobile platform that gets high market share over other mobile Operating Systems (OS) [3]. The architecture of the Android system is based on Linux, and as a result of that, the security model is based on three milestones:

- *Sandboxing*: The Android platform uses a technique called *sandboxing* to put virtual walls between applications and other software on the device. Therefore, if you download a malicious application, it cannot access data on other parts of your phone and its potential harm is drastically limited.

- *Permissions*: Android provides a permission system to help you understand the capabilities of the Apps you install, and manage your own preferences. That way, if you see a game that unnecessarily requests permission to send SMS, for example, you should not install it.

- *Malware removal*: The official Android market has a service named *Bouncer*, which provides automated scanning of Apps uploaded to Android market before being available for the users that detects potentially malicious software.

However, despite this security model, multiple types of malware embedded in Apps released in the Apps stores have been found. As Google says: ”no security approach is foolproof” [4].

In this chapter, we analyze the actors involved in the the spread of Android malware infection: Apps, official and non-official markets and users, quantifying the number of Apps in each market, which of them are malware Apps, and how the users behave when they have a smart-phone in their hands.

This chapter is organized as follows. In Section 2 we introduce the Apps characteristics and properties we are going to take into account. In Section 3 we quantify how official and non-official markets evolve, that is, we quantify the Apps and malware Apps in each market over the time, how the Apps are distributed by popularity and the mechanism of malware detection. In Section 4 we study the most important facts that determine the the user behavior in order to know if his/her smart-phone is going to be infected or not. Finally, Section 5 is devoted to conclusions.

2. The Apps

The users, with their own characteristics in their devices, access the markets and download applications (Apps) with also different characteristics. Thus, we consider two domains interacting among them:

- The markets environment.

- The users environment.

Figure 1 shows an UML (Unified Modelling Language) representation of the Apps and Clients (users). *App*, that represents the application in a given market, and *Client*, that represents the device of every user.

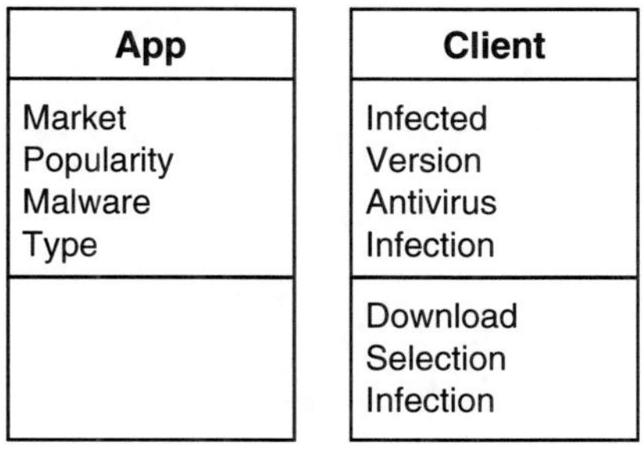

Figure 1. Agent attributes and functions.

The mobile malware spread through the Apps that are in the markets and the users download to their device. The Apps are stored in the markets and these markets can be official, as Google Play, or alternative or non-official markets. This is determined by the attribute *Market*. Every App has its own popularity that determines the probability to be downloaded and that it is stored at the attribute *Popularity*. Furthermore, malware can be classified depending on the effect they produce over the *Client* [1]:

- *Privilege Escalation*: The App gets the root privileges of the device. Depending on the Client's OS version, this kind of malware affects or not.
- *Remote Control*: Remote servers take the control of the device.
- *Financial Charge*: The App sends messages to premium accounts from the device and the money these messages cost has to be paid by the user of the smart-phone.
- *Information Collection*: The App takes private information of the device, like the contacts, agenda, SMS messages, user accounts, etc., and upload the information to a remote server.

If an App is malware and what kind of malware is, it is established by attributes Malware and Type respectively.

3. The Markets

3.1. The Official Market

The official market, also known as Google Play [5], is a repository of Apps where the users of Android smart-phones can download freely or under payment Apps, music, movies or books. We are going to focus only in Apps because they are the ones responsible of malware in the cell phones.

New Apps Entering Every Month in the Official Market

In Jul 2011, the number of applications in Google Play were $221,875$ [6]. Now, we want to describe the behavior of the official market to estimate the number of new apps every month. Some values taken in different dates from July 2011 can be seen in Table 1 [6].

Table 1. Number of Apps in the official market

Date	#Apps
July 1, 2011	221, 875
Sept 1, 2011	271, 875
Nov 1, 2011	309, 375
Jan 1, 2012	343, 750
Mar 1, 2012	400, 000
May 1, 2012	440, 645
May 20, 2012	443, 920
Feb 12, 2013	626, 865

Taking into account that the evolution is practically a linear function, we can fit $b + a, t$ with data of Table 1, obtaining the function

$$f_{OM}(t) = 225,970 + 20,740.1, t, \tag{1}$$

where t is the number of months since July 2011. Function (1) allows us to estimate the number of Apps in the official market over the next months.

New Malware Apps Entering Every Month in the Official Market

Data about malware is very difficult to find and they may not be reliable because the sources use to be antivirus developer companies. In spite of this, in order to conduct the study, we have had to trust in the few available data published in [1] and appear in Table 2.

Table 2. Number of malware Apps in the official market

Date	#Apps
July 1, 2011	86
Aug 1, 2011	86
Sept 1, 2011	103
Oct 1, 2011	200

In this case, we have less data as before and an appropriate fitting is not as good as we did above. Nevertheless, we are going to assume that the growing of malware Apps also has a linear increasing and the line that best fit the data in Table 2 is the function

$$f_{OMm}(t) = 64.9 + 35.9, t, \tag{2}$$

where t is the number of months since July 2011.

Distribution of Apps According Their Popularity

In the Android markets, Apps are classified according to their popularity as: none; less than 2.5 stars; $2.5 - 3$ stars; $3 - 3.5$ stars; $3.5 - 4$ stars; $4 - 4.5$ stars; greater than 4.5 stars. After some accesses to the distribution of the Apps by popularity website in different dates [7], we noted that there were minor changes and consequently we assume that the distribution of Apps by popularity is constant over the time. The distribution for Jul 2011 is given in Table 3.

Table 3. Distribution of Apps by popularity in Jul 2011

Popularity	None	2.5	$2.5 - 3.0$	$3.0 - 3.5$	$3.5 - 4.0$	$4.0 - 4.5$	> 4.5
#Apps	114,789	5,985	6,053	13,512	21,599	31,493	28,444
%Apps	51.74%	2.70%	2.73%	6.09%	9.73%	14.19%	12.82%

Distribution of Malware Apps According Their Popularity

The distribution of malware Apps is not uniform among popularity ratings. There is a way to create malware Apps called *repackaging* [1]. Repackaging consists of taking a popular App, introducing some malware code and upload it again. 86% of malware is repackaging [1] and we are going to assume that these malware Apps have popularity $4.0 - 4.5$ or > 4.5 distributed uniformly. Thus, in 4, we can see the distribution of the malware Apps in Jul 2011 distributed by popularity.

Table 4. Distribution of malware Apps by popularity in July 2011**, taking into account repackaging**

Popularity	None	2.5	$2.5 - 3.0$	$3.0 - 3.5$	$3.5 - 4.0$	$4.0 - 4.5$	> 4.5
#Apps	9	0	0	1	2	39	35

Malware Detection

The official Android market has a service named *Bouncer* which provides automated scanning of Apps uploaded to Android market before being available for the users that detects potentially malicious software. The admitted effectiveness of this service is around 40% [4]. This parameter will be considered in order to know the probability that the official market detects a malware App and withdraw it.

3.2. The Non–Official Market

Non-official markets are markets other than Google Play where the users can also download Android Apps. The behavior of these markets are similar to the official market, however, some differences should be taken into account because their relevance on the malware infection. First of all, we are going to assume that all the non-official markets are gathered in only one with more than $2,600,000,000$ of downloads [8].

New Apps Entering Every Month in the Non-Official Market

Non-official market [8] had $568,661$ available Apps in Jan 2012 whereas GooglePlay had, in the same month, $343,750$. Taking into account that no much more data about the number of Apps in non-official market are available, we are going to assume that the ratio 1.65 $(568,661/343,750)$ is a constant relation of the number of Apps in the official and non-official markets over the time. Therefore

$$f_{NOM}(t) = 1.65, f_{OM}(t), \tag{3}$$

describes the evolution of Apps in the non-official market, where t is the number of months since July 2011.

Malware Apps Entering Every Month in the Non-Official Market

Data about malware in the non-official market can be found in [1] and can be seen in Table 5.

Table 5. Number of malware apps in the non-official market

Date	#malware Apps
July 1, 2011	485
Aug 1, 2011	810
Sept 1, 2011	1008
Oct 1, 2011	1172

The above data can be fitted accurately using a linear function, obtaining

$$f_{NOMm}(t) = 529.9 + 225.9, t, \tag{4}$$

where t is the number of months since July 2011.

Distribution of Apps According Their Popularity

Using the same criteria as in the official market, we classify the Apps depending on their popularity in the non-official market as given in Table 6. We also consider this distribution constant over the time.

Table 6. Distribution of Apps by popularity in in July 2011 in the non-official market

Popularity	None	2.5	2.5 – 3.0	3.0 – 3.5	3.5 – 4.0	4.0 – 4.5	> 4.5
#Apps	189, 894	9, 900	10, 014	22, 353	35, 730	52, 098	47, 055
%Apps	51.74%	2.70%	2.73%	6.09%	9.73%	14.19%	12.82%

Distribution of Malware Apps According Their Popularity

Using the same criteria as in the official market, we classify the Apps depending on their popularity in the non-official market as given in Table 7. Repackaging is also considered.

Table 7. Distribution of malware Apps by popularity in July 2011 in the non-official market

Popularity	None	2.5	2.5 – 3.0	3.0 – 3.5	3.5 – 4.0	4.0 – 4.5	> 4.5
#Apps	48	3	3	6	9	219	198

Malware Detection

We do not have any information about the existence of an antivirus checking if the new Apps contain malware code in the non-official market. Therefore, we are going to assume that the non-official market does not have any control about the malware Apps.

4. The Users

The *Client* attributes determines if it is infected or not (attribute Infected), the OS version of the client's device (attribute *Version*), if the device has or not software protection (attribute *Antivirus*) and the kind of infection in case of an infected client (attribute *Infection*). The *Version* attribute is used in order to know if a *Privilege Escalation* malware affects or not the *Client*.

In 2011 there were in Spain a population of $47, 190\,493$ people [9]. 46% of them had a smartphone [10] and 50% of them had an Android terminal [11], that is, $10, 853, 813$. The population in the Community of Valencia in 2011 was $5, 117, 190$ inhabitants [9]. Applying the same rule as above we have that, in the Community of Valencia there were $1, 176, 954$ Android smartphones.

4.1. Number of Apps Downloaded Per Month

The number of Apps downloaded by a user in a month follows a Poisson distribution:

$$f(k, \lambda) = \frac{e^{-\lambda}, \lambda^k}{k!}, \tag{5}$$

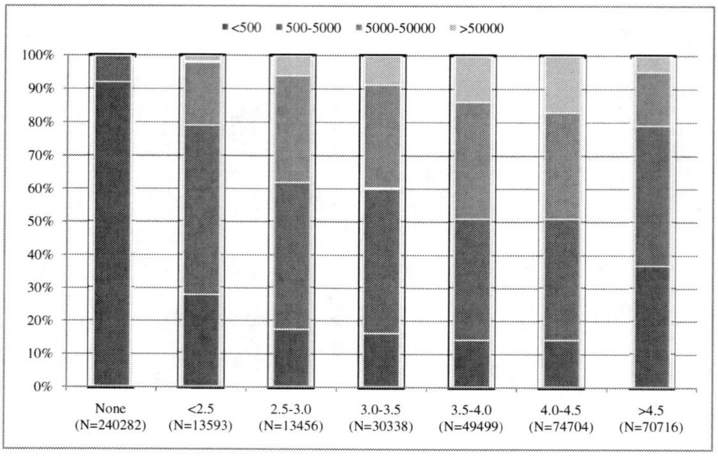

Figure 2. Downloads per popularity.

where k is the number of downloaded Apps and λ is the average number of Apps downloaded, both every month in every smart-phone. Taking into account that, in Spain, $67,293,800$ Apps were downloaded in Spain in Apr 2012 [12] by $10,853,813$ smart-phones, every user downloads an average of 6.2 Apps per month. Therefore, $\lambda = 6.2$.

4.2. App Downloads by Popularity

It is clear that, when a user is going to download an App, unless he/she wants to download a specific App, first, the user uses to have a look among the most popular Apps. Therefore, the Apps are not downloaded following a uniform distribution. In order to approach this behavior, let us consider the Figure 2, [7].

As we mentioned before, we assume that the distribution of the Apps per popularity is constant over the time. However, we do not have data about the values where the colors change in Figure 2, and we did an estimation gathered in the Table 8.

Table 8. Percentage of download distribution of Android Apps per number of downloads

Popularity/ #downloads	< 500	$500 - 5,000$	$5,000 - 50,000$	$> 50,000$
None	92%	8%	0%	0%
< 2.5	28%	51%	19%	2%
$2.5 - 3.0$	17%	45%	32%	6%
$3.0 - 3.5$	16%	44%	31%	9%
$3.5 - 4.0$	14%	37%	35%	14%
$4.0 - 4.5$	14%	37%	32%	17%
> 4.5	37%	42%	16%	5%

Let us denote by $p(i, j), i = 1, \ldots, 7, j = 1, 2, 3, 4$ the entries in Table 8. For instance, $p(2, 3) = 19\%$. Also, we call c_1, c_2, c_3 and c_4, $0 \leq c_1 < c_2 < c_3 < c_4$, the average number of downloads of Apps with less than 500 downloads, between $500 - 5,000$ downloads, between $5,000 - 50,000$ downloads and more than $50,000$ downloads, respectively. Then, taking into account that the number of Apps per popularity (h_j) are $240,282$; $13,593$; $13,456$; $30,338$; $49,499$; $74,704$ and $70,716$ (see Figure 2), the number of total downloads will be

$$\sum_{j=1}^{7}\sum_{i=1}^{4} c_i, p(j,i), h_j \tag{6}$$

where

$$\begin{array}{c} h_1 = 240,282, \quad h_2 = 13,593, \quad h_3 = 13,456, \quad h_4 = 30,338, \\ h_5 = 49,499, \quad h_6 = 74,704, \quad h_7 = 70,716. \end{array} \tag{7}$$

Simplifying the expression (6) we have

$$\frac{1}{50}\left(13,778,021, c_1 + 6,060,737, c_2 + 3,441,893, c_3 + 1,348,749, c_4\right).$$

If we were trying to find out the average number of downloads (c_i, $i = 1, 2, 3, 4$) for all the people over the world, we would have to assume that $c_1 < 500$, $500 \leq c_2 < 5,000$, $5,000 \leq c_3 < 50,000$ and $c_4 > 50,000$. However, we are going to restrict the downloads to the Community of Valencia and consequently, c_i, $i = 1, 2, 3, 4$ do not have to satisfy the above restrictions. In fact, they will be much lower. Thus, taking into account that $67,293,800$ Apps were downloaded in Spain in Apr 2012 [12] (closest data available to Jul 2012), the population in Spain in Apr 2012 was $46,185,697$ inhabitants and in the Community of Valencia is $5,009,635$ [9], we are going to assume that the number of Apps downloaded in the Community of Valencia in Apr 2012 was

$$67,293,800, \frac{5,009,635}{46,185,697} = 7,299,173, \text{Apps.} \tag{8}$$

Consequently, for the Community of Valencia we have that the following equality should be satisfied

$$\frac{1}{50}\left(13,778,021, c_1 + 6,060,737, c_2 + 3,441,893, c_3 + 1,348,779, c_4\right) = 7,299,173. \tag{9}$$

Isolating c_1, we have

$$c_1 = \frac{887,150,988,850,000}{33,491,973,850,823} - \frac{6,060,737}{13,778,021}, c_2 - \frac{3,441,893}{13,778,021}, c_3 - \frac{1,348,7497}{13,778,021}, c_4. \tag{10}$$

Taking into account that $0 \leq c_1 < c_2 < c_3 < c_4$, c_4 will take its maximum value when $c_1 = c_2 = c_3 = 0$ and, in this case, we have that

$$\frac{887,150,988,850,000}{33,491,973,850,823} - \frac{1,348,7497}{13,778,021}, c_4 = 0, \tag{11}$$

and the maximum value that c_4 can reach is 270.59. Summarizing the above reasoning, if we call d_1, d_2, d_3 and d_4 the probabilities a user in the Community of Valencia downloads an App which number of downloads in all the world are less than 500, between $500-5,000$, between $5,000-50,000$ or more than $50,000$ downloads, respectively, is

$$d_i = \frac{c_i}{C}, \quad i = 1, 2, 3, 4, \tag{12}$$

where $C = c_1 + c_2 + c_3 + c_4$ and c_i, $i = 1, 2, 3, 4$ should satisfy that

$$c_1 = \frac{887,150,988,850,000}{33,491,973,850,823} - \frac{6,060,737}{13,778,021}, c_2 - \frac{3,441,893}{13,778,021}, c_3 - \frac{1,348,7497}{13,778,021}, c_4, \tag{13}$$

where

$$0 \leq c_1 < c_2 < c_3 < c_4 < 270.59. \tag{14}$$

4.3. OS Version Evolution and Infection by *Privilege Escalation* Malware

The OS version is an important parameter in order to estimate the infection by *Privilege Escalation* malware. We assume the evolution of the OS version installed on the smart-phones as given in Table 9 [13].

Table 9. Distribution of OS version in Android smart-phones from July 2011 **until Feb** 2013

Version	Affected	July 2011	Oct 2011	Feb 2012	June 2012	Oct 212	Feb 2013
1.5	Cupcake	1.40%	0.90%	0.40%	0.20%	0.10%	0.00%
1.6	Donut	2.20%	1.40%	0.80%	0.50%	0.30%	0.20%
2.1	Eclair	17.50%	10.70%	6.60%	4.70%	3.10%	1.90%
2.2	Froyo	59.40%	40.70%	25.30%	17.30%	12.00%	7.50%
2.3	Gingerbread	18.60%	44.40%	62.00%	64.00%	54.20%	44.10%
3.0	Honeycomb	0.90%	1.90%	3.30%	2.40%	1.80%	1.20%
4.0	Icecream	0.00%	0.00%	1.60%	10.90%	25.80%	28.60%
4.1	Jelly	0.00%	0.00%	0.00%	0.00%	2.70%	16.50%

The percentage of devices that can be affected by the most common Android privilege escalation vulnerabilities is given in Table 10 [14].

4.4. Users with Antivirus Installed in Their Devices

The number of users with antivirus installed in their devices is 33% [15]. The admitted effectiveness of these antivirus software is between 20.2% and 79.6% [1].

4.5. Conditions for a User to Be Infected by Malware App

We will know if a downloaded malware App infects the client if one of the following conditions are met:

- Malware App (Privilege escalation) + Vulnerable OS + None antivirus installed.
- Malware App (Remote Control, Financial Charge or Information Collection) + not antivirus installed.

Table 10. Percentage of devices that can be affected by the most common privilege escalation vulnerabilities, depending on the Android OS version

Version	Name	Affected
1.5	Cupcake	100%
1.6	Donut	100%
2.1	Eclair	96.70%
2.2	Froyo	98.80%
2.3	Gingerbread	100%
3.0	Honeycomb	0.00%
4.0	Icecream	31.00%
4.1	Jelly	0.00%

- Malware App (Privilege escalation) + Vulnerable OS + Probability of no detection by antivirus installed.
- Malware App (Remote Control, Financial Charge or Information Collection) + Probability of no detection by antivirus installed.

4.6. Probability a User Detects His/Her Smart-Phone Is Infected and Repair It

We assume that a user only detects and repairs infections caused by Financial Charge malware. The detection is made monthly when the user receives the mobile bill. Other cases are difficult to estimate but we consider also that a smart-phone user changes his/her smart-phone, as average, every 12 months.

5. Conclusion

In this chapter we described how the Apps and the malware Apps evolve over the time in the Official market and the non-official market. Also, we included all the relevant characteristics about the behavior of a user, that is, what a user does when downloading an App, the probability the Apps is malware, the probability it infects the smart-phone and what the user does if finally it is infected.

Our idea, in a future work, is to use these data to build a dynamic model to introduce the malware propagation on mobile devices via applications. Then, the results of the model will give us the population affected by malware and the type of infection can be quantified and qualified. Thus, given a population, the number of infected devices would be estimated over the time and the kind of infection would be determined. Moreover, it would be possible to analyze the critical part of the smart-phones business related to malware: if the Privilege Escalation malware would be prevalent, that would mean that the vulnerabilities of the Android platform permiting the rooted of the device would be critical and more effort on patching these vulnerabilities should be addressed. Nevertheless, if the prevalent malware would be Remote Control, Financial Charge or Information Collection, that would mean that the users, giving indiscriminate permissions to the Apps or not protecting their devices with mobile anti-virus software, would be primarily responsible for the infection of the devices. Furthermore, with the results, it would be possible to estimate the cost the users should afford in case of malware that causes financial charges.

References

[1] Zhou, Y; Jiang, X. (2012). Dissecting Android Malware: Characterization and Evolution. *Proceedings of the 33rd IEEE Symposium on Security and Privacy*, Oakland 2012, San Francisco, CA.

[2] http://www.malgenomeproject.org (accessed at 10 October, 2013).

[3] http://en.wikipedia.org/wiki/Mobile-operating-system (accessed at 10 October, 2013).

[4] http://googlemobile.blogspot.com.es/2012/02/android-and-security.html (accessed at 10 October, 2013).

[5] https://play.google.com/store (accessed at 10 October, 2013).

[6] http://www.appbrain.com/stats/number-of-android-apps (accessed at 10 October, 2013).

[7] http://www.appbrain.com/stats/android-app-ratings.

[8] http://www.getjar.com.

[9] Spanish National Institute of Statistics. http://www.ine.es (accessed at 10 October, 2013).

[10] TomiAhonen Consulting Analysis December 2011, based on raw data from Google/Ipsos, *the Netsize Guide/Informa*, and TomiAhonen Almanac 2011 reported data.

[11] http://en.wikipedia.org/wiki/Android (operating system) (accessed at 10 October, 2012).

[12] http://xyologic.com (accessed at 10 October, 2013).

[13] http://en.wikipedia.org/wiki/Android (operating system) (accessed at 10 October, 2013).

[14] Webinar: *X-Ray Results- Mobile Device Vulnerabilities, Duo Security*, www.duosecurity.com, October 2012.

[15] Mylonas, A; Kastania, A; Gritzalis, D. (2013). Delegate the smartphone user? Security awareness in smartphone platforms, *Computers & Security*, 34, 47–66.

In: Mathematical Modeling in Social Sciences … ISBN: 978-1-63117-335-6
Editors: J. C. Cortés López et al.

Chapter 11

A Stochastic Agent-Based Approach to Interregional Migration in Quantitative Sociodynamics

Minoru Tabata[1,*], Nobuoki Eshima[2], Keiko Kanenoo[1] and Ichiro Takagi[3]
[1]Department of Mathematical Sciences,
Osaka Prefecture University, Sakai, Osaka, Japan
[2]Department of Statistics, Oita University, Oita, Japan
[3]Department of Business Management, School of Business Studies,
Tokai University, Kumamoto, Japan

Abstract

In order to describe interregional migration we construct an agent-based model whose agents relocate stochastically to obtain higher utility in a bounded discrete domain. By making use of numerical solutions of the discrete master equation, we will describe the behavior of the agent-based model in order to study how the concentration of agents changes with the cost incurred in moving.

Keywords: Migration; sociodynamics; agent-based model

1. Introduction

There are recently pros and cons over joining TPP (Trans-Pacific Strategic Economic Partnership Agreement) in Japan. By the way, many free trade areas have been established since 1990s such as CEFTA ('93), AFTA ('93), NAFTA ('94), SADC ('94), ANCOM ('95), and FTAA (2005). Goods and services are freely traded within such trade areas, but population mobility is restricted in general. However, there is a new move to abolish the

[*] E-mail address: mnrtabata@luck.ocn.ne.jp, Fax: +81 72 254 9916.

restriction. For example, the EU has been moving toward an economic integration of 27 countries, and it is trying to remove the restriction in the near future. If the restriction is abolished entirely, then population will move to more desirable locations within the EU countries. Such a socioeconomic phenomenon is an example of *interregional migration*, and has been fully studied in quantitative sociodynamics (see, e.g., [1], [7] and [8]). We will construct an agent-based model to describe the phenomenon. We assume that each agent represents an individual and relocates to obtain higher *utility*, where the utility denotes a quantity representing socioeconomic desirability. In the real world we often observe that the socioeconomic desirability depends on the population density. Hence, we reasonably assume that the utility is a function of the density of agents. The socioeconomic desirability is evaluated by each individual according to his/her own preference. Nonetheless, preferences may vary very greatly, and it is impossible to explicitly know preference of each individual. In addition, various unpredictable socioeconomic events may occur. Therefore, it is necessary to assume that the utility function contains a *random variable*. Hence, because of this stochastic property, it is very difficult to observe the behavior of the agent-based model. However, it is proved that the behavior of the agent-based model is approximated by solutions of *the discrete master equation* (see, e.g., [1-6]). Hence, we will overcome such a difficulty by making use of this result to describe the behavior of the agent-based model.

2. The Agent-based Model

By D we denote a rectangle of the form, $D := [0,a)\times[0,b)$,where a and b are positive constants. We divide D into small disjoint rectangles as follows:

$$D = \cup_{i,j=1,\dots,N}[(i-1)a/N, ia/N)\times[(j-1)b/N, jb/N), \tag{1}$$

where N is an integer. We call these small rectangles *sections*. We number the sections from 1 to N^2, and we denote them by d_j, $j = 1,\dots,N^2$. We denote the total number of agents by R. We assume that each agent has *the size*, γ/R, where γ is a positive constant. Assuming that agents are distributed uniformly in each section, we define *the density of agents* $\mathbf{f} = \mathbf{f}(t,x)$ as follows:

$$\mathbf{f}(t,x) := (\gamma/R)R_j(t)/(ab/N^2) \text{ for each } (t,x)\in[0,+\infty)\times d_j,\ j = 1,\dots,N^2, \tag{2}$$

where $R_j = R_j(t)$ denotes the number of agents located in d_j at time $t \geq 0$, $j = 1,\dots,N^2$. Note that the factor ab/N^2 is equal to the area of each section. We regard $R_j = R_j(t)$, $j = 1,\dots,N^2$, as random variables depending on the time variable $t \geq 0$. Because R is not a random variable but a positive constant, we obtain *the conservation law of total number of agents*,

$$R = \sum_{j=1}^{N^2} R_j(t), \text{ for each } t \geq 0. \tag{3}$$

We assume that each agent can relocate at each time $t = n\Delta t$, $n\in\mathrm{N}$, where we denote the set of all natural numbers by N, and Δt is a small positive constant that represents the least unit of time variable. We see that $\mathbf{f} = \mathbf{f}(t,x)$ is constant in each subset of the form, $(n\Delta t,(n+1)\Delta t)\times d_j$, where $j = 1,...,N^2$, and $n \in \mathrm{N}\cup\{0\}$. We assume that the utility depends only on $\mathbf{f} = \mathbf{f}(t,x)$. We will make the following assumption in order that the utility changes stochastically:

Assumption 1. The utility has the following form:

$$\mathbf{U} = \mathbf{U}(t,j) := U(\mathbf{f}(t,X_j)) + S(t,j),\ j = 1,...,N^2, \tag{4}$$

where we denote the utility of a section d_j at time $t = n\Delta t$, $n\in\mathrm{N}\cup\{0\}$, by $\mathbf{U} = \mathbf{U}(t,j)$, $j = 1,...,N^2$, $U = U(\mathbf{f})$ is a smooth given function of $\mathbf{f}\in [0,+\infty)$, and $S = S(t,j)$ is a nonnegative-valued random variable depending on $(t,j) \in \{t = n\Delta t;\ n \in \mathrm{N}\cup\{0\}\}\times\{1,...,N^2\}$. We denote the center of d_j by X_j, $j = 1,...,N^2$.

We assume that the random variables $S = S(t,j)$, $t = n\Delta t$, $n\in \mathrm{N}\cup\{0\}$, $j = 1,...,N^2$, are independent of each other, and that their density functions are the same. We denote the density function by $\rho = \rho(S)$. In the same way as [6], we assume that

$$\rho(S) := \exp(-S) \text{ for each } S \geq 0. \tag{5}$$

If an individual moves from one section to another in the real world, then he/she needs to bear *the cost of moving*, which increases with the distance between the sections. Hence we assume that the cost incurred in moving from d_j to d_i is equal to $C = C(|X_i - X_j|)$ for each $i,j = 1,...,N^2$, where $C = C(r)$ is a sufficiently smooth increasing nonnegative-valued given function of $r \geq 0$.We will impose the following assumption on the relocation of agents:

Assumption 2. At each time $t = n\Delta t$, $n \in \mathrm{N}\cup\{0\}$, each agent contained in d_j decides whether or not to move to the chosen section d_i by comparing the utility of d_j with that of d_i. If

$$\mathbf{U}(t,i) - \{\mathbf{U}(t,j) + C(|X_i - X_j|)\} > A, \tag{6}$$

where A is a positive constant, then the agent moves from d_j to d_i. If not, then the agent stays in d_j in the time interval $[n\Delta t,(n+1)\Delta t)$.

3. The Discrete Master Equation

By **M** we denote the agent-based model constructed above. Let us construct the corresponding continuous model, which is described by the discrete master equation,

$$df(t,X_i)/dt =$$

$$-w(f(t,\cdot);X_i)f(t,X_i) + \sum_{j=1}^{N^2} W(f(t,\cdot);X_i|X_j)f(t,X_j)(ab/N^2),\ i = 1,...,N^2, \tag{7}$$

where

$$W(f(t,\cdot);x|y) := \nu\exp\{U(f(t,x)) - U(f(t,y)) - C(|x-y|)\},$$

$$w(f(t,\cdot);x) := \sum_{j=1}^{N^2} W(f(t,\cdot);X_j|x)(ab/N^2),$$

$f = f(t,x)$ denotes an unknown function of $(t,x) \in [0,+\infty)\times D$ and, ν is a positive constant called the *flexibility*, which describes the activity of population. It is proved in [2] and [5]that the initial value problem has a unique global solution. Therefore, we see that the discrete master equation can define a continuous model. By M we denote the continuous model thus defined. The following theorem is proved in [6].

Theorem. If the total number of agents R tends to infinity and if the least unit of time variable Δt converges to 0+0, then the stochastic agent-based model **M** converges to the continuous model M with a probability converging to 1.

Hence, if we describe interregional migration in terms of **M**, then the description is almost the same as done by M when R is sufficiently large and $\Delta t > 0$ is sufficiently small. Hence, in order to describe the behavior of **M** we will do numerical simulations of M. We assume that the initial density of agents is equal to the following function:

$$f(0,x) = 1/2 + (1/2)\sin(9\pi^2 x_1 x_2/4), \tag{8}$$

where we denote the i-th component of x by x_i, $i = 1,2$, i.e., $x = (x_1,x_2)$. Figure1 describes the graph of (8).

We perform simulations when

$$a, b = 1, C(z) = cz, U(w) = w, N = 33, \nu = 1, \tag{9}$$

where c denotes a nonnegative constant. We do numerical simulations of M in exactly the same way as [3]. Figures2-5 describe numerical simulations when $c = 0$, i.e., when the effort incurred in moving is identically equal to 0. Inspecting these figures, we see that the evolution of the density of agents has two different stages. Figures 2 and 3 and Figures 4 and 5 describe the first stage and the second stage, respectively. The density of agents exhibits self-organization in the first stage. However, the spatial structure thus self-organized comes to crumble quickly in the second stage.

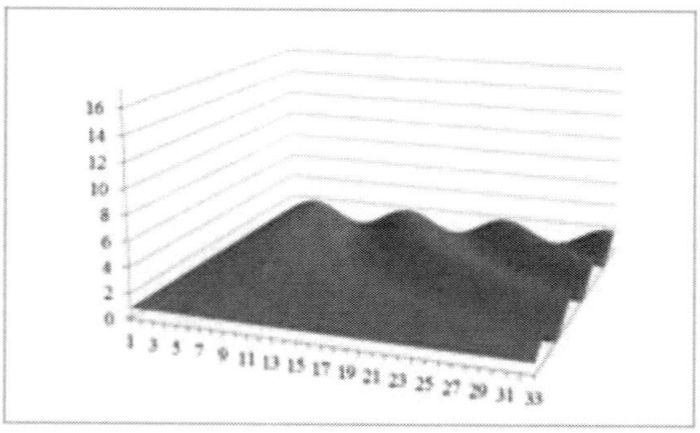

Figure 1. The initial density of agents.

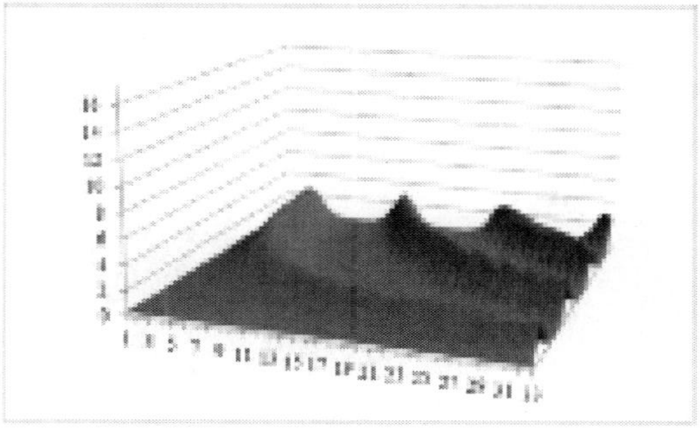

Figure 2. The density of agents at t = 0.85 when c = 0.

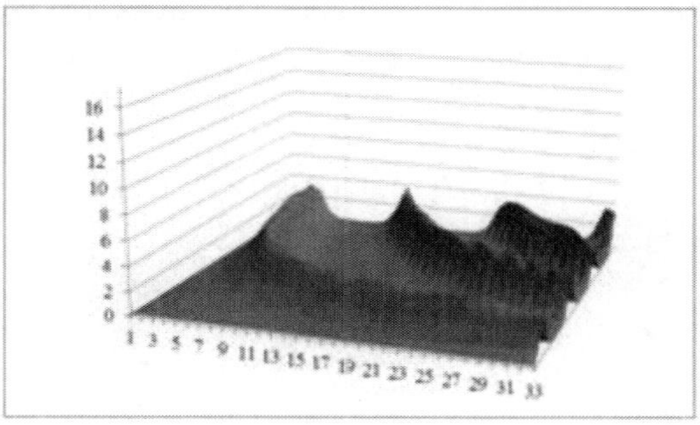

Figure 3. The density of agents at t = 1.00 when c = 0.

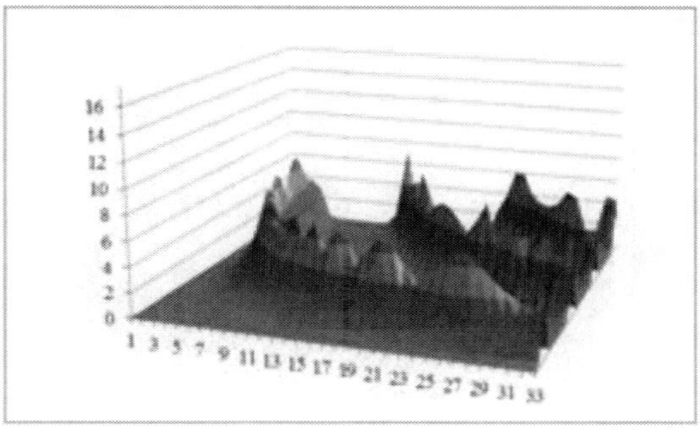

Figure 4. The density of agents at t = 1.15 when c = 0.

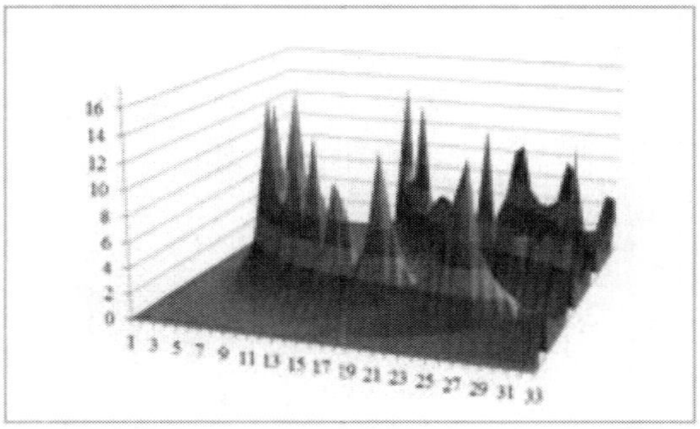

Figure 5. The density of agents at t = 1.17 when c = 0.

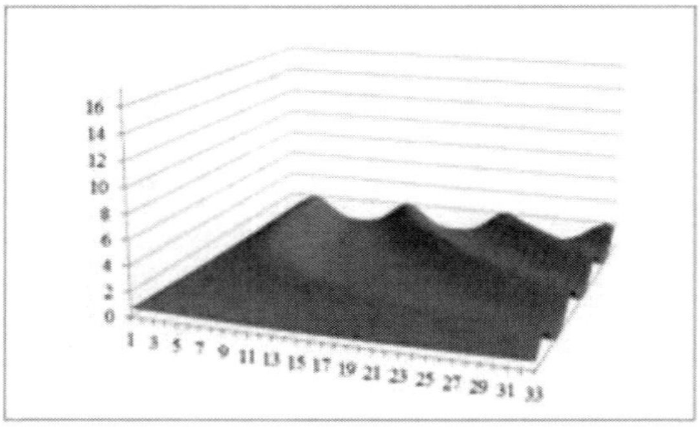

Figure 6. The density of agents at t = 4.25 when c = 5.

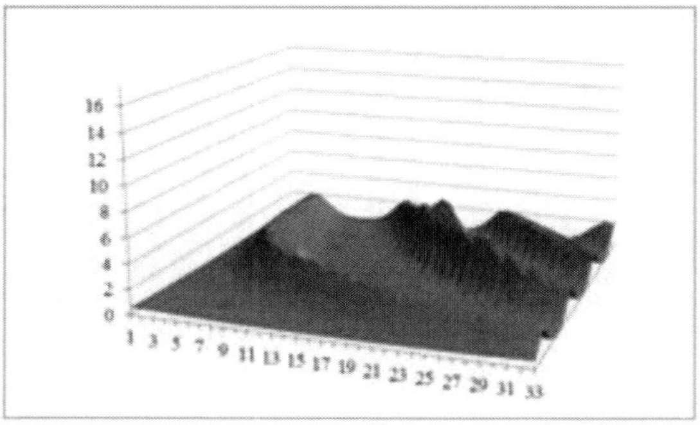

Figure 7. The density of agents at t = 6.0 when c = 5.

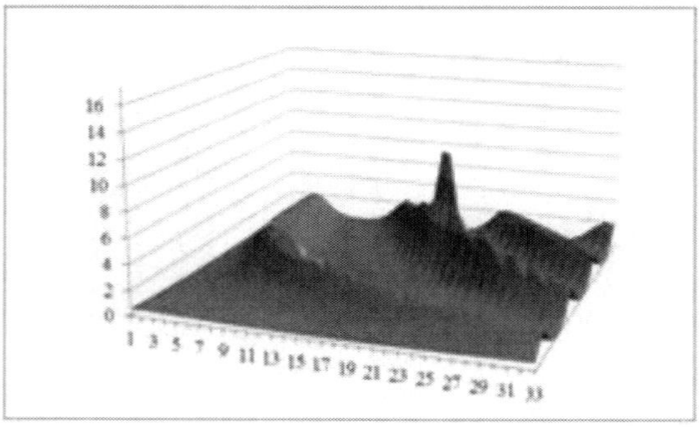

Figure 8. The density of agents at t = 6.1 when c = 5.

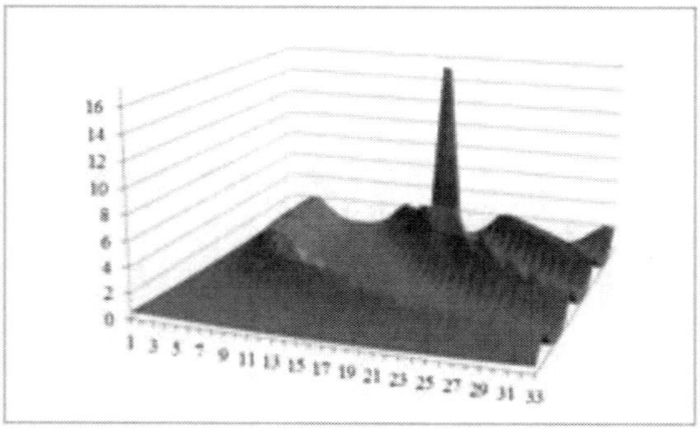

Figure 9. The density of agents at t = 6.11 when c = 5.

Figures 6-9 describe numerical simulations when $c = 5$. Inspecting these figures, we see that the density of agents slowly exhibits self-organization as the time variable increases. Inspecting Figures 7-9, we see that the spatial structure thus self-organized is pulled strongly in the neighborhood of the central part of the domain D as the time variable increases.

Conclusion

If the moving cost is large, then the center of the domain gravitates agents strongly. If not, then the center of the domain cannot gravitate agents.

References

[1] Helbing, D. *Quantitative Sociodynamics*; Springer: Heidelberg, 2010.

[2] Tabata, M. and Eshima, N. The behavior of solutions to the Cauchy problem for the master equation, *Appl. Math. Comput.* 2000, vol. 112, No. 1, pp 79-98.

[3] Tabata, M., Eshima, N., and Takagi, I.The master equation approach to self-organization in labor mobility, In *Evolutionary Controversy in Economics Towards a New Method in Preference of Trans-discipline*; Aruka, Y.; Ed.; Springer-Verlag: Tokyo, 2001.

[4] Tabata, M., Eshima, N., and Takagi, I. An infinite continuous model that derives from a finite discrete model describing the time evolution of the density of firms, *Appl. Math. Comput.* 2002, vol. 125, pp 105-132.

[5] Tabata, M. and Eshima, N. The Cauchy problem for the nonlinear integro-partial differential equation in quantitative sociodynamics, *Appl. Math. Comput.* 2002, vol. 132, No. 2-3, pp 537-552.

[6] Tabata, M. and Eshima, N. The behavior of stochastic agent-based models when the number of agents and the time variable tend to infinity, *Appl. Math. Comput.* 2004, vol. 152, No.1 pp 47-70.

[7] Weidlich, W. and Haag, G. *Interregional Migration*; Springer: Berlin, 1988.

[8] Weidlich, W. *Sociodynamics*; Harwood Academic Publishers; Amsterdam, 2000.

In: Mathematical Modeling in Social Sciences ...
Editors: J. C. Cortés López et al.
ISBN: 978-1-63117-335-6

Chapter 12

A BAYESIAN MATHEMATICAL MODEL TO ANALYSE RELIGIOUS BEHAVIOR IN SPAIN

***R. Cervelló-Royo*[1,*], *A. Sánchez-Sánchez*[2,†], *F. Guerrero*[3,‡], *F. J. Santonja*[4,§] *and R. J. Villanueva*[2,ϕ]**

[1]Departamento de Economía y Ciencias Sociales.
Universitat Politècnica de València, Spain
[2]Instituto de Matemática Multidisciplinar,
Universitat Politècnica de València, Spain
[3]Departament de Matemàtica Aplicada,
Universitat de València, Spain
[4]Departament d'Estadística i Investigació Operativa,
Universitat de València, Spain

Abstract

In order to study religious behaviors in Spain, two conceptions are studied by a mathematical modelling approach: (a) the religious ideas are transmitted (by social contacts) versus (b) the *agent* transmitted are the non-believers ideas. A mathematical model based on ordinary differential equations is presented to understand it. Approximate Bayesian Computation scheme (ABC scheme) for parameters estimation and model selection is used. Prediction of the evolution of catholics, non-believers and believers of other religions are presented for the next few years.

Keywords: Social Behaviours; Religion; Modeling; Parameters Estimation; Model Selection

*E-mail address: rocerro@esp.upv.es
†E-mail address: alsncsnc@posgrado.upv.es
‡E-mail address: guecor@uv.es
§E-mail address: francisco.santonja@uv.es
ϕE-mail address: rjvillan@imm.upv.es

1. Introduction

In order to study religious behaviors using mathematical models, two conceptions are considered and studied: (a) the religious ideas are transmitted (by social contacts) or (b) the *agent* transmitted are the non-believers ideas [2]. Therefore, to understand religious behavior in Spain, we construct two mathematical models (model 1 and model 2). Model 1 is constructed considering religious ideas as the agent transmitted through social contact from one person to other. Model 2 is constructed considering the transmission of ideas of agnosticism or atheism. The consideration that religious ideas or non-believers ideas are trasmitted through social contact from one person to other allows us to model using epidemiological mathematical models based on a system of differential equations [3]. After that, we use an approximate bayesian computation (ABC) scheme [10] to compare the predicted evolution of the subpopulations of each model with real data (Table 1) and, then, select the model that best describes the real evolution. In addition, the ABC scheme also provides the parameters estimation for the selected model. Solving this model using the estimation obtained for the parametes, we will be able to define credible intervals in the predicted evolution of the subpopulations for the next few years.

To perform our study we have taken into account the evolution of catholics (C), non-believers (N) and believers of other religions (O) population. See Table 1. Data are from the Spanish Centre for Sociological Research (2000-2010) for Spain, for the over-18s [8].

Table 1. Evolution of catholics (C), non-believers (N) and believers of other religions (O) population for the period 2000-2010 [8]

	C	N	O
2000	0.853	0.130	0.017
2001	0.832	0.148	0.020
2002	0.820	0.160	0.019
2003	0.825	0.159	0.016
2004	0.801	0.183	0.016
2005	0.802	0.179	0.018
2006	0.787	0.196	0.017
2007	0.789	0.193	0.017
2008	0.779	0.202	0.019
2009	0.774	0.205	0.020
2010	0.760	0.221	0.019

The paper is organized as follows. Section 2 presents the two models considered and a briefly description of the procedure used to select and estimate the parameters of the best model. Section 3 shows the results of the model selection, the parameters estimation and the prediction given by the selected model. Finally, Section 4 presents the main conclusions derived from this work.

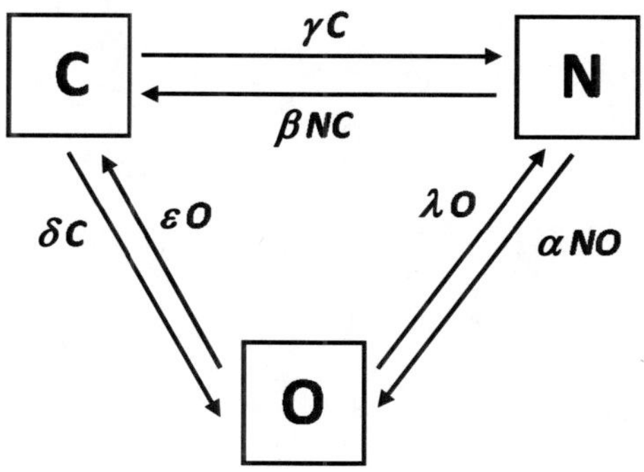

Figure 1. Flow diagram of model 1 where the contagion terms are due to the transmission of religious beliefs. The boxes represent the subpopulations and the arrows represent the transmission between the subpopulations. Arrows are labeled by their corresponding model tranmission terms.

2. Method

2.1. Mathematical Models

As it has been said above, we want to contrast two main hypotheses. This selection will allow us to understand religious behavior scenario in Spain. The first hypothesis is that religious ideas are spread through social contact. This means that the idea of converting into a religion is transmitted from one person to another. The second hypothesis is that the ideas that are transmitted from person to person are the ideas of agnosticism or atheism.

According to this, we propose two epidemiological models to study the evolution in Spanish population (older than 18 years old, and with Spanish nationality) of the subpopulations of catholics (C), non-believers (N) and believers of other religions (O). The information about these subpopulation for the period 2000-2010 is shown in Table 1.

The first hypothesis leads to the model 1 described below, and the second hypothesis leads to the model 2. These two models will be compared with data by means of a Bayesian scheme for model selection [10] to decide which one best explains the observed data (Table 1). Both models are constructed assuming constant population. This is correct because data are obtained only from people with Spanish nationality and this population has been approximately constant in the period under study according to the official census of Spain [9].

The transitions between the subpopulations C, N and O according to the model 1 are shown in Figure 1. We have not included the flows corresponding to births and deaths to simplify the figure. We have assumed constant population and that the rates of birth and death are equal for the three subpopulations. So, we think that it is not necessary to add and subtract the same term in every box.

The parameters of the model 1 are defined as follows:

- α, transmission rate due to social contacts to non-believers to adopt other religions.

- β, transmission rate due to social contacts to non-believers to adopt Catholicism.
- γ, rate at which catholics become non-believers.
- δ, rate at which catholics become believers of other religions.
- ϵ, rate at which believers of other religions become catholics.
- λ, rate at which believers of other religions become non-believers.

The model 1 shown in Figure 1 is described by the following system of differential equations:

$$\begin{aligned} C'(t) &= \beta N(t)C(t) - \gamma C(t) - \delta C(t) + \epsilon O(t) \\ N'(t) &= \gamma C(t) + \lambda O(t) - \beta N(t)C(t) - \alpha N(t)O(t) \\ O'(t) &= \delta C(t) + \alpha N(t)O(t) - \epsilon O(t) - \lambda O(t) \end{aligned} \tag{1}$$

In order to build the model the following assumptions are considered:

1. Let us assume homogeneous population mixing, i.e. each individual may transmit the religious ideas, or ideas of agnosticism or atheism, to any other one [6].
2. The transitions between the different subpopulations are determined as follows. An individual in $N(t)$ transits to $C(t)$ because people in C transmit religious ideas by social contact at rate β. Therefore, this is a non-linear term modelled by $\beta N(t)C(t)$. We also admit this consideration to model the transition from $N(t)$ to $O(t)$. An individual in $N(t)$ transits to $O(t)$ because people in $N(t)$ transmit religious ideas by social contact at rate α. In this case, the non-linear term considered is $\alpha N(t)O(t)$. The remainder transits are governed by the terms proportional to the sizes of the subpopulations:
    - $\gamma C(t)$ to transit from $C(t)$ to $N(t)$,
    - $\delta C(t)$ to transit from $C(t)$ to $O(t)$,
    - $\epsilon O(t)$ to transit from $O(t)$ to $C(t)$ and
    - $\lambda O(t)$ to transit from $O(t)$ to $N(t)$.

Data in Table 1 are in percentages meanwhile model (1) is referred to number of individuals. It leads us to transform (scaling) the model into the same units as data. To do that, we follow the techniques developed in [4, 5]. Here, we are not going to show the process and the scaled model because it is a technical transformation, the resulting equations are more complex and longer and does not provide extra information about the model. Moreover, the scaled model has the same parameters as the non-scaled model with the same meaning. In order to avoid introducing new notation, we are going to consider that the subpopulations $C(t)$, $N(t)$ and $O(t)$ correspond to the percentage catholics, non-believers and believers of other religions, respectively.

According to the second hypothesis, the transmission of ideas of agnosticism and atheism, we have constructed the model 2. The assumptions to build the model and the conditions of constant population and equal birth and death rates are the same as those for model

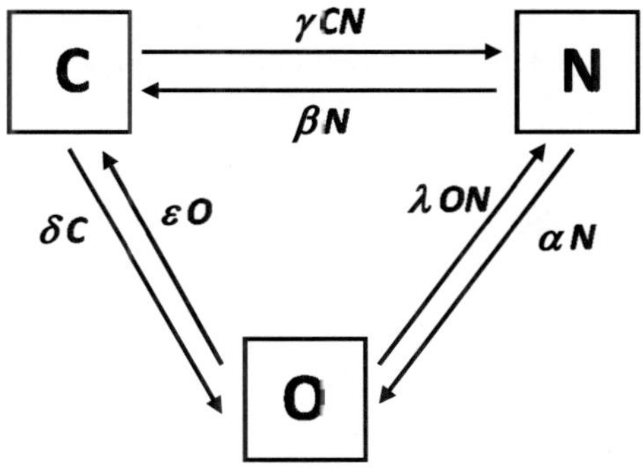

Figure 2. Flow diagram of model 2 where the contagion terms are due to the transmission of non-believers ideas.

1. In this case, the model has also been scaled. The transitions between the subpopulations for model 2 are shown in Figure 2.

The parameters of the model 2 are defined as follows:

- α, rate at which non-believers become believers of other religions.
- β, rate at which non-believers become catholics.
- γ, transmission rate due to social contacts to leave Catholicism and become non-believers.
- δ, rate at which catholics become believers of other religions.
- ϵ, rate at which believers of other religions become catholics.
- λ, transmission rate due to social contacts to leave other religions and become non-believers.

The model 2 is described by the following system of differential equations:

$$
\begin{aligned}
C'(t) &= \beta N(t) - \gamma C(t)N(t) - \delta C(t) + \epsilon O(t) \\
N'(t) &= \gamma C(t)N(t) + \lambda O(t)N(t) - \beta N(t) - \alpha N(t) \\
O'(t) &= \delta C(t) + \alpha N(t) - \epsilon O(t) - \lambda O(t)N(t)
\end{aligned} \quad (2)
$$

2.2. Model Selection and Parameter Estimation

We have already presented the two models and now we have to decide which one is the best to describe the evolution of the different subpopulations. To do this, we are going to use the approximate bayesian computation sequential Monte-Carlo approach (ABC SMC) proposed by T. Toni in [10].

Our objective is to obtain a set of N vectors $\theta(m)$, with $m = 1$ or 2 (model 1 and model 2), distributed between the two models, that satisfy the final condition $d(x^*, x_0) \leq \epsilon_T$. This condition means that the prediction given by model m with values of the parameters $\theta(m)$, is separated from the observed data by a distance less than ϵ_T. The model with the highest number of parameters vectors assigned will be selected.

To obtain a final estimation of the parameters, we are obtaining intermediate estimations, that is sets of N vectors $\theta(m)$, by refining the values of the maximum distance permitted ϵ_t in each iteration. This means that ϵ_t has to satisfy that $\epsilon_1 > \epsilon_2 > ... > \epsilon_T$.

The ABC SMC algorithm for model selection proceeds as follows (more details in [10]):

Step 1. Initialize $\epsilon_1, \epsilon_2, \cdots, \epsilon_T$.
Set the population indicator $t = 1$ (t varies from 1 to T).

Step 2.

Step 2.0. Set the particle indicator $i = 1$ (i varies from 1 to N).

Step 2.1. Sample m^* from $\pi(m)$. If $t = 0$, sample θ^{**} from $\pi(\theta(m^*))$.If $t > 0$, sample θ^* from the previous population of the parameters $\{\theta(m^*)_{t-1}\}$ with weights $w(m^*)_{t-1}$ and perturb θ^* to obtain $\theta^{**} \sim K_t(\theta|\theta^*)$. K_t is the perturbed kernel.
If $\pi(\theta^{**}) = 0$ return to Step 2.1.
Simulate a candidate dataset $x^* \sim M(x|\theta^{**}, m^*)$, where $M(X|\theta^{**}, m^*)$ is the dynamic model (1) if $m^* = 1$ and the dynamic model (2) if $m^* = 2$.
If $d(x^*, x_0) \geq \epsilon_t$, return to Step 2.1. Where x_0 are the observed data.

Step 2.2. Set $m_t^{(i)} = m^*$ and add θ^{**} to the population of particles $\{\theta(m^*)_t\}$, and calculate its weight as:

$$w_t^{(i)} = \begin{cases} 1, & \text{if } t = 0 \\ \frac{\pi(\theta^{**})}{\sum_{j=1}^{N} w_{t-1}^{(j)} K_t(\theta_{t-1}^{(j)}, \theta^{**})}, & \text{if } t > 0 \end{cases}$$

If $i < N$, set $i = i + 1$, go to Step 2.1

Step 3. For every m, normalize the weights. If $t < T$, set $t = t + 1$ and go to Step 2.0

The outputs of the algorithm are the approximations of the marginal posterior distribution of the model parameter $P(m|Data)$ and the marginal posterior distributions of each parameter of each model $P(\theta_i|Data, m)$. In our case, $m = 1$ for model 1 and $m = 2$ for model 2.

The parameter estimation for each model is calculated simultaneously with the model selection. The model with the highest posterior probability will have the greater number of parameters vectors $\theta(m)$. This ensures a good estimation of its parameters.

If one of the models gives a poor description of data, then that model can eventually dissapear, as we introduce decreasing values of ϵ_t, in the sense that all the parameters vectors belong to the other model.

In our case, the algorithm shown above, works as follows.

We have two models and each of them has a set of parameters:

$$\theta(m = 1) = (\alpha_1, \beta_1, \gamma_1, \delta_1, \epsilon_1, \lambda_1) \tag{3}$$

$$\theta(m=2) = (\alpha_2, \beta_2, \gamma_2, \delta_2, \epsilon_2, \lambda_2) \quad (4)$$

Where $m = 1$ is for model 1 and $m = 2$ is for model 2.

In Step 2.1. of the algorithm we obtain a random value for m from a uniform distribution $\pi(m)$ with only two possible values $m = 1, 2$. This means that for the first iteration of the N parameters vectors we will have more or less half of the parameters vectors $\theta(m)$ from model 1 and half from model 2. This step allows us to select a model.

In the first iteration ($t = 0$) we obtain a set of values for the parameters of the selected model m from a uniform distribution for each of them and we verify that the prediction given by this set of values satisfies the condition for the distance $d(x^*, x_0) \leq \epsilon_1$. Then, in Step 2.2 we asign weights for this set of parameters.

When we have obtained the N vectors $\theta(m)$, with a set of parameters for each one, we normalize the weights to unity.

In the following iterations, after choosing a value for m, we choose a vector of parameters (θ^*) from the previous population of vectors $\theta(m)$ according to its weights. We perturb the values θ^* to obtain a new set of values θ^{**} (close to θ^*, but different) according to $\theta^{**} \sim K_t(\theta|\theta^*)$. For more details about $K_t(\theta|\theta^*)$, see the Appendix.

The rest of the process is equal to the first iteration. First, we have to check that the set θ^{**} satisfies the condition for the distance, and finally we have to normalize to unity the weights obtained. This process is repeated again and again until the prediction reaches the minimum required distance ϵ_T.

3. Results

3.1. Results of the Model Selection

In our case, we have applied the ABC SMC algorithm for model selection shown above, for model 1, eqs. (1), and for model 2, eqs. (2).

The values of ϵ_t that we have used to ensure the transition from the a priori distributions for the parameters to the posterior distributions are $\epsilon_1 = 0.05$, $\epsilon_2 = 0.025$, $\epsilon_3 = 0.012$ and $\epsilon_4 = 0.0055$. Then, $T = 4$, and, in addition we have considered $N = 1000$ for the number of parameters vectors $\theta(m)$. The distance function $d(\cdot;\cdot)$ is defined by the root mean square error. The value ($\epsilon_{T=4}$) is considered taking into account the deterministic estimation of the models in the mean square sense. The lowest distance to be reached is expected to be close to this numbers and we choose the tolerance level ϵ_T accordingly. Note that we choose e_1, e_2, e_3 and e_4 in decreasing sense and we select the values to ensure that the distribution gradually evolves towards the posterior one, i.e., the distrution defined by $\epsilon_{T=4}$.

The parameters considered are: $\alpha, \beta, \gamma, \delta, \epsilon$ and λ for each of the models. And also m is considered a parameter with allowed values 1 and 2 to represent the two models under study.

The prior distributions for the parameters of the model are taken to be uniform and the intervals of definition are shown in Table 2 for model 1 and in Table 3 for model 2.

At this point we have to make a comment. The limits of the intervals are taken according to some references found in the literature for the rates of conversion of religions [1, 7]. In both cases, the order of magnitude of the estimated rates of conversion per year is around

0.01. To allow more flexibility to our estimation, and since we are describing a different country, we have proposed an upper limit of 0.05. Taking into account this value and the structure of the flows given in each model we have obtained the limits of Table 2 and Table 3.

Table 2. Prior definition for the parameters of model 1. These values are for the period 2000-2010 and are obtained from [1, 7]

Model 1	Min	Max
α	a_α=0.0	b_α=1.0
β	a_β= 0.0	b_β=0.075
γ	a_γ=0.0	b_γ=0.05
δ	a_δ= 0.0	b_δ= 0.05
ϵ	a_ϵ=0.0	b_ϵ=0.05
λ	a_λ=0.0	b_λ= 0.05

Table 3. Prior definition for the parameters of model 2. These values are for the period 2000-2010 and are obtained from [1, 7]

Model 2	Min	Max
α	a_α=0.0	b_α=0.05
β	a_β= 0.0	b_β=0.05
γ	a_γ=0.0	b_γ=0.25
δ	a_δ= 0.0	b_δ= 0.05
ϵ	a_ϵ=0.0	b_ϵ=0.05
λ	a_λ=0.0	b_λ= 0.25

Figure 3 shows the distributions obtained for the parameter $\theta(m)$ for each iteration $t = 1, 2, 3, 4$ according to the four values of ϵ_t. We can see how the number of times that the algorithm selects the model 2 is decreasing as ϵ_t is decreasing. Finally, only model 1 survives. So, model 1 is the one that best describes the evolution of the subpopulations of catholics, non-believers and believers of other religions.

The algorithm also provides the posterior distributions for the parameters of model 1 (the best model). The distribution is defined by the N=1,000 values of $\theta(m = 1)$. Table 4 summarizes the posterior probability distributions for the parameters. It shows the 90% credible interval taking into account percentile 5 and 95.

3.2. Prediction of the Evolution of the Different Subpopulations

Considering the posterior distribution for the parameters based on N=1,000 values for each one (Table 4), we solve numerically the system of differential equations corresponding to

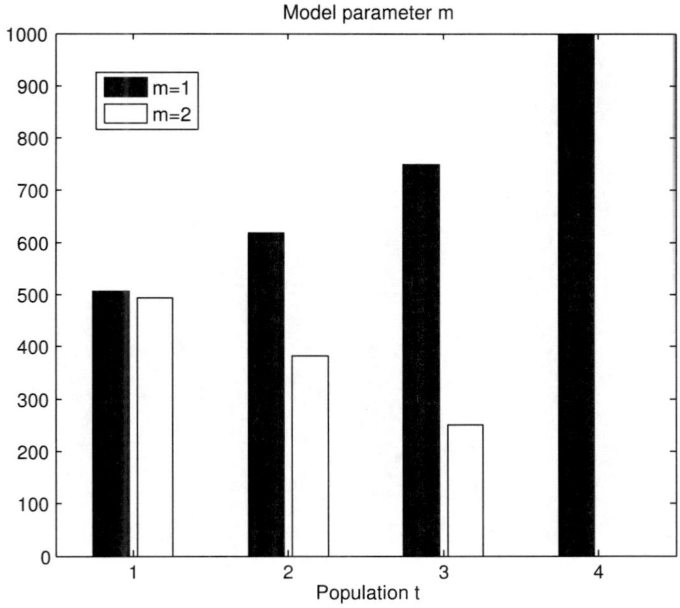

Figure 3. Evolution of the number of parameters vectors $\theta(m)$ corresponding to each model in each iteration $t = 1, 2, 3, 4$.

the model 1, and we obtain the predicted evolution of the subpopulations of catholics, non-believers and believers of other religions. Figures 4, 5 and 6 show this predicted evolution for the period 2000-2015 by 90% credible intervals.

In the years 2000-2010, we can see how our model 1 fits perfectly with the observed data. This allows us to make a short term prediction for the years 2011-2015 showing how the proportions of each subpopulation will evolve.

In Table 5, 6 and 7 we show the prediction given by the model 1 for the period 2011-2015. We see how the proportion of catholics is decreasing and the proportion of non-believers is increasing correspondingly. The proportion of believers of other religions is more or less constant but with a significant increasing in the amplitude of the credible interval. This means that the uncertainty in the evolution of this subpopulation is growing. Taking into account the mean values, we predict a decreasing of 2.4 points in the proportion of catholics, and increasing of 2.2 points in the proportion of non-believers and a small increase of 0.2 points in the proportion of believers of other religions.

We have restricted our predictions to the period 2011-2015 because we can not assume that the probability distributions of the parameters will remain the same in the long term. However, we have evaluated, what the evolution will be if that condition could be assumed. The result is shown in Figure 8. The decrease in the proportion of catholics is slowing down until the equilibrium value around 0.5926. Non-believers and believers of other religions increase until equilibrium around 0.2652 and 0.1422, respectively. These evolutions have been calculated using the mean values of the estimation of each parameter. The values of the equilibrium point are obtained setting equal to zero eqs. (1).

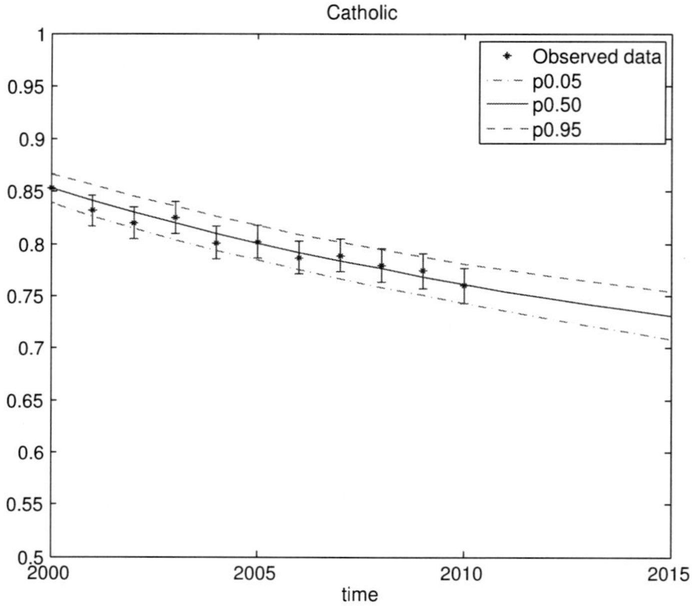

Figure 4. Evolution of the proportion of catholics taken into account the distribution of the parameters shown in Table 4.

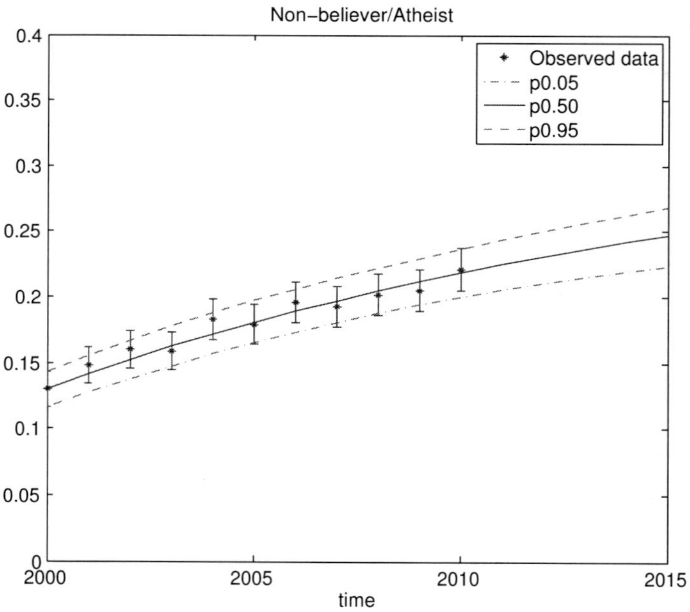

Figure 5. Evolution of the proportion of catholics taken into account the distribution of the parameters shown in Table 4.

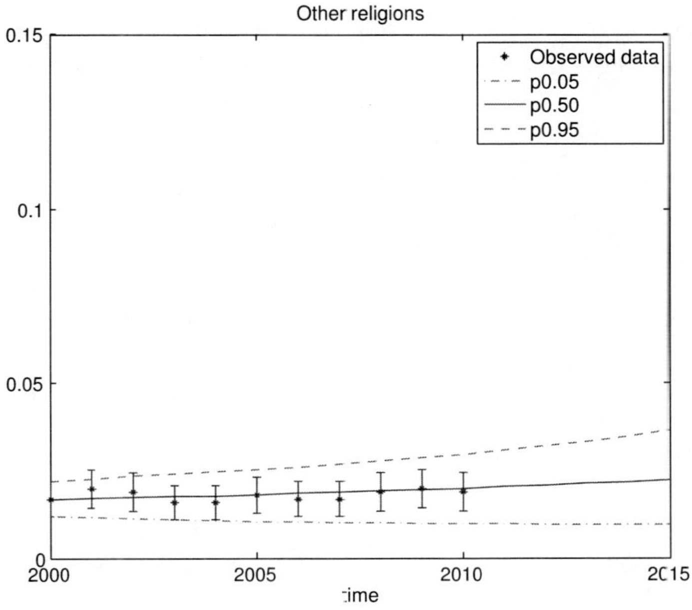

Figure 6. Evolution of the proportion of catholics taken into account the distribution of the parameters shown in Table 4.

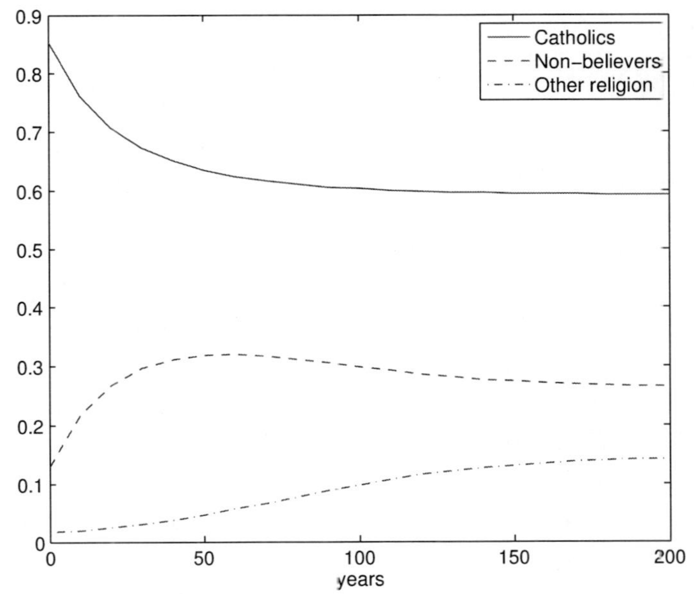

Figure 7. Asymptotic evolution towards the equilibrium point predicted by the model 1.

Table 4. Summary of the posterior probability distributions for the parameters of the model 1 after applying ABC SMC. The 90% credible interval is computed taking into account percentile 5 and 95.

Parameter	Mean	Median	90% Credible Interval
α	0.213777	0.191574	(0.019326; 0.477202)
β	0.055480	0.057903	(0.030229; 0.073108)
γ	0.021110	0.021427	(0.016985; 0.024264)
δ	0.000831	0.000716	(0.000081; 0.001913)
ϵ	0.029916	0.031593	(0.005467; 0.048865)
λ	0.029477	0.031838	(0.003931; 0.048191)

Table 5. Evolution of the proportion of catholics for different years. The 90% credible interval is computed by percentiles 5 and 95

	Mean predicted	90% Credible interval
2011	0.754528	(0.735875; 0.774859)
2012	0.748113	(0.728704; 0.769157)
2013	0.742112	(0.721823; 0.763754)
2014	0.736379	(0.715018; 0.758643)
2015	0.730953	(0.708416; 0.753947)

Conclusion

In this work we present an application of a Bayesian computation scheme (ABC scheme) for model selection (and estimation) for the case of the evolution of religious beliefs in Spanish population.

We have shown that the dynamics of the religious behavior in a society can be described by means of an epidemiological model based on a system of differential equations with random parameters. Moreover, we have been able to show that the model that best explains the observed proportions of catholics, non-believers and believers of other religions is the one constructed under the hypothesis of the contagion of religious ideas.

In addition, the ABC scheme provides us with an approximation to the posterior probability distributions for the parameters of the selected model. This fact allows us to predict the evolution of the subpopulations in the near future for the Spanish population. Solving the mathematical model (1,000 times) we have calculated the prediction for the proportions of catholics, non-believers and believers of other religions by 90% credible intervals.

The prediction for the period 2011-2015 shows a continuous increasing in the proportion of non-believers at the expense of the proportion of catholics. The proportion of believers of other religions remains approximately constant but with a significant increase of its uncertainty.

Table 6. Evolution of the proportion of non-believers for different years. The 90% credible interval is computed by percentiles 5 and 95

	Mean predicted	90% Credible interval
2011	0.224983	(0.205618; 0.243463)
2012	0.230935	(0.210715; 0.250126)
2013	0.236510	(0.215423; 0.256408)
2014	0.241800	(0.219616; 0.262689)
2015	0.246924	(0.223429; 0.268852)

Table 7. Evolution of the proportion of believers of other religions for different years. The 90% credible interval is computed by percentiles 5 and 95

	Mean predicted	90% Credible interval
2011	0.020471	(0.009747; 0.030819)
2012	0.020955	(0.009700; 0.032028)
2013	0.021375	(0.009646; 0.033405)
2014	0.021821	(0.009541; 0.034940)
2015	0.022214	(0.009461; 0.036635)

Although this prediction can not be extended to the long term, because we can not assume that the probability distribution will remain the same, we have calculated the mean evolution of the subpopulations to see whether the decreasing in catholics stops or not. We found that finally the three subpopulations reach an equilibrium point at $C = 0.59$, $N = 0.27$ and $O = 0.14$.

To finish, we want to emphasize the fact that we have applied these mathematical and statistical techniques to a specific population and to its real observed data. This means that the ABC scheme for model selection and the epidemiological models are powerful tools to explain and predict social behaviors in real populations.

In our opinion, the application of these techniques is a promising area of research in social sciences.

Appendix

In Step 2.1 of the ABC SMC scheme, we consider independence in the parameters. Therefore, we define $K_t(\theta|\theta^*)$ for each parameter. For example, for parameter α, we have the following definitions

$$K_t(\alpha|\alpha^*) \quad = \quad Uniform(\alpha^* - \sigma_{t-1}^{\alpha}, \alpha^* + \sigma_{t-1}^{\alpha}) \tag{5}$$

where

$$\sigma_{t-1}^{\alpha} = \frac{1}{2}[max(\alpha_{t-1}^{(i)}) - min(\alpha_{t-1}^{(i)})] \tag{6}$$

In Step 2.2, we consider

$$K_t(\alpha_{t-1}^{(j)}, \alpha_t^{(i)}) = \frac{1}{Min(b_\alpha, \alpha_{t-1}^{(j)} + \sigma_{t-1}^{\alpha}) - Max(a_\alpha, \alpha_{t-1}^{(j)} - \sigma_{t-1}^{\alpha}))} \tag{7}$$

(8)

Analogously for the rest of parameters: $\beta, \gamma, \delta, \epsilon$ and λ.

References

[1] Barro, R.J. and Hwang, J. (2009): *Religious conversion in 40 countries*, National Bureau of Economic Research working paper series. No. 13689. http://www.nber.org/papers/w13689. Accessed 20 November 2011.

[2] Geertz, A.W. and Markusson, G.I. (2010): Religion is natural, atheism is not: On why everybody is both right and wrong, *Religion* 40, 152-165.

[3] Ma, Z. and Li, J. (2009): *Dynamical modeling and analysis of epidemics*. World Scientific.

[4] Martcheva, M. and Castillo-Chavez, C. (2003): *Diseases with chronic stage in a population with varying size*, *Mathematical Biosciences* 182, 1-25.

[5] Mena-Lorca and J., H.W. Hethcote, H.W. (1992): Dynamic models of infectious diseases as regulators of population sizes, *Journal of Mathematical Biology* 30, 693-716.

[6] Murray, J.D. (2002): *Mathematical biology*. Springer.

[7] Shy, O. (2007): Dynamic models of religious conformity and conversion: Theory and calibrations, *European Economic Review* 51, 1127-1153.

[8] Spanish Centre for Sociological Research CIS (2010):. *Barometer* (2000-2010). http://www.cis.es/. Accessed 20 November 2011.

[9] Spanish Statistic Institute. *Demography and population.* http://www.ine.es. Accessed 20 november 2011.

[10] Toni, T., Welch, D., Strelkowa, N., Ipsen, A., Stumpf, M.P.H. (2009): *Approximate bayesian computation scheme for parameter inference and model selection in dynamical systems*, *Journal of the Royal Society Interface* 6, 187-202.

In: Mathematical Modeling in Social Sciences …
Editors: J. C. Cortés López et al.
ISBN: 978-1-63117-335-6

Chapter 13

MODEL OF PROBLEMS CLEANING IN EDUCATION

Jan M. Myszewski**, *Malgorzata Gromek*, and *Joanna Oczkowicz

Department of Quality Management,
Kozminski University, Warsaw, Poland

Abstract

An educational system of secondary schools is considered. Focus is given to two processes: educational process, which involves students and a teacher in lesson; improvement of the educational process, which involves teachers and administration of the school. It is shown that both can be represented as special cases of a process of eliminating multiple problems, referred to as "problems cleaning". Consequently, the effectiveness of the processes depends on characteristics: of amount and quality of resources used in the process as well as those of control of the resources.

Considerations are illustrated by a case study of improvement initiated by the evaluation procedure in a school. It demonstrates: the relationship of model parameters with characteristics of the educational system and that change of the characteristics depends on the very involvement of the administration in the improvement processes.

Keywords: Improvement, resources, effectiveness, evaluation

Introduction

The specificity of the "problems cleaning" process is the number of improvement tasks, higher than in generic improvement processes. Education is an environment, where the streams of problems intersect constantly. They include: learning difficulties encountered by students; efforts of teachers to ensure the effectiveness of the teaching; management efforts to meet the expectations of various stakeholders of the school.

The chapter is composed of four parts. It begins with a description of the model of problems cleaning. Its mathematical complexity is moderate: refers to the Poisson rate

* E-mail address: myszewski@wspiz.edu.pl.

functions involved in a linear ordinary differential equation with constant coefficients. Its advantage is the ability to represent many important phenomena contributing to the performance of the elimination of numerous problems.

The next section outlines the educational processes seen from the perspective of the "problems cleaning". On this basis, a procedure is shown that can be used for analyzing and improving the "problems cleaning" in education. The functions: effectiveness A and efficiency E allow diagnosing the improvement performance in a school. The presentation is illustrated by some case study. It shows the context of improvement, the importance of the model parameters and can be an inspiration to the application of the model in other environments than education.

1. Model of the Process of Problems Cleaning

1.1. Prerequisites of the Model

We denote by "*problems cleaning*" a general procedure to eliminate the numerous failure states of definite organizational function and their consequences. This concept represents composition of various activities conducted in organization, which are to get rid of unwanted states of organization's functions. In one reference model it combines all possible routes that can be used and all operations that are necessary to ensure effective improvement.

Idea of the Problems Cleaning Model

At the input of the problems cleaning there is a time series which represents tasks expected to be done while eliminating failure states of the organizational functions and their consequences (see the Figure 2).

At the output there is a time series, which represents tasks not completed by problems cleaning. The output time series is combined with the input time series synchronously. This represents an accumulation of current tasks when the cleaning is ineffective.

Resources in the Problems Cleaning Process

Problems cleaning involves definite resources and control standards. The resources may be shared with other functions of organization. The part, which is dedicated exclusively to cleaning operations is referred to as *improvement resources*.

Category 'improvement resources' may include such elements as e.g.: personnel (individuals or some groups of people who are assigned to eliminate problems), infrastructure (rooms, testing equipment, computers etc used for purpose of improvement work), explicit knowledge resources (including databases, which can be used in the process), financial capital to complement the resources and services necessary to achieve improvement.

Improvement resources are the key cleaning agent. The indispensable amount is proportional to number of tasks to be handled. However, some resources are lost due to ineffective use of them and some are inhibited by actions to eliminate impacts of problems. If the amount of available resources is adequate, then the cleaning process is able to cope with them and number of outstanding problems lowers to zero. Otherwise some tasks remain unaccomplished and feed a list of tasks to be done.

Assumptions of the Model of Problems Cleaning

Assumption A (ideal homogeneity of resources):

- A1. Resources are perfectly homogeneous: each individual has the same knowledge and skills, all material items have the same performance characteristics.
- A2. Effectiveness of improvement resources in the cleaning process is uniform with respect to time and amount of resources: the longer a fixed amount of resources can be used, the bigger portion of improvement tasks can be accomplished or alternatively, the more resources can be used, the more tasks can be completed.

Assumption B: We assume constant conditions in organization within a period of observation [0, T).

1.2. Equation of Problems Cleaning

Variables in the Model of Problems Cleaning

Amount of resources available in the interval [t, t+h) is a random variable with average that can be estimated by formula $R = R(t, h) = r(t) \cdot h$, for positive relatively small h. Stream of problems, which enter the cleaning process is represented by number of respective improvement tasks in the interval [t, t+h), which is a random variable with average that can be estimated by formula $P = P(t, h) = p(t) \cdot h$. Some tasks may be left uncompleted and are returned back to the queue of tasks to be served again. Number of unaccomplished tasks in the interval [t, t+h) is a random variable with average that can be estimated by formula $Q = Q(t, h) = q(t) \cdot h$.

The functions r, p and q are Poisson process rate functions [1]. By the Assumption B, r and p are constant, q is assumed piecewise differentiable over some interval [0, T).

Parameters in the Model of Problems Cleaning

By the assumption A, there is a random variable with average x such that $x \cdot P$ is the average amount of resources, which are necessary to have an number P of tasks eliminated by the time h. We call the parameter x a *complexity* of problems represented by average P. We assume that the same complexity x can be also used to express resources consumption when eliminating unsolved problems represented by the average Q. There is a random variable with average u such that $u \cdot (x \cdot P)$ represents amount of resources, which are blocked by ad hoc actions to suppress impacts of problems represented by average P (see the formula 5-2). We call the parameter u an *onerousness* of problems represented by average P. We assume that the same onerousness u can be used to express loss of resources related to outstanding problems Q. There is a random variable with average c such that $(1 - c) \cdot R$ represents amount of resources that can be lost because of ineffectiveness in handling improvement resources (see the formula 5-3). Parameter c is called an *efficiency* of use of resources.

Equation of the Problems Cleaning

Below we present a quantitative model of "problems cleaning". The model represents relationship between average number of tasks related to new problems (represented by variable P), and average number of outstanding tasks as a result of ineffective cleaning (represented by variable Q). Expected function of the cleaning process is to reduce number of unaccomplished tasks, when the average number of tasks related to new problems at the input is constant. An idea of the "problems cleaning" is shown in the Figure 1.

Statement 1. The average number of outstanding problems satisfies the equation:

$$Q' = \alpha\cdot(Q+P) - \beta\cdot R \tag{1}$$

with $\alpha := (1+u)$, $\beta := c/x$, h is set a unit length on the t-axis.

Solution to the Equation of Problems Cleaning

Statement 2. The average number of outstanding problems can be estimated by the formula:

$$P(t) = Q(0) + A\cdot(1 - \exp(\alpha t)) \tag{2}$$

$$A: = E\cdot R - (P + Q(0)). \tag{3}$$

$$E: = c / \{x\cdot(1+u)\} \tag{4}$$

Statement 3. There are three states of process of the problems cleaning. They can be distinguished by a sign of function A. The three states of problem cleaning are:

- $A > 0 \Leftrightarrow$ problems cleaning is 'effective': improvement resources are big enough,
- $A < 0 \Leftrightarrow$ problems cleaning is 'blocked': improvement resources are too small,
- $A = 0 \Leftrightarrow$ problems cleaning is 'idle': improvement resources are at the neutral level.

1.3. Performance Characteristics of the Problems Cleaning

Statement 4. The function E defined by (4) is a measure of overall efficiency of the problems cleaning process. E·R approximates average amount of tasks, which can be completed with the available resources by the unit time h.

Sketch of the proof. For u close to zero, $R/(1+u) \approx R\cdot(1 - u)$. Therefore, $E\cdot R \approx (c\cdot R/x)\cdot(1 - u)$ approximates (from above) the average amount of tasks, which can be completed, regarding all losses related to impacts of particular factors.

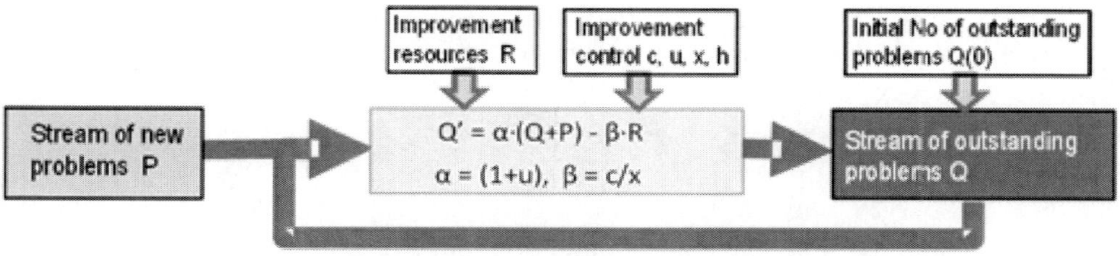

Source: Authors.

Figure 1. Structure of factors influencing organization's ability to eliminate problems.

Statement 5. The function A defined by (3) is a measure of effectiveness of the problems cleaning process. For any average number of problems P+Q(0) and any positive overall efficiency of the cleaning process E, there exists an average amount of resources R, for which improvement process can be effective (i.e. A > 0) under some fixed conditions.

Statement 6. By increasing the efficiency E of the problems clearing, it is possible to reduce the amount of resources R, whilst ensuring appropriate effectiveness A.

Remark. Decisions on assigning resources to the cleaning process, on establishing measures to stimulate the growth of efficiency and how to make use of the growth, belong to the strategic level management of organization.

2. Process of Problems Cleaning at School

The basic educational process is realized at operational level (lesson). It can be shown that it is a process of problems cleaning. It is to close the gaps concerning knowledge, skills, and students' attitude. Stiff lesson timing makes that in cleaning process there is no time reserve that enables explaining every doubt of a student after a given lesson.

Problems observed at a lesson level can be the results of the problems that belong to tactical level of the course. Therefore, at the level of course the actions that increase efficiency and efficacy of the lesson should be expected.

It is done as a result of cleaning at the course level by means of a suitable timetable modification or changing the methods or didactic tools.

At a school level the decisions concerning contribution of sources are made. They include improvement sources, the regulations are established, impulses of improvement are generated.

The actions of school management should express support for improvement processes at operational and tactical level. Owing to limited timing for presentation we will focus on a tactical level.

Tactical Level-Course

The course level comprises the actions that belong to planning e.g. transformation the requirements related to school curriculum content into plans of particular lesson units including their factual content, the way of conduct and didactic tools. At this level the following aspects are arranged: plans of particular lesson units from the point of view of their factual content, working methods, order and applied didactic tools.

Problems at Tactical Level

Problems at tactical level are called the faults of course program. They involve accepting in the program "solutions", that make a thorough program realization in given conditions more risky than accepted, as far as program implementation is concerned. The cause of the faults can be, among others, false assumption concerning the conditions of realization or lack of experience in planning. The results of faults can involve among others difficulties in course program realization during a lesson and as a result gaps in course realization, identified during tests. It is assumed that they are evenly set in the program and appear gradually in the course of semester.

Problems Cleaning at Tactical Level

Problems cleaning involves:

- identification and removing potential threats before accepting the program for implementation (prevention actions)
- making suitable alterations in the program while detected (corrective actions) or
- introduction corrections to the course of program realization without program alteration (immediate actions at the course level)
- correction of lesson content that comprises focusing the attention on more difficult part of the material at the expense of the others (immediate actions at lesson level).

Outstanding Problems at Tactical Level

Outstanding problems are determined as program faults, that were not eliminated or corrected at a given time.

Time Resources at Tactical Level

Time of teacher's work in accordance with the regulations: W = 40 hours per week. Each lesson unit is related to various amount of time that is used for supportive actions, e.g. preparation for classes.

If it is assumed that 18 h is devoted to direct work with students, 18 h to checking assignments and preparation for classes), 2 h for administrative duties and others ordered by management; then the teacher has time in the amount of $R = 40 - (18 + 18 + 2) = 2$ h /week at his/her disposal. The said amount of time constitutes potential improvement sources at a course level.

Time Sources Management at Tactical Level

h = 1 week (time unit size to which average values are related);
P = 1 (average number of problems appearing per week);

$x = 2$ h (average time amount needed to eliminate one problem per week),

Time resources are becoming lower and lower because of interruptions independent of the teacher such as: technical damage, administration faults and others

$1–c = 0,1$ (average time fraction of wasted time as a result of interruptions independent of a teacher)

$u = 0,2$ (average time fraction of time wasted for removing the effects of the problem);

Formulas (2) and(3) result in

$$A = E \cdot R - (P+Q(0)) > 0 \Leftrightarrow R > (P+Q(0))/E \quad (5)$$

$$E = 0,9/(2 \cdot (1+0,2)) = 0,375\ [1/h],\ A = 2 \cdot 0,375 - 1 = -0,25 < 0,$$

$$R\_min = P/E = 1/0,375 = 2,67.$$

Notice. As a result of waste (c, u) real need for sources 2,67 is higher than nominal: $1 \cdot 2 = 2,00$.

Support of Cleaning Process by School Management Staff at Tactical Level

Support of management staff involves:

- assuring indispensable amount of R improvement sources by due care for task delegation;
- increasing c, by improving efficiency of teachers' work organization;
- decreasing x, by creating conditions and encouragement system to make teachers' work more effective;
- decreasing P and u, by introduction of prevention actions to program planning;

Notice. The significance of the problems cleaning scheme involves highlighting attributes of the process of problems elimination, which have an influence on efficiency of elimination process and on accuracy of resources assumed for cleaning. The model of problems cleaning focuses on the significance of management support for educational process as well as encouraging teachers and students to improve efficiency of the basic resource: time devoted to education and more effective time management.

3. Methodology of Testing Opportunities of Function Improvement – Case Study

The case study below is the background to the presentation of methods aimed at checking the opportunities of school improvement by means of evaluation.

The selected school function: assessment of student educational progress, subject: English language. The data is fictious, but all similarities between real phenomena are not accidental.

3.1. Description of the Problem to Solve

After parents' complaint concerning the problem of unfair grades the school headmaster decided to look at the scale of the problem thoroughly. It appeared that on average every 20th negative grade was the subject of complaints. General overview of the controversial assignment showed that there were formal inconsistencies, so that the headmaster started evaluation of the assessment process at school.

3.2. Evaluation of Assessment Function at a Level of the Course

Step 0. Establishment of evaluation team.

The team consisted of deputy headmaster responsible for School Grading System and two other teachers. Discussion started when the teacher, whose grade was the subject of complaint, was appointed to the team. While opening the reunion the headmaster explained that the team's task is to examine the circumstances that are the subject of questioning as well as mechanism of the problem, excluding looking for guilty persons. A person who encountered difficulties can share precious experience and evoke the origins of a problem.

Step 1. Compilation and review of the documentation with grade system requirements and procedures.

The basic purpose of this phase of evaluation is to state if there is evidence of application of the set requirements. Evaluation was introduced as a result of the reliability of a given grade, so that the team decided to focus on such inconsistency. While reviewing the documents only lack of completion of some entries was noticed.

Step 2. The team started to look for the answer to the question: "Why evaluation can be unreliable?"

In order to find and register the potential causes the diagram of the cause tree was used [3]. On the diagram the potential mechanisms of unreliability were presented. Among others the "special" mechanisms were distinguished: "liking/antipathy to some students", "grades given in a hurry". The most "common" causes were: "momentary emotions" (e.g. tiredness, nervousness), and "common mistakes". It was stated that "ambiguity of criteria" as a system fault will be analyzed separately.

Step 3. The team started to look for actions that could eliminate special precautions.

After a short discussion it was approved that both mechanisms concern more school problems: communication and employee's workload. It was suggested that both issues should be a matter of special teachers' meeting during winter holidays. The suggestion was not enthusiastically accepted by teachers as it coincided with their various personal plans.

Step 4. Analysis of short-term actions, when a given grade can be a matter of question.

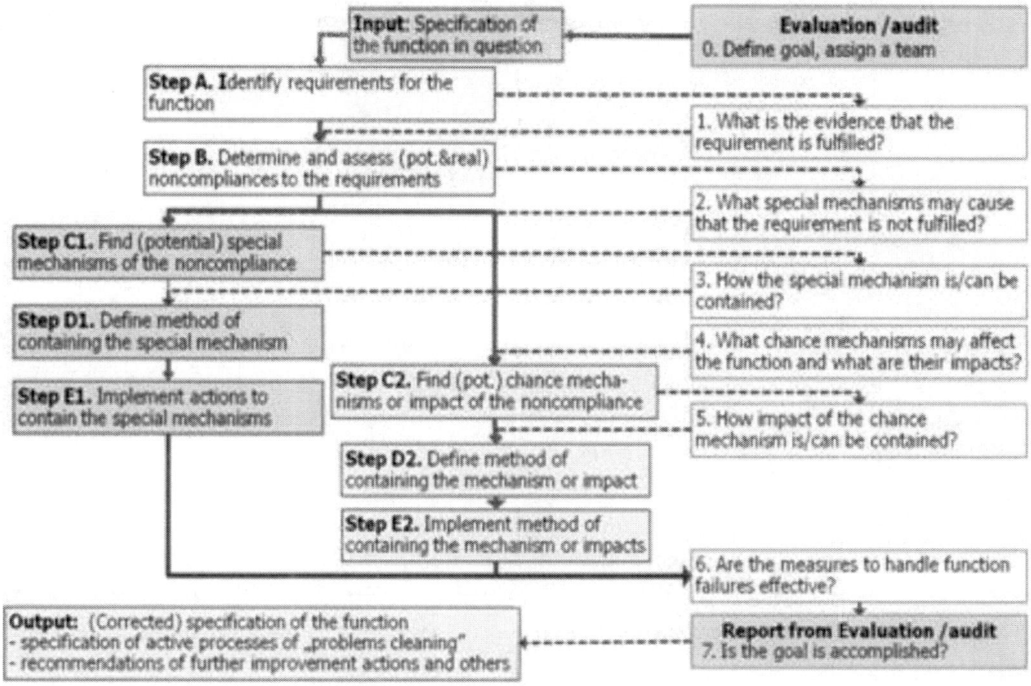

Figure 2. Scheme of the process of function evaluation.

Vice-headmaster suggested that until the correction of grade system the rule of "special precaution" should be obeyed, in the case a grade could disqualify a student assignment. The teacher will be obliged to additional and independent verification of grade, giving enclosure with the list of mistakes as well as method of assessment. Moreover the teacher has to conduct a conversation with a student and give justification to negative grades. In justified cases the teacher is obliged to ask for signature of the guardians under the grade justification.

There was a critical remark that the procedure is time consuming. The vice-headmaster gave arguments that when the opportunity occurs, it is possible to raise self control of the teachers also with other grades. It was decided that the procedure will be taken for a trial period (one month) preceding the period of final tests' heap. The remarks will be presented during the methodological meeting to be held out of school.

Step 5. The team started assessment of efficiency of applied and planned actions aimed at diminishing the number of complaints about the grades given by teachers.

Formal accuracy of procedure was not questioned. The doubts were evoked by its labor intensity. The preliminary estimation of the amount of time indispensable to its efficient application was made. The results are presented in the next section.

3.3. Analysis of Adequacy of Resources Planned in the Procedure of "Special Precaution"

In resource analysis the following data was taken into consideration:

Tasks 1 and 2 are respectively: verification of assignments assessed negatively and conversation with a student.

h = 1 week (size of time unit to which the average values are related)

P = 10 (average number of negative grades per week for written assignments, given by one teacher)

x_i = average amount of time indispensable for task fulfillment in the course of week;

R_i = average amount of time designed for task fulfillment in the course of the week;

$1\text{-}c_i$ = average fraction of time wasted by week, out of the amount designed for task fulfillment, owing to causes independent of the teacher.

u_i = (average fraction of time designed for assessment, wasted for discussions e.g. with students who have not deserved a higher grade).

It is assumed that there are not outstanding matters: Q(0)=0. The procedure will take at least 6,13 + 7,52 = 13,65 [h/week] of teacher work, which equals approximately 34% of weekly workload. It can be predicted that the procedure is too charging for teachers to be applied effectively.

Table 1. Assessment of minimal weekly workload related to the procedure of "special caution"

Element	x_i [h/week]	c_i	u_i	E_i [1/h]	R_i_min [h/week]
Task 1	0,5 h	0,9	0,1	0,9/(0,5·(1+0,1))=0,9/0,55 = 1,63	10/1,63 = 6,13
Task 2	0,5 h	0,8	0,2	0,8/(0,5·(1+0,2))=0,8/0,6 = 1,33	10/1,33 = 7,52
Total	1,0 h	0,85	0,15	0,85/(1·(1+0,15)) = 0,74	10/0,74 = 13,51

4. Methodology of Function Improvement (Algorithm of Assessment of Function Improvement)

The case presented in the chapter 3 is the background for the analysis of methods of function improvement.

4.1. Description of a Situation

After a parents' complaint, there was the evaluation of the School Grading System initiated. The procedure of "special precaution" was introduced as a temporary measure. It was decided to discuss the attempt to eliminate special causes on a special Teachers' Council meeting when school semester finished.

4.2. Session – the Function Development on a School Level

Step 0. The headmaster made a decision to organize Teachers' Council meeting at the last weekend of winter holidays and sent the meeting schedule.

It was decided to have an obligatory meeting for all the teachers outside school area. The schedule included team works in a few problem sessions alternatively to plenary discussions of the results. The headmaster was responsible for the supervision on the organization of the meeting. The teachers with long term experience and good communication skills were asked to lead the teams.

Step 1. The risks associated with each particular mechanism were analyzed in teams.

The analysis concerned the mechanisms which were indicated by the evaluation team, among which there were severity of the results of the mechanisms as well as their frequency and possibilities to resist them immediately when they occurred [3]. On the basis on the analysis, the priorities for each solution to the causes were assessed. During plenary session, the results were discussed without crucial controversy.

Step 2. Teams started to analyze the causes of each particular mechanism.

The cause diagram was used as a tool for registration and to analyze the causes [3]. The results were taken under discussion on plenary session. It was much more turbulent then the previous one and it brought many changes to the diagrams. There were weights assigned for each particular elementary cause and root cause for each particular mechanism.

Step 3. The teams started to search for some actions to eliminate the special causes.

The brainstorm method was used[3]. There were many suggestions. It was required to analyze some of them in details. The results were presented on plenary session when correction to particular assessment was made. There were also new solutions suggested, which were supposed to bring significant effects inspite of poor financial conditions of a school. The analysis of the causes was of significant value as it was much more detailed than the one during the evaluation.

Step 4. Common mechanisms were analyzed separately.

There were not any new ideas apart from those which came as a result of the evaluation. The team, which task was to monitor the procedure of "special precaution", prepared the report. There were not as many appeals to negative grades as it was before. Moreover, there were fewer negative grades. In doubtful situations, teachers started to give better grades in order to avoid time consuming procedure. It was an important sign of that it could be much more difficult to eliminate the results of the problems (mistaken grades) than to prevent from the problems. It became obvious that the crucial factor of improvement was to prepare homogeneous criteria of the grade system.

Consequently, the headmaster prepared an experiment where the team of experts made a copy of some tests completed by weak students. The task for some teachers was to check the tests anonymously in a particular amount of time. The grades were verified by the experts. The results showed that some of the criteria in the grade system could be misleading so that it might be the cause of some teachers' mistakes. Furthermore, the teachers did not obey those criteria which were clear to understand.

Step 5. Teams started to make a list of actions to decrease the risk of giving mistaken grades.

There were two tasks: correction of grade system criteria and to encourage teachers to use the criteria. The second task caused a turbulent discussion while some of the participants supported the idea to punish for making mistakes, others were the supporters of giving prizes to teachers who work without mistakes. Finally, it was assumed that there should be two tests for teachers twice a year and the results should be analyzed with all the participants. The significant improvement of grade system in a particular school team of teachers should be prized (lower average number of mistakes during 3, 4 years). It was decided to support the cooperation, long-term effects and to encourage improving the grade criteria.

Step 6. The assessment of the effectiveness of planned actions.

The ideas, encouraging the team to improve the grade system, seemed to be better solutions than the procedure of "special precaution". Similarly to a potential effectiveness assessment, there was necessary to analyze the resources. The results are in the next section.

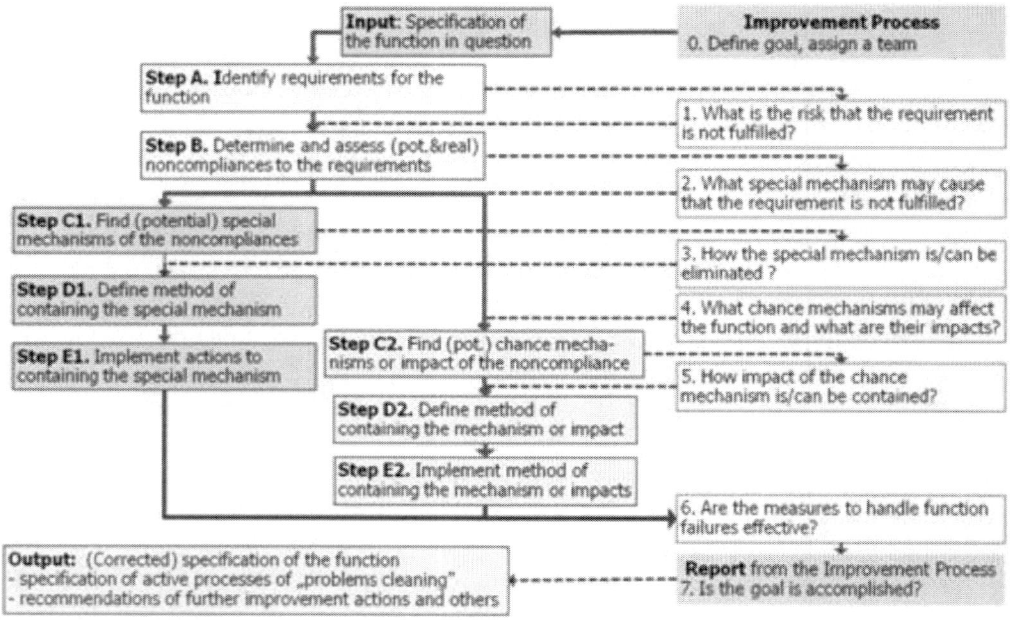

Figure 3. The scheme of the function improvement process.

At the end of the meeting, the participants admitted that they had never expected such significant results in such a short length of time and they declared their willingness to take part in a similar meeting about the improvement of the basic school functions in the future.

4.3. The Analysis of Adequacy of the Resources Planned in the "Correctness of Grades" Test

The subject to the analysis was the adequacy of time resources of the experts who were expected to participate in test procedure. It was assumed that in order to realize the procedure it was necessary to plan and do such tasks as:

1. Preparation of a set of tests for all of the teachers of each particular subject
2. Taking a test in similar circumstances
3. Analysis of the results of the test (number and type of mistakes, and their association with the criteria for a particular school subject)
4. Writing conclusions and recommendations concerning teachers and the grade system
5. Presentation of results and the analysis of recommendations with interested people.

It was assumed that the realization of each task takes 1 day (8 hours). There was one expert chosen to one school subject. While the test was prepared and carried out, the teachers were not given other tasks, so Q(0) = 0.

P=5 (the average amount of teachers, participating in a test)

X_i= average amount of time, necessary for an expert to complete the task for one teacher)

R_i= average amount of time for an expert to complete a particular stage of work

1-c_i = average fraction of time, wasted by the interference independent on an expert, e.g. "other important" tasks or emergency

u_i= average fraction of time, wasted by secondary problems, e.g. explanation, cooling emotions.

The result of the calculation above shows that the tasks 1 and 5 are in danger of being unrealized. They are sensitive to interference. If it is successful to plan the test realization by the teachers in such a way that they take tests simultaneously, it will be possible to have some time reserve after the phase, which could be used to verify the tests and to start talks with participants in the last phase.

Table 2. Assessment of minimal weekly workloads related to the "correctness of grades" test

Element	Task 1	Task 2	Task 3	Task 4	Task 5	Total
x_i [h/day]	1	1	1	1	1	5 [h/week]
c_i	0,7	0,7	0,8	0,9	0,7	0,76
u_i	0,2	0,1	0,1	0,1	0,3	0,16
E_i	0,7/(1+0,2) =0,58	0,7/(1+0,1) = 0,64	0,8/(1+0,1) = 0,73	0,9/(1+0,1) = 0,82	0,7/(1+0,3)=0,54	0,76/(5*(1+0,16) = 0,13
R_{i}_min	5/0,58 = 10,34	5/0,64 = 7,81	5/0,73 = 6,85	5/0,82 = 6,10	5/0,54 = 9,26	5/0,13 = 38,46

Conclusion

The model of "problems cleaning" shows that the conditions, which determine the effectiveness of eliminating problems are fixed before the process begins. Process parameters describe the basic features of the environment in which problems are eliminated. The effectiveness A and efficiency E of improvement process depend on the management decisions that were made much earlier, see the equations (1)-(4).

The information that enables to diagnose and determine the necessary actions is available to teachers implementing the cleaning process at the levels of the lesson and course. Instruments and decisions, which are necessary to accomplish the change, are in the hands of the management of the school. The effectiveness of the action depends on the efficiency of information use. The key factor is communication in the school. The vertical one, allows managers to improve the use of available resources and opportunities for improvement. The horizontal one, allows to search for optimal solutions to common problems of teachers. A record of the vertical and horizontal communication is provided in the case study in Sections 3 and 4.

The use of right measures at the right time and in the right way are the expected attributes of improvement process. In constantly changing environment, they are not likely to be achieved without close interaction between managers and teachers. The effect is the involvement of teachers and leadership exercised by managers. Examples of leadership are reported in the case study in Sections 3 and 4.

Support of school leaders for improvement of the educational process is essential. One of its manifestations may be patronage of projects, which encourage teachers and students to improve the efficiency use of the basic resource - time spent on learning and that of improving the use of time.

> "Specifically, the relationship between leadership and improvement capacity is best described as one of mutual influence or reciprocity. (..) Leadership may diffuse through the organization, transforming from an individual characteristic (e.g., the principal) to an attribute of a team, and finally into an organizational property. (..) Strong learning directed, collaborative leadership appeared to be an important factor, for change in the capacity of the (..) school to improve" (in [2] p. 22).

References

[1] Cox, D. R.; Miller, H. D. *The Theory of Stochastic Processes;* Chapman & Hall/CRC: London-Boca Raton, 1977.

[2] Hallinger, P.; Heck, R. H. Exploring the journey of school improvement: classifying and analyzing patterns of change in school improvement processes and learning outcomes, *SCH. EFF. SCH. IMPROV.* 2011, 22 (1), 1–27.

[3] Myszewski, J.M. *Simply the quality* (In Polish), WAIP: Warsaw, 2009.

In: Mathematical Modeling in Social Sciences ...
Editors: J. C. Cortés López et al.
ISBN: 978-1-63117-335-6

Chapter 14

DOES VAT GROWTH IMPACT COMPULSIVE SHOPPING IN SPAIN?

***E. de la Poza*[1,*], *I. García*[2,†], *L. Jódar*[3,‡] *and P. Merello*[3,§]**
[1]Departamento de Economía y Ciencias Sociales,
Universitat Politècnica de València, Spain
[2] Departamento de Comunicación Audiovisual y Publicidad,
Universidad del País Vasco, Spain
[3]Instituto Universitario de Matemática Multidisciplinar,
Universitat Politècnica de València, Spain

Abstract

Compulsive buying is a mental disorder based on gratifying dissatisfaction through excessive and unplanned purchase behaviours. Thus, a priori, the number of compulsive buyers should be influenced by those factors that enhance or reduce their compulsive behaviour. This chapter uses a dynamic approach by difference equations to model the influence of two events of opposite nature on compulsive shopping in Spain; on one hand the influence of VAT growth and, on the other, the stress the Spaniards suffer as consequence of the bad economic perspectives. The parameters associated with the event will be estimated using the algorithm of Nelder -Mead. The results show that the 41.85 % of the rational buyers with emotional distress but with good economic expectations increase their level of compulsion becoming excessive buyers. The influence of VAT (decrease of the percentage of addicts) was lower than expected since compulsive buying does not respond to rational behaviour.

Keywords: Event study, compulsive shopping, mathematical modeling

1. Introduction

The consumption of goods and services is the rational way to satisfy human needs such as hunger in the case of food or safety when renting or buying a house, what is assumed as ra-

*E-mail address: elpopla@esp.upv.es
†E-mail address: irene.garcia@ehu.es
‡E-mail address: ljodar@mat.upv.es
§E-mail address: palomamerello@outlook.com

tional behavior. However, at the present time in western societies individual's consumption is more a response to an attempt to improve individual's level of self-esteem and well-being and the desire to repair self-thread [1]. One of the mechanisms used by consumers in the shopping process is the imitation whether a person claims to be experienced in connection with the purchase concrete to be carried out [2], bandwagon [3] or in the absence of that person concerning the imitation of the buying habits of most people [4].

The addiction to purchase differs from the normal purchase in quantitative terms (the compulsive shopper has to buy items incessantly), and is based on looking for immediate gratification through the acquisition of material goods. The compulsive shopper is not capable to control their purchase appetite what impacts negatively on their economic situation, as well as on their family relationships [5]. The birth of such compulsive buying behaviour is influenced primarily by cultural factors and socialization, as well as family and peers [6], while also psychobiological factors that make people have a natural inclination towards this type of behaviour [5].

Compulsive shopping is considered as a mental disorder by psychologists. However, it has not been studied extensively, but it is thought to be an impulse control disorder [7].

Compulsive buying disorder occurs mainly in developed countries due to cultural and social factors causing or promoting the disorder. Elements which appear necessary for the development of compulsive buying disorder include the presence of a market-based economy, the availability of a wide variety of goods, easily obtained credit, disposable income, and significant leisure time [8].

Thus, the number of compulsive buyers increases during the expansion stages of the economic cycles due to their consumption by credit [9] what reinforces their compulsive behaviour but also they become highly indebtedness [10].

For the particular case of Spain, the economic blooming occurred during the period 2002-2007, however from 2008 until the present time, Spain is submerged in its worst economic crisis with high and stable unemployment annual rates (over 20%) combined with the loss of trust on the political system of Spanish citizenship. The main economic measure taken by Spanish Government so far consists of increasing taxes (VAT, income taxation), and cut of wages, reducing the household's disposable income.

In this context, a decline in the number of compulsive buyers would be expected due to the fiscal policy carried out by the Spanish Government, combined with the difficulty of households to access to credit.

In this chapter, following the model developed by [11], we propose a population discrete mathematical model represented by a system of difference equations that considers the increase of the Spanish VAT to estimate the prevalence rate of compulsive buyers in Spain in 2013 comparing our results with the previous ones obtained by [11] in which the fiscal policy was not considered.

The increase of the Spanish VAT rate by the Government is modeled following the event study methodology [12], [13], based on the effect that a new event in our case the increase of taxes has on the consumers' behaviour. We applied the event study methodology modeling the impact of the VAT increase on consumers' behaviour and comparing our results with the previous study [11].

Our approach is epidemiological dealing with populations instead of individuals whose behavior may be erratic while aggregate behaviour is often quite predictable [13]. The foundations of such approach are based on the mimetic human behaviour [14], human herding [4] and social contagion [6].

This type of population approach has been recently employed in the study of sociological problems such as anxiolytic consumption dependency [15], workaholism [16], eating addiction [17]. The study of the propagation of compulsive buyers is relevant for curbing its causes and also because of its social and helthcare implications.

The population object of study is the Spanish citizen aged from 18 to 65 years old. This population is split in three subpopulations according to their purchase behavior. Firstly, we employed the Compulsive Buying Scale proposed by [18]. Then, a cluster analysis (with the K-means algorithm) was performed to classify the individuals among three categories: N (rational shopper), S (over-shopper) and A (compulsive shopper) [19].

Traditionally Event Study methodology [20], [21], has based on adjusted linear regression models, assuming that a given event would modify the response variable of the model. Since in the present study a model of simple linear regression with a single response variable would not be able to model the evolution of the three buyers subpopulations analyzed and, also, a dynamical approach by difference equations was considered, the parameters associated with the event will be estimated in this study using the algorithm of Nelder-Mead.

The inter-subpopulations transits are modeled according to demographic, sociological, economic and fiscal factors that will allow the construction of a system of difference equations whose solution will permit forecast the prevalence of compulsive buyers in the next years.

The chapter is organized as follows, section 2 introduces the method, section 3 shows the model, and results of the application of the event methodology through the VAT impact. Finally, section 4 includes the conclusions and recommendations of the chapter.

2. Method

A survey was performed in April 2013 in the province of Vizcaya by telematic means, replicating the one performed in May 2010 [11], consisting of the Compulsive Buying Scale [18] and as demographic variables the gender and the age.

The database was cleaned of outliers and sex and age ratios of the Spanish population were replicated. The final sample used in the analysis was composed by 275 individuals.

3. Results

3.1. Mathematical Model of Compulsive Shopping

The shopping addiction has been significantly related to self-esteem and satisfaction of emotional instability [22]. Therefore, in the mathematical modeling of addiction we considered as causes of transition to higher levels of compulsive buying, those causes which may cause distress or capable of inducing the individual to use shopping as a tool to evade personal conflicts: such as a bad economic situation or the influence of compulsive buyers from his/her social environment (social contagion [6]).

The Spanish population of buyers was divided according to their compulsive behaviour, thus three subpopultions were identified [11]: rational shoppers (N), over-shoppers (S) and compulsive shoppers (A).

Thus, the total population size (P) at any time n is given by [11]:

$$P_n = N_n + S_n + A_n \,. \tag{1}$$

The indicated population dynamics can be described by the following system of equations (n, time in months):

$$\left.\begin{aligned} N_{n+1} - N_n &= \mu P_n - dN_n - \beta_1 \tfrac{N_n A_n}{P_n} \\ &- \beta_2 N_n + \epsilon_2 A_n, \\ S_{n+1} - S_n &= \beta_1 \tfrac{N_n A_n}{P_n} + \beta_2 N_N - dS_n \\ &- \gamma_1 \tfrac{S_n A_n}{P_n} - \gamma_2 S_n + \epsilon_1 A_n, \\ A_{n+1} - A_n &= \gamma_1 \tfrac{S_n A_n}{P_n} + \gamma_2 S_n \\ &- \epsilon_2 A_n - dA_n - \epsilon_1 A_n. \end{aligned}\right\} \tag{2}$$

where the state population vector $PS_n = (N_n, S_n, A_n)$, gives the number of rational buyers, excessive buyers and addicts at time n.

In the dynamic of the addiction the following is assumed:

- A buyer requires a period of adaptation to move from one state to another, so individuals transit level by level, i.e., $N \to S$, $S \to A$,
- All new buyers (Spanish population with consumption capacity, i.e. over 16 years old) enter to the model as rational buyers subpopulation (N).

The values of all parameters were estimated from different sources of information and hipotheses (detailed below), with the exception of contagious rates γ_1 and β_1, which were adjusted by the least squares Nelder-Mead algorithm as specified in [11].

The parameters were estimated as follows:

- μ, birth rate in Spain. $\mu = 0.000833$ months^{-1}, Spanish average birth rate between the years 2002 to 2009 [23].
- d, death rate in Spain. $d = 0.000666$ months^{-1}, Spanish average death rate between the years 2002 to 2009 [23].
- β_1, social contagion rate (rational buyers $\to$ excessive buyers). $\beta_1 = 0.002453$, estimated by Nelder-Mead.
- β_2, transition rate from rational buyer (N) to escessive buyer (S) related to the economic situation. $\beta_2 = ICC_n \times (0.25 \times 0.026)$ months^{-1}. Where ICC [24] represents

the proportion of the population that has an optimistic view of the economic situation in the month n. This parameter is estimated assuming that 25% population behave on impulse to shopping (with a non-pathological behaviour) [25] and the 2.6% of the population suffers from emotional instability [26], [27].

- γ_1, social contagion rate (excessive buyer $\rightarrow$ addict). $\gamma_1 = 0.0048$, estimated by Nelder-Mead.

- γ_2, transition rate from excessive buyer (S) to addict (A) related to the economic situation. $\gamma_2 = ICC_n \times (0.25 \times 0.026) + k$ months^{-1}. It is assumed that $\gamma_2 \geq \beta_2$, $\gamma_2 = \beta_2 + k$, i.e., is admitted that the economic situation affects more to excessive buyers (S) than rational buyers (N). $k = 0.00013$, estimated by Nelder-Mead.

- ϵ_1, self-induced recovery rate (from addict to excessive buyer). It is assumed that the 24.1% of the addicts self-stop their compulsive behaviour [28], as an addict needs 10 years for the recognition of the pathological behaviour [29], thus $\epsilon_1 = 0.241 \times \frac{1}{10 \times 12}$. $\epsilon_1 = 0.0020$ months^{-1}.

- ϵ_2, recovery rate by therapy (from addict to rational buyer). $\epsilon_2 = 0.0035 \times 0.5 \times (1/3) = 0.0005833$ months^{-1}. This parameter has been estimated assuming:

  * The percentage of addicts who begin therapy each year (0.35%) [30].
  * The average success rate for treatment programs (50%) [31].
  * Average duration of therapy is 12 weeks (3 months aprox.) [31].

3.1.1. Subpopulations Estimated by the Model (April 2013)

Based on the actual data available for the ICC for the period May 2010 to April 2013 [24], and starting from the known data of the consumer population for May 2010 ($N = 44.14\%$, $S = 39.1\%$ and $A = 16.76\%$), solving the equations of the model the percentages of each subpopulation (N, S and A) for the Spanish population in April 2013 were estimated (Table 1).

Table 1. Forecast of buyers by subpopulation (April 2013)

Year	Subpopulation	Percentage
April 2013	N	41.84%
	S	38.87%
	A	19.29%

3.2. Event Study

To estimate the prevalence of addictive buying, performing a k-means cluster analysis on data obtained from the survey variables of the Compulsive Buying Scale. The result of the

Table 2. Results of k-means cluster analysis (April 2013)

Year	Subpopulation	Percentage
April 2013	N	39.64%
	S	41.82%
	A	18.54%

cluster analysis has determined the percentages of each subpopulation (N, S and A) in the Spanish population, the results are listed in Table 2.

The model originally proposed for modeling the dynamics of the consumer population in Spain has shifted in the prediction, primarily for excessive (S) and rational buyers (N).

The traditional event study methodology [20, 21] is based on fitting a linear regression model assuming that a particular event (modeled as different values of a certain variable) can modify the response variable of the model.

In our case, a simple linear regression model with a single response variable is not able to model the evolution of the three subpopulations present in the consumer population. Thus, in this work we estimate the parameters associated with the events by the Nelder-Mead algorithm [11, 15, 16].

It is reasonable to expect that due to a stress situation or event as a bad economic forecast, a percentage of rational consumers modify their behaviour purchasing as a self-compensation attitude trying to reduce their anxiety. So, this event would be modeled within the parameter β_2 from the original model (rate at which a rational buyer transits to the excessive buyer's subpopulation because of the effect of a bad economic situation). Furthermore, it is known that the crisis has increased emotional distress by 30% [27], [32], so the percentage of the population with emotional distress will now be 2.6%. Thus, the new parameter $\beta_2^{'}$ will be defined as follows:

$$\beta_2^{'} = ICC \times 0.026 \times \rho$$

Where ρ represents the proportion of people that behave on impulse to shopping. This parameter is estimated by the Nelder-Mead algorithm.

On the other hand, a possible reducing effect is expected caused by the upward Spanish tax policies that have produced an increase of the VAT in recent years. This event will be modeled within the parameter γ_2 of the original model (rate at which an excessive buyer transits to the addictive buyer's subpopulation because of the effect of a bad economic situation). Given that the percentage of people with emotional distress has risen to 2.6% and assuming that the effect of the increase in VAT on consumption extends over 12 months, the new parameter ($\gamma_2^{'}$) would take the following form:

$$\gamma_2^{'} = [(ICC \times 0.026 \times 0.25) + 0.00013] - \sigma(IVA_{n+1} - IVA_{n-11})$$

Where σ is a parameter estimated by the Nelder-Mead algorithm.

The adjusted parameters are estimated by fitting the model to real data (surveys performed in Vizcaya, May 2010 and April 2013). Taking as the initial conditions of the model (year 2010-May, i.e., $n = 0$), $N_0 = 0.4414$, $S_0 = 0.391$ and $A_0 = 0.1676$ (Prevalence study for 2010) and the final conditions of the model (year 2013-April, i.e., $n = 35$) $N_{35} = 0.3964$, $S_{35} = 0.4182$ and $A_{35} = 0.1854$ (Prevalence study for 2013), the parameters ρ and σ have been estimated by fitting the scaled model. The values of ρ and σ that fit the model with real data are: $\rho = 0.4185$ y $\sigma = 0.0471$.

Regarding the value estimated for the stress event (ρ), the parameter value indicates that a 41.85% of the population with emotional distress that felt good economic prospects transit from rational buyer's subpopulation (N) to excessive buyer's subpopulation (S), compared to 25% that was considered in the original model.

Regarding the value estimated for the event "VAT increase" (σ), the fitted parameter indicates that for every 1% (0.01) increase of the VAT, the monthly transit proportion from S to A is reduced by 0.000471. Thus, the VAT increase has an effect on the addictive buyers, but much less than a priori expected. Contrary to what one might initially intuit, the tax increase has less effect than expected as a result of that addiction does not respond to rational behaviour.

This result is consistent with several studies, on one hand with those which have contrasted that the rate increase in addictive products does not reduce addiction levels [33], [34], and on the other hand with economic works in the OECD countries that have confirmed that during crisis time and high levels of public debt consumers do not substantially alter their consumption as a response to potential future tax changes [35].

Conclusion

The model proposed in this chapter applying the event study approach improves the previous one by [11]; the results obtained by our model adjust precisely the Spanish rate of compulsive buyers estimated throughout the Survey in April 2013.

The proposed model introduces two events of opposite nature; the first event is based on the stress the Spaniards suffer caused by the bad economic perspectives while the second one is related to the level of VAT. The Nelder-Mead algorithm adjusted both events.

The results show, on one side that the 41.85% of the population with emotional distress but with good economic expectations transit from rational buyer's (N) to excessive buyer's subpopulation (S), while in the original model [11] was the 25%. On the other hand, we found a decrease of the compulsive buyers (A) associated with increases of the VAT rates. However, this was smaller than it would have been expected since compulsive buying does not respond to rational behaviour.

Compulsive buying disorder is not included in DSM-IV [36], however there is not a clear conclusion about its nature, if it should be considered an addictive disorder, an obsessive-compulsive disorder or a mood disorders. There is evidence of connection between compulsive buying and other disorders such as bipolar disorder; the maniac part of a bipolar diagnosed person can consist of buying compulsively during their episodes of good mood.

The importance of this disorder requires public health action, throughout prevention, mainly pursued by education in schools. Also it is relevant to diagnose as soon as possible any kind of mental disorder affecting the life of a person; when the person is diagnosed cognitive behavioural techniques can be helpful, although more research is necessary to determine what types of therapy are effective for whom.

References

[1] Sivanathan, N.; Pettit, N. C. Protecting the self through consumption: Status goods as affirmational commodities *Journal Exp Soc Psychol*. 2010, 46, 564-570.

[2] Sheth, J. N. How Adults Learn Brand Preference *J Advertising Res*. 1968, 8, september, 25-36.

[3] Kastanakis, M. N.; Balabanis, G. Between the mass and the class: Antecedents of the bandwagon luxury consumption behavior *J Bus Res*. 2012, 65 (10), 1399-1407.

[4] Raafat, R .M.; Chater, N.; Frith, C. Herding in humans *Trends Cogn Sci*. 2009, 13 (10), 420-428.

[5] Rodríguez-Villarino, R.; Rodríguez-Castro, R. La adiccion a la compra: revisión y necesidad de estudio en la realidad española *Estudios sobre consumo*. 2000, 52, 75-98.

[6] Christakis, N. A.; Fowler, J. H. *Connected: The Surprising Power of Our Social Networks and How They Shape Our Lives*; Back Bay Books; Little Brown and Company: New York City, US, 2009.

[7] Johnson, B. A. *Addiction Medicine: Science and Practice*; Springer: New York, US, 2010; Vol 1.

[8] Black, DW. Compulsive buying disorder: definition, assessment, epidemiology and clinical management *CNS Drugs*. 2001, 15, 17-27.

[9] Petit, N. C.; Sivanathan, N. The Plastic Trap: Self-Threat Drives Credit Usage and Status Consumption *Soc Psychol Person*. 2011, 2 (2), 146-153.

[10] Christenson, G. A.; Faber, R. J.; Mitchell, J. E. Compulsive buying: Descriptive characteristics and psychiatric comorbidity *J Clin Psychiat*. 1994, 55 (12), 545-546.

[11] García, I.; Jódar L.; Merello, P.; Santonja, F.J. A discrete mathematical model for addictive buying: Predicting the affected population evolution *Math Comput Model*. 2011, 54 (7-8), 1634-1637.

[12] Fama, E. F.; Fisher, L.; Jensen, M. C.; Roll, R. The adjustment of stock prices to new information *Int Econ Rev*. 1969, 10, 1-21.

[13] MacCluer, C. R. *Industrial Mathematics: Modeling in Industry, Science, and Government*; Prentice Hall: Upper Saddle River, US, 2000.

[14] Girard, R. *Mimesis and Theory: Essays on Literature and Criticism*, 1953-2005. Stanford University Press: CA, USA, 2008.

[15] De la Poza, E.; Guadalajara, N.; Jódar, L.; Merello, P. Modeling Spanish anxiolytic consumption: Economic, demographic and behavioral influences *Math Comput Model*. 2013, 57 (7-8), 1619-1624.

[16] De la Poza, E.; Del Libano, M.; García, I.; Jódar, L.; Merello, P. Predicting workaholism in Spain: a discrete mathematical model *Int J Comput Math*. 2013, In press.

[17] Santonja, F. J.; Villanueva, R. J.; Jódar, L.; González-Parra, G. Mathematical modelling of social obesity epidemic in the region of Valencia, Spain *Math Comp Model Dyn*. 2010, 16 (1), 23-34.

[18] Valence, G.; D'Astous, A.; Fortier, L. (1988). Compulsive buying: concept and measurement *J Consume Policy*. 1988, 11, 419-433.

[19] Olabarri, E.; García, I. La compra por impulso y la adicción al consumo en el País Vasco *Estudios Sobre Consumo*. 2003, 65, 86-109.

[20] Mackinlay, A. C. Event Studies in Economics and Finance *J Econ Lit*. 1997, XXXV, 13-39.

[21] Kukar-Kinney, M.; Ridgway, N. M.; Monroe, K. B. The Role of Price in the Behavior and Purchase Decisions of Compulsive Buyers *J Retailing*. 2012, 88, 63-71.

[22] Rodríguez-Villarino, R.; González-Lorenzo, M.; Fernández-González, A.; Lameiras-Fernández, M.; Foltz, M.L. Individual factors associated with buying addiction: an empirical study *Addict Res Theory*. 2006, 14 (5), 511-525.

[23] INE. (2010). *Indicadores demográficos basicos* [statistical data]. Available at: Instituto Nacional de Estadstica. [http://www.ine.es/jaxiBD/menu.do?L=0&divi=IDB&his=0&type=db] [accesed on 30/5/2013].

[24] Centro de Investigaciones Sociológicas. (2013). Indicador de Confianza de los Consumidores [statistical data]. Available at:

[http://www.cis.es/cis/opencms/ES/13Indicadores/Indicadores/ICC/index.jsp] [accesed on 30/5/2013].

[25] Garcés, J. (2000). Experiencias de trabajo en la prevencion y tratamiento de la adiccion al consumo, 2000. Available at:

[http://webs.uvigo.es/consumoetico/textos/consumo/experiencias.pdf.]

[26] Departamento Valenciano de Salud. (2005). Encuesta de salud, 2005. Available at: [http://www.san.gav.es/cas/prof/homeprof.html].

[27] El Economista. (2013). El "malestar psiquíco" de la población ha crecido un 30% por la crisis [newspaper article]. *El Economista*. Available at: [http://www.eleconomista.es/publicidad/acierto-abril/espana/noticias/4570612/02/13/El-malestar-psiquico-de-la-población-ha-crecido-un-30-por-la-crisis.html] [accesed on 18/06/2013].

[28] Toneatto, T.; Nett, J. C. Natural recovery from problem gambling. In *Promoting Self-Change from Addictive Behaviors. Practical Implications for Policy, Prevention, and Treatment*; Klingemann, H.; Sobell, L.C.; Ed.; Springer: Berlin, DE, 2007.

[29] McElroy, S. L.; Keck, P. E. Jr.; Pope, H. G. Jr.; Smith, M. R.; Strakowski, S.M. Compulsive buying: a report of 20 cases. *J Clin Psychiatry*. 1994, 55, 242-248.

[30] Rodríguez, R.; González, M.; Fernández, A.; Lameiras, M. Explorando la relación de la adicción a la compra con otros comportamientos excesivos: un estudio piloto. *Adicciones*. 2005, 17, 231-240.

[31] Mitchell, J. E.; Burgard, M.; Faber, R.; Crosby, R. D.; de Zwaan, M. Cognitive behavioral therapy for compulsive buying disorder *Behav Res Ther*. 2006, 44, 1859-1865.

[32] AEN; FADSP. (2011). Crisis económica y repercusión sobre la salud. Manifiesto de la Asociación Española de Neuropsiquiatría (AEN) y la Federación de Asociaciónes para la Defensa de la Sanidad Pública (FADSP), November. Available at: [http://www.aen.es/index.php?option=com-content&view=article&id=489:crisis-economica-y-repercusion-sobre-la-salud&catid=417:comunicados-aen&Itemid=135]

[33] Callison, K.; Kaestner, R. Do higher tobacco taxes reduce adult smoking? New evidence of the effect of recent cigarette tax increases on adult smoking *Econ Inq*. 2013, Article in Press.

[34] Thomas, D.P.; Ferguson, M.; Johnston, V.; Brimblecombe, J. Impact and perceptions of tobacco tax increase in remote Australian aboriginal communities *Nicotine Tob Res*. 2013, 15 (6), 1099-1106.

[35] Bhattacharya, R.; Mukherjee, S. Non-Keynesian effects of fiscal policy in OECD economies: An empirical study *Appl Econ*. 2013, 45 (29), 4122-4136.

[36] American Psychiatric Association. *Diagnostic and statistical manual of mental disorders*; American Psychiatric Press: Washington, US, 1980; 3rd ed.

In: Mathematical Modeling in Social Sciences ...
Editors: J. C. Cortés López et al.
ISBN: 978-1-63117-335-6

Chapter 15

IS FITNESS ACTIVITY AN EMERGENT BUSINESS? ECONOMIC INFLUENCES AND CONSEQUENCES OF MALE FITNESS PRACTICE

***M. S. S. Alkasadi*[1], *E. De la Poza*[2,*] *and L. Jódar*[1]**
[1]Instituto Universitario de Matemática Multidisciplinar,
Universitat Politécnica de Valéncia, Valencia, Spain
[2]Facultad de Administración y Dirección de Empresas,
Universitat Politécnica de Valéncia, Valencia, Spain

Abstract

Males under forty years old increase their gym practice to improve their self-esteem and sexual appeal through their body image. In this chapter we develop a discrete mathematical model to forecast the future loyal fitness customer rate in Spain over the next years. For this purpose, economic, emotional and social contagion variables are taken into account in order to quantify the dynamic behavior of men gym users. Economic consequences of gym customers' behavior are studied and also future possibilities of business are suggested. Also, the model can be exported without additional effort to other countries where data is available.

Keywords: loyal fitness customer, fitness business, economic influences, contagion effect

1. Introduction

Our society is concerned with people's physical appearance and the ideal body image [1] [2]. As a result, physical fitness practice has grown considerably which leads to health benefits such as the reduction of obesity, increase of personal self-esteem, and also healthy spare time activities. Apart from this, it is clear that a possible increase of the fitness practice business involves a parallel increase of the related economic sectors such as sports clothing, energy drinks and sports equipment [3].

*E-mail address: elpopla@esp.upv.es; Tel: +34963877032; Fax: +34963877032

The Spanish population who exercise regularly has evolved from 27% in 2005 to 35% in 2010 [4]. Also, in this context the role of the media and fashion industry encourages people to take care of their image, influencing society's behavior [5].

The purpose of this chapter is to develop a discrete mathematical model to forecast the future loyal fitness customer rate in Spain over the next five years. For this purpose, economic, emotional and social propagation variables [6] [7] [8] are taken into account in order to quantify the dynamic behavior of men gym users. Also, personal and social consequences of this behavior are studied and future possibilities of business are suggested.

This chapter is organized as follows: Section 2 Deals with the model construction throughout a discrete system of difference equations representing the population of interest. In section 3 Computations and simulations are carried out after assuming several possible economic scenarios for the next coming years. Section 4 Shows the conclusions of the study.

2. Mathematical Model Construction

2.1. Data Collection and Sampling

The population of the study is composed by the Spanish males who attend the gym among the age interval $[18, 40]$. We classified the gym users into three categories according to their score obtained in the ten questions abbreviated Adonis Complex Questionnaire (ACQ) [9]. The questionnaire measures both the level of gym attendance and also the individuals' psychological dependence to the gym practice.

According to the score obtained the three categories are:

- $I(n)$: defined as incidental gym users; those men whose score was equal or lower than 2 points in the questionnaire and the number of working out sessions is one's or twice per week at year n .
- $F(n)$: Frequent gym users were those whose scores were 3 or 4 points in the questionnaire and attend the gym at least three times per week at year n .
- $R(n)$: Regular gym users were those whose score was higher than 5 points in the questionnaire and also attend the gym more than three times per week at year n.

2.2. Mathematical Model

The dynamic behavior of the gym users is based on their transition among subpopulations explained by coefficients that need to be found according to economic, socio-demographic and social propagation hypotheses. Our attention is focused on forecasting the number of loyal customers to fitness centers for the period $2012 - 2015$.

Data was collected from two samplings: first one taken in 2011 and the second one in 2012 at the public gym of the Polytechnic University of Valencia, and at a private multi-located gym; then, we classified the Spanish gym users by subpopulations after adjusting statistically using data from the [10].

Then, the variation between subpopulations I, F and R was estimated for the interval $[n, n+1]$.

2.2.1. Hypothesis of the Model

1. The individuals fall into the model under two hypotheses:

    a. Male gym users older than 17 years old. It is assumed that the increase of I and F users at year n is equal to the Spanish birth rate at that year, while it is 0 for R users.

    b. In case of an economic improvement occurs, population invests on sportive practices, increasing I subpopulation.

2. The gym users leave the model at year n, due to possible scenarios:

    a. The gym user becomes older than 40 years old or he passes away at year n.

    b. As a result of the economic crisis, a proportion of the I users emigrate abroad, giving up on their gym practices.

    c. Due to a possible economy deterioration, I users decrease their sport practices which F and R users do not.

3. The gym users can only transit jumping one category or subpopulation [6] [7]. Also, it is assumed a possible recovery transit from $F \longrightarrow I$.

Hence, the transits among subpopulations are due to:

- In an attempt to recover their self-esteem, those individuals who suffered from traumas such as bullying during childhood [11]; also looking for improving their sex-appeal [12] [13]; or rebuilding their personal life [14] defined by α_e .

- The combination of the emotional factors and the economic worsening influences is compiled by coefficient α_t.

- The influence of personal relationships, especially with regular consumers, is a determinant of the people's habits and behaviors' as their diet, measured by γ_1 for the transit from $I \longrightarrow F$, and γ_2 for the transit from $F \longrightarrow R$ [6] [7].

- The influence of an economic improvement in case of occurrence combined with the rebuilt of the personal life can produce a backward transit $F \longrightarrow I$ subpopulation, measured by α_2 [15].

  The dynamics of the model is shown in the Figure 1.

Thus, the dynamic model of the gym consumer's propagation can be modeled by the following equations:

$$P(n) = I(n) - F(n) + R(n). \tag{1}$$

$$\left.\begin{aligned} I(n+1) - I(n) &= b_1 I(n) - d_1 I(n) + \alpha_2(n)F(n) \\ &- \alpha_e I(n) - \gamma_1 I(n) - \alpha_f I(n) + \alpha_\rho(n) I(n), \\ \\ F(n+1) - F(n) &= b_2 F(n) - d_2 F(n) + \alpha_e I(n) \\ &- \alpha_2(n)F(n) + \gamma_1 I(n) - \alpha_t(n)F(n) - \gamma_2 F(n), \\ \\ R(n+1) - R(n) &= -d_3 R(n) + \alpha_t(n)F(n) + \gamma_2 F(n). \end{aligned}\right\} \tag{2}$$

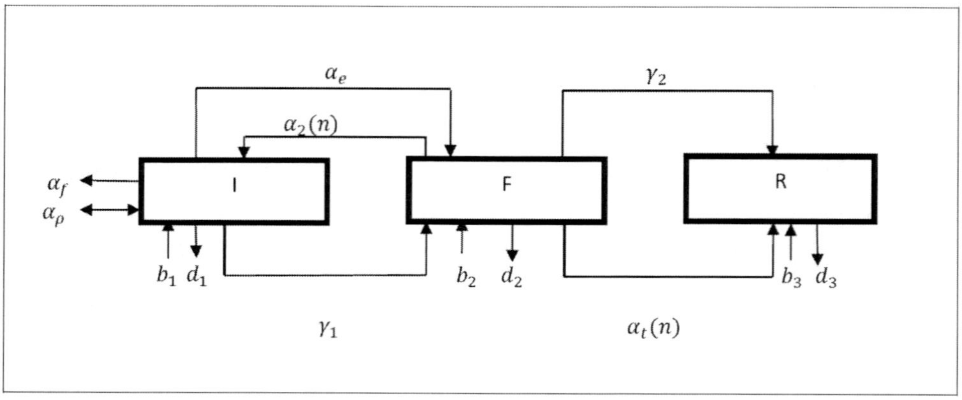

Figure 1. Dynamics of the population.

2.2.2. Computation and Estimation of the Parameters

The values of all parameters were estimated or computed from different sources of information and hypotheses as follows:

- b_i: Birth rate of the population in 2011 by categories $(i = 1, 2)$. We consider the birth rate forecasted by [10] for the period $2011 - 2015$ remains constant [16]. The birth rate is distributed among subpopulations as follows: $b_1 = 0.8 \times (10.66/1000) = 0.0085$; $b_2 = 0.2 \times (10.66/1000) = 0.0021$.

- d_i: Spanish mortality rate in 2011 by categories of gym consumption $(i = 1, 2, 3)$. We assume this rate remains constant for the period $2011 - 2015$, where $d_i = (8.8/1000)/3 = 0.00293$.

- $\rho_{(n)}$: Unemployment rate at the year n. For the year 2011 the unemployment rate proceeds from the [10]. For the next years 2012 and 2013, the economic forecast is taken from [17]and [18]. For 2014 and 2015 we assumed four possible scenarios from the [17] and [18] unemployment forecast and also two more assuming an optimistic and a pessimistic scenario.

- α_f: It is defined as the emigration rate caused by unemployment in Spain. A total of 150 000 people left Spain in 2011 looking for a job abroad, of this number, a 5% [4] were gym users; thus, $\alpha_f = (0.05 \times 150\,000)/537\,064 = 0.014$. We assume the population that emigrates is incidental gym users (I); This rate remains constant for the period $2011 - 2015$ [16].

- $\alpha_\rho(n)$: Economic influence over I subpopulation. From the observed data in previous years from [10] and [4] it is assumed that for each 1% decrease of the unemployment rate the I category increases 0.4% at year n; also for each 1% increase of unemployment rate, the I category decreases 0.2% at year n. Then,

$$\alpha_\rho(n) = \begin{cases} -0.004 \times \left(\rho_{(n+1)} - \rho_{(n)}\right) & \text{if,} \quad \rho_{(n+1)} < \rho_{(n)}, \\ \\ -0.002 \times \left(\rho_{(n+1)} - \rho_{(n)}\right) & \text{if,} \quad \rho_{(n+1)} > \rho_{(n)}. \end{cases} \tag{3}$$

- $\alpha_2(n)$: Rate of recovery of the F subpopulation due to two components: an employment recovery (α_{21}) and/or emotional recovery (α_{22}).
  $\alpha_{21}(n)$: The gym user reduces his gym activity when becoming employed. Thus, the 80% of the jobs produced by the economic recovery [10] are absorbed by the population aged in the interval $[18, 40]$ from whom a 5% are gym users [4]. Then,

$$\alpha_{21}(n) = 0.05 \times \left(\rho_{(n+1)} - \rho_{(n)}\right)^{-} \times 0.8. \tag{4}$$

$$\left(\rho_{(n+1)} - \rho_{(n)}\right)^{-} = \begin{cases} 0 & \text{if,} \quad \rho_{(n+1)} \geq \rho_{(n)}, \\ \rho_{(n)} - \rho_{(n+1)}, & \rho_{(n+1)} < \rho_{(n)} \quad \text{if the economy recovers.} \end{cases} \tag{5}$$

- The second component that determines the transit from $F \rightarrow I$ is due to a personal life rebuilt. We estimate α_{22} as the proportion of gym users over the total Spanish population aged $[18, 40]$ that rebuild their life, what is estimated in terms of the marriage rate in Spain in 2008 [10].

  Hence, $\alpha_{22} = 0.05 \times 0.7796 = 0.0389$. We assume this rate remains constant for the period of study [4].

- α_e : Rate of emotional impact is estimated as the weighted average of two addends resulted of dividing the population into two age intervals:

  α_{e1}: Estimated as the proportion of Spanish people aged in the interval $[18, 25]$, 9.06% [10], that are gym users (5%), and have lower level of self-esteem as consequence of their past experiences such as a childhood trauma (4%) [11], or they search for improving their physical attractiveness (8%), [19] [12] [13] [20] but also other causes 1%(emotional drawbacks). In all cases the individuals are pushed to the gym practice.

  Then, α_{e1} is weighted as $\frac{1}{3}$ of $\alpha_e(\frac{9.06\%}{25.63\%} = \frac{1}{3} = 0.33)$, where 9.06% is the Spanish rate of men aged $[18, 25]$ years old in 2011 [10], 25.63% Spanish rate of men aged $[26, 40]$ years old [10].

$$\alpha_{e1} = 0.33 \times (0.0906 \times 0.05 \times 0.13) = 0.0001943.$$

  α_{e2} : is the proportion of Spanish people aged in the interval $[26, 40]$, that are gym users (5%), and experience any of the following status: divorce (35%), childhood trauma (4%) [11], or they look for improving their physical attractiveness (8%), [19] [12] [13] [20].

$$\alpha_{e2} = 0.66 \times [(0.2563 \times 0.05) \times (0.13 + 0.35)] = 0.004059.$$

- $\alpha_t(n)$: Is the rate of transit $F \rightarrow R$ as consequence of the combination of opposite effects, the emotional impact rate (α_e) and an economic recovery (α_{21}) in case it happens.

$$\alpha_t(n) = \alpha_e - \alpha_{21} = 0.0042 - (0.05 \times \left(\rho_{(n+1)} - \rho_{(n)}\right)^{-} \times 0.8).$$

- γ_1: Is the propagation effect derived from the contact between R and I user, then $I \rightarrow F$. This value was adjusted from the model, using the data collected by the two samples 2.2 $\gamma_1 = 0.08415$ [6] [7].
- γ_2: Is the propagation effect produced by contact between R and I users $F \rightarrow R$; γ_2, is assumed constant for all years. After matching data the found value was $\gamma_2 = 2 \times \gamma_1$ [6] [7].

3. Results and Simulations

3.1. Computation of the Expected Gym Users

The mathematical model allows us to compute the subpopulations $I(n)$, $F(n)$ and $R(n)$ at any year n.

We assumed the economic forecast from [17] and [18] from 2012 until 2015. Apart from these economic scenarios, we introduced two others one optimistic and one pessimistic for the total period of study; thus, any possible economic future situation will be enclosed in the range of variation of our scenarios,(table 1).

Table 1. Economic forecast of the Spanish unemployment rate

	OECD	FUNCAS	Optimistic	Pessimistic
2011	0.23	0.23	0.23	0.23
2012	0.2502	0.2502	0.2502	0.2502
2013	0.269	0.272	0.24	0.28
2014	0.25	0.26	0.22	0.29
2015	0.24	0.25	0.21	0.3

The following table 2 collects the results performed by the computation of the system expressed in volume of users. Thus, the percentages of F and R increase over time independently of the economic scenario, with scarce differences between them, which confirms the robustness of our model. The F subpopulation evolves from 5.97% in 2011 to the interval $[29.15\%, 29.53\%]$ in 2015 depending on the economic scenario; while the R subpopulation increases from 1% in 2011 to the interval $[11.52\%, 11.70\%]$ in 2015 depending on the economic scenario.

3.2. Sensibility Analysis of the Proportionality Ratio of the Contagion Parameters

In the model construction we assumed that the propagation parameter γ_2 was two times the value of γ_1 , knowing that $\gamma_2 > \gamma_1$. As it is uncertain how much bigger is γ_2 than γ_1 , it is advisable to perform a sensitivity analysis of this proportionality factor between both parameters. We simulate the number of regular consumers under a variation of the proportionality factor, considering the interval $[1, 3]$, this is from the situation $\gamma_2 = \gamma_1$ to $\gamma_2 = 3 \times \gamma_1$.

Table 2. Subpopulations forecasts in volume of gym users according to the simulated scenarios

		OECD	FUNCAS	Optimistic	Pessimistic
	I	537,092	537,092	537,092	537,092
2011	F	34,466	34,466	34,466	34,466
	R	5,744	5,744	5,744	5,744
	I	484,292	484,292	484,292	484,292
2012	F	80,573	80,573	80,573	80,573
	R	11,674	11,674	11,674	11,674
	I	438,747	438,457	441,556	437,682
2013	F	120,179	120,179	120,179	120,179
	R	25,543	25,543	25,543	25,543
	I	404,218	402,760	406,949	398,310
2014	F	154,274	154,205	154,479	154,040
	R	46,296	46,263	46,301	46,157
	I	372,671	371,345	375,156	364,239
2015	F	183,931	183,736	184,369	183,061
	R	72,842	72,796	72,882	72,539

Results are shown in Figure 2. Note that for a fluctuation of the social propagation parameter γ_2 between the interval $[1, 3]$ the prevalence rate of regular fluctuates between the interval $[6.70\%, 15.87\%]$.

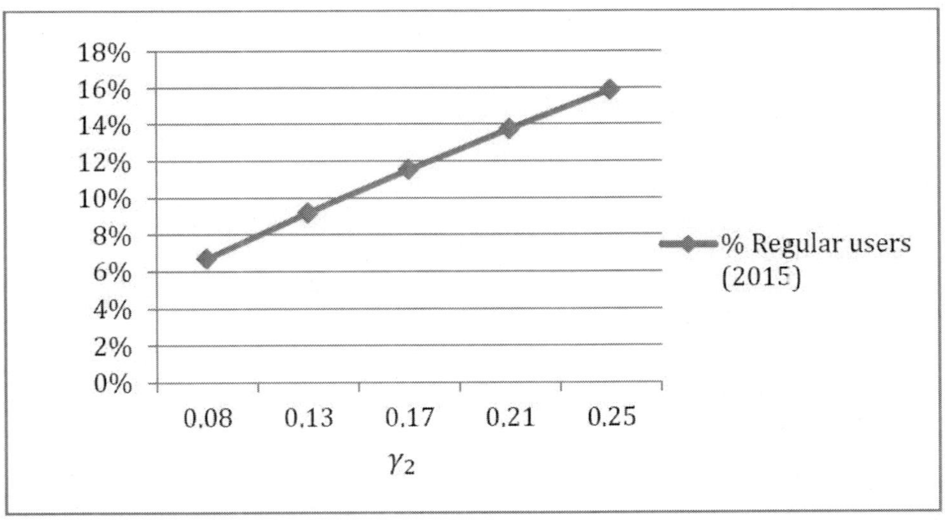

Figure 2. Sensitivity analysis of γ_2 for regular gym users.

Conclusion

The gym consumption of frequent and regular users in Spain is going to grow around 10% in next four years almost independently of the behavior of the economy. Our forecast does not take into account the marketing strategies of the current fitness centers in the present economic turmoil.

We can suggest the general recommendation of promoting the young male sportive practices; mainly the team sports activities that prevent from social isolation of the gym users allowing their self-esteem improvement and physical appearance.

The proposed model becomes a tool to estimate not only the potential gym users but also the potential consumption of correlated markets such as sportswear, sports equipment, energy drinks and foods. Also, the model can be exported to any other western country when data will be available.

References

[1] Hamermesh, D. S. *Beauty Pays: Why Attractive People Are More Successful*; Princeton University Press: Princeton, NJ ; Oxford, 2011.

[2] Hakim, C. *Erotic Capital: The Power of Attraction in the Boardroom and the Bedroom*; Basic Books: New York, 2011.

[3] French, S. A.; Story, M.; Downes, B.; Resnick, M. D.; Blum, R. W. Frequent dieting among adolescents: psychosocial and health behavior correlates *Am J Public Health*. 1995, 85(5), 695-701.

[4] Ferrando, M.; Goig, R. *Ideal democrático y bienestar personal. Encuesta sobre los hábitos deportivos en España*. Consejo Superior de Deportes, Centro de Investigaciones Sociológicas, [in Spanish]: Madrid, 2011.

[5] Ricciardelli, L. A.; McCabe, M. P. Sociocultural and individual influences on muscle gain and weight loss strategies among adolescent boys and girls *Psychology in the Schools*. 2003, 40(2), 209-224.

[6] Christakis, N. A.; Fowler, J. H. *Connected: The Surprising Power of Our Social Networks and How They Shape Our Lives*; Back Bay Books, Little, Brown and Company: New York, 2009.

[7] Raafat, R. M.; Chater, N; Frith, C. Herding in humans *Trends Cogn Sci*. 2009, 13(10), 420-428.

[8] Kastanakis, M. N.; Balabanis, G. Between the mass and the class: Antecedents of the "bandwagon" luxury consumption behavior *J Bus Res*. 2012, 65(10), 1399-1407.

[9] Pope, H. G. Jr.; Gruber, A. J.; Mangweth, B.; Bureau, B.; deCol, C.; Jouvent, R.; Hudson, J. I. Body image perception among men in three countries *Am J Psychiatry*. 2000, 157(8), 1297-1301.

[10] Spanish Statistical Institute(INE).(2012). Available at: http://www.ine.es/ [last access: August 2012].

[11] Wolke, D.; Sapouna, M. Big men feeling small: Childhood bullying experience, muscle dysmorphia and other mental health problems in bodybuilders *Psychology of Sport and Exercise*. 2008, 9(5), 595-604.

[12] Brown, J.; Graham, D. Body satisfaction in Gym-active males: An exploration of sexuality, gender, and narcissism *Sex Roles*. 2008, 59, 94-106.

[13] Pompper, D. Masculinities, the metrosexual, and media images: Across dimensions of age and ethnicity *Sex Roles*. 2010, 63(9-10), 682- 696.

[14] Duato, R.; Jódar, L. Mathematical modeling of the spread of divorce in Spain *Math Comput Modell*. 2013, 57(7-8), 1732-1737.

[15] Popkin, B. M. The nutrition transition in the developing world *Dev Policy Rev*. 2003, 21(5-6), 581- 597.

[16] Arango, J. *Después del gran boom: la inmigración en la bisagra del cambio;* in Aja, E.; Arango, J.; Alonso, J. O. (Eds.); *La inmigración en tiempos de crisis, Anuario de la Inmigración en España*, CIDOB Edicions: Barcelona, 2010, (pp.52-73), [in Spanish].

[17] The Organisation for Economic Co-operation and Development (OECD).(2012). Available at: http://www.oecd.org/economy/ spaineconomicforecastsummary.htm [last access: August 2012].

[18] Foundation of Savings Banks (FUNCAS).(2012). Available at: http://www.funcas.es/, [in Spanish], [last access: August 2012].

[19] Hönekopp, J.; Rudolph, U.; Beier, L.; Liebert, A.; Müller, C. Physical attractiveness of face and body as indicators of physical fitness in men *Evol Hum Behav*. 2007, 28(2), 106-111.

[20] Varangis, E.; Lanzieri, N.; Hildebrandt, T.; Feldman, M. Gay male attraction toward muscular men: Does mating context matter? *Body Image*. 2012, 9(2), 270-278.

In: Mathematical Modeling in Social Sciences ...
Editors: J. C. Cortés López et al.
ISBN: 978-1-63117-335-6

Chapter 16

POPULAR SUPPORT TO TERRORIST ORGANIZATIONS: A SHORT-TERM PREDICTION BASED ON A DYNAMIC MODEL APPLIED TO A REAL CASE

***Matthias Ehrhardt*[1,*], *Miguel Peco*[2,†], *Ana C. Tarazona*[3,‡], *Rafael J. Villanueva*[3,§] *and Javier Villanueva-Oller*[4,ϕ]**
[1]Lehrstuhl für Angewandte Mathematik und Numerische Analysis, Fachbereich C – Mathematik und Naturwissenschaften, Bergische Universität Wuppertal, Wuppertal, Germany
[2]PhD in International Security. Independent Researcher.
[3]Instituto Universitario de Matemática Multidisciplinar, Universitat Politècnica de València, Spain
[4]Centro de Estudios Superiores Felipe II, Aranjuez, Spain

Keywords: Modelling, Dynamical Systems, Extremism, Support to Terrorism

1. Introduction

Popular support is an important enabler for radical violent organizations and it may be crucial for their survival. At the same time, extremist groups have also an impact in the societies where they are inserted, especially if those groups are engaged in violent activities. Social and behavioral scientists try to find clues about how that interaction may affect those people, either at the group or at the individual level, in order to foresee subsequent dynamics [1, 2, 3, 4].

Out of the social and behavioral fields, the model presented by Castillo-Chavez and Song [5] deals with similar processes from a mathematical modeling perspective. In that

*E-mail address: ehrhardt@math.uni-wuppertal.de
†E-mail address: mpeco.research@gmail.com
‡E-mail address: actarazona@asic.upv.es
§E-mail address: rjvillan@imm.upv.es
ϕE-mail address: jvillanueva@pdi.ucm.es

paper, the authors divide the total population into what they call the core, i.e., people belonging or supporting an extremist organization, objective or idea, and the non-core, usually larger than the previous one. At the same time, the core is divided into the people who are not fully committed yet, what they call the semi-fanatic population, and the fully fanatic people. They also assume that an individual may become more fanatic by contacting with people more fanatic than him/her and, at the same time, that individuals in the core may leave the group at a certain rate. With these assumptions in mind, they present a continuous model and its long term analysis. Other authors [6, 7, 8] consider a network version of the Castillo-Chavez and Song's model based upon a system of ordinary differential equations and also study its long-term dynamics.

In this chapter we apply the Castillo-Chavez and Song's model to the Basque Country citizens' attitude towards the terrorist organization ETA (Basque Fatherland and Liberty) after that organization declared the cease of its violent activity, in 2011 [9]. Of course, what we apply here is the version of the model related to people supporting the organization and not to the version related to people belonging to it. Our objective is to analyze any short term dynamics appearing after that event. To do so, we take data from the *Euskobarometro* survey [10, Table 20], one of the best-known independent opinion polls in the region, as well as demographic index. Then, according to those data, we divide the population into the sub populations appearing in the Castillo-Chavez and Song's model, and we fit the model parameters by least square techniques. After that, we are able to predict in the short term the quantitative evolution of the full-supporting population, which in turn might constitute in our opinion an estimation of the bulk of people able to become new ETA members in upcoming years.

This chapter is organized as follows. In Section 2, we retrieve and prepare the necessary data from *Euskobarometro*. In Section 3, we recall the Castillo-Chavez and Song's model, scale it in order to adapt to the data magnitudes and assign values to the demographic parameters. In Section 4 we fit the model with the data and predict the evolution of the sub populations over the next few years. Finally, Section 5 is devoted to conclusions.

2. The Data for the Model

The *Euskobarometro* [10] ("Basque-barometer") is a sociological statistical survey in the Basque Country. It is conducted by the Department of Political Science of the University of the Basque Country and it is based on personal interviews at home, asking questions about the sociological current issues, including ETA. In particular, the question #20 asks about the attitude of the Basque population towards the ETA and divide the population depending on their answer into eight sub populations: Total support; Justification with criticism; Goals yes / Means no; Before yes / Not now; Indifferent; ETA scares; Total rejection; No answer. In order to fit these eight sub populations with the four Castillo-Chavez and Song's ones, we group them into the following ones:

- Total support towards the ETA.
- Attitude of justification with criticism.
- Remote justification attitude.

- Remaining attitudes (indifference, rejection, etc.).

In Table 1, we show the percentages for every sub population since January 2011, when the ETA declared the cease of violent activities [9]. Note that the first *Euskobarometro* after January 2011 was issued in May 2011.

Table 1. Percentage of Basque people in each sub population, classified depending on their attitude towards the ETA

Date	Total support	Justification with criticism	Remote justification	Remaining attitudes
May 2011	1	3	21	75
Dec 2011	1	2	25	72
May 2012	1	4	29	66
Dec 2012	1	2	25	72
May 2013	1	2	28	69

3. The Model

First, we recall the Castillo-Chavez and Song's model [5]. This model is given by the following nonlinear system of ordinary differential equations

$$\begin{aligned}
G'(t) &= \Lambda T(t) - \beta_1 G(t)\tfrac{C(t)}{T(t)} + \gamma_1 S(t) + \gamma_2 E(t) + \gamma_3 F(t) - \mu G(t),\\
S'(t) &= \beta_1 G(t)\tfrac{C(t)}{T(t)} - \beta_2 S(t)\tfrac{E(t)+F(t)}{C(t)} - \gamma_1 S(t) - \mu S(t),\\
E'(t) &= \beta_2 S(t)\tfrac{E(t)+F(t)}{C(t)} - \beta_3 E(t)\tfrac{F(t)}{C(t)} - \gamma_2 E(t) - \mu E(t),\\
F'(t) &= \beta_3 E(t)\tfrac{F(t)}{C(t)} - \gamma_3 F(t) - \mu F(t),\\
T(t) &= G(t) + C(t),\\
C(t) &= S(t) + E(t) + F(t).
\end{aligned} \tag{1}$$

In (1), $G(t)$ is the non-core population, $C(t)$, in turn, is the core population, which includes $S(t)$, $E(t)$ and $F(t)$,

- $S(t) + E(t)$ is the semi-fanatic sub population,
- $F(t)$ is the fanatic sub population, which includes individuals who are completely committed.

$T(t)$ encompasses the total population. Finally, Λ is the constant birth rate, μ is the constant death rate, β_i, $i = 1, 2, 3$ are the transmission rates and γ_i, $i = 1, 2, 3$ are the transition backward rates.

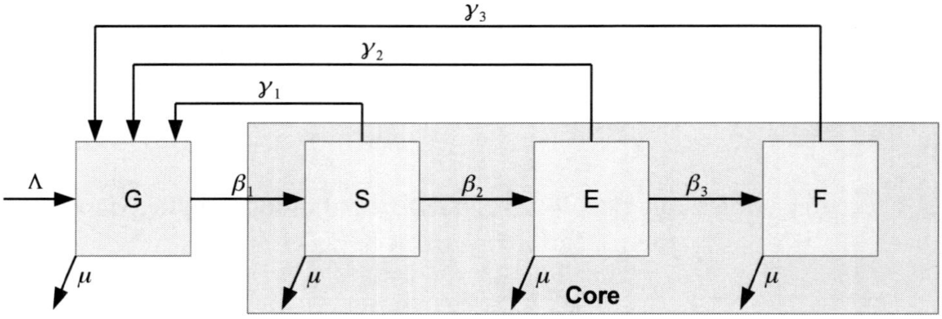

Figure 1. Model flow diagram. The arrows indicate the flow labelled by the corresponding parameters. It is an adaptation of Castillo-Chavez & Song model [5] for our purposes.

In Figure 1, we can see a flow diagram of the model.

As said before, we identify *Euskobarometro* populations (see Table 1) with the model populations. Then,

- $F(t)$ will be those who have a total support attitude towards ETA.
- $E(t)$ will be the ones with an attitude of justification with criticism.
- $S(t)$ are those with an attitude of remote justification.
- $G(t)$ will be the remaining population.

Taking into account that the data in Table 1 is related to percentages meanwhile the model (1) is referred to number of individuals, we transform (scaling) the model into the same units as the data, because one of our objectives is to fit the data with the model in the next section.

Hence, following the ideas developed in the papers [11, 12] about how to scale models where the population is varying in size, we use the code described in [13] to scale the model. This process is very technical and does not provide relevant information, therefore, we are not going to describe here in detail and refer the interested reader to the references [11, 12, 13]. Furthermore, in the following, we are going to consider the populations $F(t)$, $E(t)$, $S(t)$ and $G(t)$ as the scaled ones.

In [14] we can find that the birth rate in the Basque Country in 2011 is $\Lambda = 0.00969$ and the mortality rate in the same year is $\mu = 0.00908$. We also consider the birth and death rates over the next years the same as the ones in 2011. The remaining model parameters $\beta_i, \gamma_i, i = 1, 2, 3$ are fitted with the data in Table 1.

4. Model Fitting and Prediction over the Next Few Years

In order to compute the best fitting, we carried out computations with *Mathematica* [15] and we implemented the function

$$\begin{array}{rccc} \mathbb{F}\ : & \mathbb{R}^6 & \rightarrow & \mathbb{R} \\ & (\beta_1, \beta_2, \beta_3, \gamma_1, \gamma_2, \gamma_3) & \mapsto & \mathbb{F}(\beta_1, \beta_2, \beta_3, \gamma_1, \gamma_2, \gamma_3) \end{array}$$

such that:

1. Solve numerically (using *Mathematica* command `NDSolve[]`) the system of differential equations (1) with initial values given by the first row of Table 1,
2. For $t =$ May 2011, Nov 2011, May 2012, Nov 2012, May 2013, evaluate the computed numerical solution for each sub population $F(t)$, $E(t)$, $S(t)$ and $G(t)$.
3. Compute the mean square error between the values obtained in Step 2 and the data in Table 1.

The function $\mathbb{F}$ takes values in $\mathbb{R}^6$ and returns a positive real number. Hence, we minimize this function using the Nelder-Mead algorithm [16, 17], that does not need the computation of any derivative or gradient, which is impossible to know in this case. Thus, the values of β_1, β_2, β_3, γ_1, γ_2, γ_3 (all of them positive) that minimize the objective function $\mathbb{F}$ are

$$\begin{array}{lll} \beta_1 = 6.39902, & \beta_2 = 1.03593 \times 10^{-9}, & \beta_3 = 0.36436, \\ \gamma_1 = 4.98113, & \gamma_2 = 0.13285, & \gamma_3 = 6.30120 \times 10^{-7}. \end{array} \qquad (2)$$

The obtained model parameters (2) indicates that there is a large flow, entering and exiting, between populations G (Remaining attitudes) and S (Remote justification). Furthermore, the transition from S (Remote justification) to E (Justification with criticism) is very difficult. Also, it is very difficult for the strongest supporters (F, the more "fanatics") to reconsider their position.

We substitute the fitted model parameters (2) into the scaled version of the model (1) and we calculate the output until May 2017. In Figure 2 we can see the prediction for the evolution of all the populations over the next few years. Numerical values in the dates of the coming eight *Euskobarometro* surveys are shown in Table 2.

The prediction figures indicate stabilization in the evolution of the attitudes towards the ETA over the next few years, and therefore stabilization in a hypothetical pool of candidates willing to join the organization in upcoming years.

Conclusion

In this chapter we applied the Castillo-Chavez and Song's model to a real situation where there is a significant impact of violent activities into the public opinion and vice-versa.

To do so, we have divided the Basque population depending on their support attitude towards the ETA, by using data series of the *Euskobarometro*, since January 2011, when the ETA declared the cease of its violent activity. By using these data, we have developed

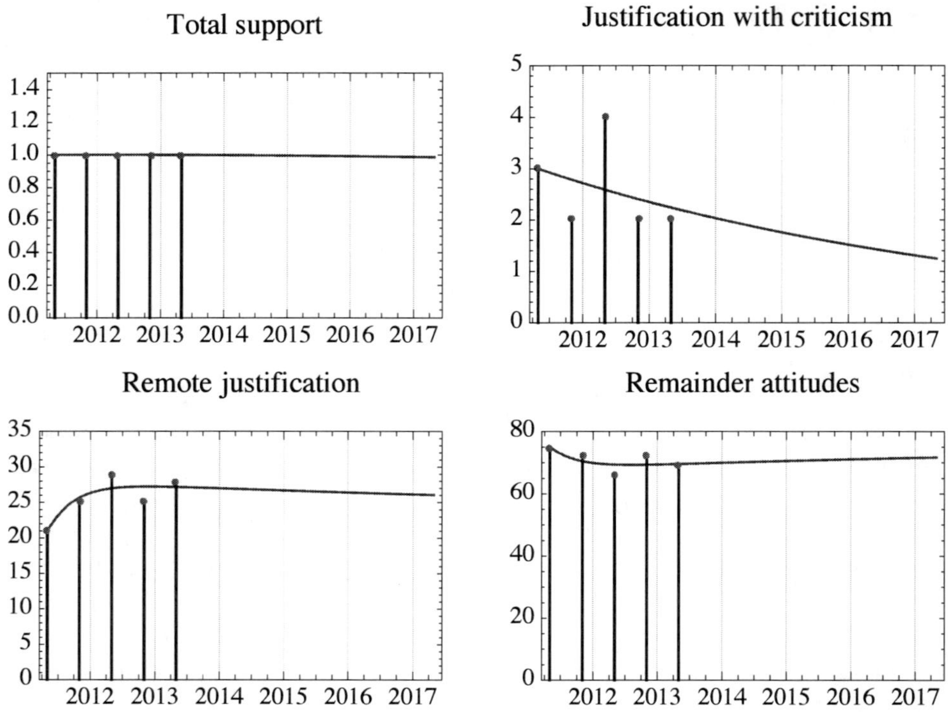

Figure 2. Model fitting (from May 2011 to May 2013) and prediction (from November 2013 to May 2017). Points are data in Table 1. The continuous line is the model output. Units are in percentages. Note that the scales for every graph are different. The decreasing in the population "Justification with criticism" is less than 2% from May 2011 to May 2017. The prediction is very stable over the next four years.

Table 2. Predicted percentage of Basque people in each sub population for the next eight *Euskobarometro* surveys, from May 2014 until May 2017. The predictions show a stable situation. The predicted variations over the next four years in each population are less than 1%

Date	Total support	Justification with criticism	Remote justification	Remaining attitudes
Nov 2013	0.999	2.080	27.050	69.870
May 2014	0.997	1.935	26.900	70.170
Nov 2014	0.996	1.797	26.750	70.460
May 2015	0.994	1.672	26.600	70.730
Nov 2015	0.992	1.553	26.460	71.000
May 2016	0.990	1.444	26.320	71.240
Nov 2016	0.988	1.342	26.190	71.480
May 2017	0.986	1.248	26.060	71.700

an algorithm to find the model parameters that best fit the model with the data. Once the model has been calibrated, we use the obtained model parameters to predict the evolution of the different populations in the Basque Country over the next four years. As a result, the presented prediction states that the popular support to the ETA will remain stable, if and when the current scenario does not change.

However, as an epilogue, this might not be the case. In fact, the Spanish Ministry of Internal Affairs announced recently (Oct 27th, 2013) [18] that the Application no. 42750/09 of the European Court of Human Rights (Oct 21st, 2013) [19] will allow to release 50 members of the ETA from prison in two or three months. This notice constitutes an undoubted change in the present scenario, and therefore may provoke an impact in the above conclusions.

References

[1] M. Peco, *Aproximacion funcional a los movimientos radicales en el ejercicio de la violencia politica* (A Functional Approach to Radical Movements Engaged in Political Violence), Ph.D. Dissertation, UNED, 2011.

[2] L.L. Cavalli-Sforza and M.W. Feldman, *Cultural Transmission and Evolution: A Quantitative Approach*, Princeton University Press, Princeton, NJ, 1981.

[3] C. Lumsden and E.O. Wilson, *Genes, Mind and Culture: The Coevolutionary Process*, Harvard University Press, Cambridge, MA, 1981.

[4] F.J. Santonja, A.C. Tarazona and R.J. Villanueva, A mathematical model of the pressure of an extreme ideology on a society. *Computers and Mathematics with Applications* 56 (2008), 836–846.

[5] C. Castillo-Chávez and B. Song, Models for the transmission dynamics of fanatic behaviors, in *Bioterrorism: Mathematical Modeling Applications in Homeland Security*, SIAM Frontiers in Applied Mathematics, ed.: H.T. Banks and C. Castillo-Chávez, SIAM, Philadelphia, 28 (2003), 155–172.

[6] A. Cherif, H. Yoshioka, W. Ni and P. Bose, *Terrorism: Mechanism of Radicalization Process, Control of Contagion and Counter-Terrorist Measures*. Preprint 2010.

[7] D. Stauffer and M. Sahimi, Discrete simulation of the dynamics of spread of extreme opinions in a society. *Physica A* 364 (2006), 537–543.

[8] D. Stauffer and M. Sahimi, Can a few fanatics influence the opinion of a large segment of a society? *The European Physical Journal B* 57 (2007), 147–152.

[9] ETA communiqué on 08/01/2011. Video clip issued in GARA. Retrieved from: http://www.gara.net/bideoak/110108_video/. A transcription (in Spanish) is available at: http://www.elpais.com/elpaismedia/ultimahora/media/201101/10/espana/20110110elpepunac_1_Pes_PDF.pdf.

[10] Euskobarometro data series. Available at http://www.ehu.es/euskobarometro/.

[11] M. Martcheva and C. Castillo-Chavez, Diseases with chronic stage in a population with varying size. *Mathematical Biosciences* 182 (2003), 1–25.

[12] J. Mena-Lorca and H.W. Hethcote, Dynamic models of infectious diseases as regulators of population sizes. *Journal of Mathematical Biology* 30 (1992), 693–716.

[13] http://scaling.imm.upv.es

[14] http://www.ine.es

[15] http://www.wolfram.com/mathematica

[16] J.A. Nelder and R. Mead, A simplex method for function minimization. *Computer Journal* 7 (1964), 308–313.

[17] W.H. Press, B.P. Flannery, S.A. Teukolsky and W. Vetterling, *Numerical Recipes: The Art of Scientific Computing*, Cambridge University Press, New York (1986).

[18] http://www.abc.es/espana/20131027/abcp-entrevista-minist
ro-interior-20131027.html (in Spanish)

[19] http://hudoc.echr.coe.int/sites/eng/pages/search.aspx?i=001-127697

In: Mathematical Modeling in Social Sciences ... ISBN: 978-1-63117-335-6
Editors: J. C. Cortés López et al.

Chapter 17

Mathematical Modelling of the Consumption of High-Invasive Plastic Surgery: Economic Influences and Consequences

***M. S. S. Alkasadi*[1]*, E. De la Poza*[2,*] *and L. Jódar*[1]
[1]Instituto Universitario de Matemática Multidisciplinar,
Universitat Politécnica de Valéncia, Valencia, Spain
[2]Facultad de Administración y Dirección de Empresas,
Universitat Politécnica de Valéncia, Valencia, Spain

Abstract

Plastic surgery among women grows continuously in Western countries due to body image dissatisfaction, the aging anxiety and ideal body image propagated by media. The consumption growth is so important that the plastic surgery is becoming a normal practice among women, like other cosmetic products, with the risk of suffering psychopathology disorders in the sense that plastic surgery could be regarded as an instrument to recover personal self-esteem, or even happiness.

In this chapter we develop a mathematical model to forecast the High-Invasive Plastic Surgery (HIPS) consumption in Spain. We simulate possible economic scenarios. Our results show an increasing trend of occasional and regular women HIPS consumers independently of the economic situation.

Keywords: Plastic Surgery, contagion effect, economy, satisfaction, mathematical modelling

1. Introduction

Cosmetic procedures can be classified into two categories surgical and non-surgical. While non-surgical procedures are low-invasive treatments such as botox, chemical peelings, prevelle, rosacea treatments and vampire facelift. The surgical ones are more aggressive from

*E-mail address: elpopla@esp.upv.es; Tel:+34963877032; Fax: +34963877032 (Corresponding author)

a medical point of view what is associated to greater healthcare risks and require hospitalization [1].

The difference in the level of medical-invasion between both categories (Low-Invasive Plastic Surgery (LIPS) and High-Invasive Plastic Surgery, (HIPS)) explains the LIPS are cheaper and as consequence more affordable than HIPS which are more expensive.

The HIPS category embraces procedures such as breast augmentation, breast reduction, rhinoplasty and liposuction. The consumption of HIPS has been traditionally related to women in western societies [2]. However, recent studies show an increase of men consumption [3].

The nature of the drivers that explain the consumption of HIPS is different. We can group them into three kinds: economic [4] psychological [5] and contagion effect [6]. As any other consumption good, HIPS consumption is affected by the economic-cycle; thus, there is positive relation between the real net income and the number of HIPS performed [7] but also the HIPS demand is influenced by the access to the credit what depends on the financial markets stability [4].

At the present time, Spain is impacted by a ferocious economic and financial crisis with stable unemployment rates over 21%. In this context, it would be expected a decrease of HIPS consumption, however two opposite forces emerge: women looking for a physical improvement throughout HIPS procedures as a tool for achieving professional success [8]; also, any economic crisis increase unequal income distribution [9] producing the expansion of demand from wealthy women, (mainly HIPS consumers) while consumption from middle class decreases, (more oriented to LIPS consumption, [10]). Psychological effects that drive women to practice HIPS can be explained as a mechanism to recover their well-being and personal satisfaction with their physical appearance [11].

Also, the contagion effect promoted by media, (TV, marketing, advertising.....etc) that spreads the message of perfect bodies (diets [12], muscular [13] and breast boobs [14]) but also by the interactions among women that practice HIPS regularly with those that do not, producing the propagation of the consumption of this product.

The aim of this chapter is to develop a mathematical model to forecast the future consumption rate of high-invasive plastic surgery in Spain over the next five years. To our knowledge there is none questionnaire in the literature measuring the level practice of HIPS consumption or practice, neither any study that models and predicts the level of consumption of this sector of economic activity.

The chapter is organized as follows: Section 2 deals with the model construction throughout a discrete system of different. In Section 3 computations and simulations are carried out after assuming several possible economic scenarios for the next coming years. Section 4 shows the conclusions.

2. Mathematical Model Construction

2.1. Data Collection and Sampling

The population of the study consists of Spanish women who underwent HIPS aged among the interval $[16, 60]$.

We classified the population into three categories depending on their level of activity measured throughout a survey carried out for this purpose.

The three categories were defined according to HIPS women consumption as follows:

- $P(n)$: Defined as rational women when their level of consumption was equal to 0 times at year n.
- $O(n)$: Defined as occasional-consumers when they practiced HIPS just 1 time at year n.
- $R(n)$: Is defined as regular-consumers when their HIPS practice was higher than 1 time at year n.

2.2. Mathematical Model

The dynamic behavior of the HIPS procedure is based on their transition among subpopulations explained by coefficients that need to be found according to economic, psychological and social propagation hypotheses. Our attention is focused on forecasting the number of HIPS consumption for the period $2012 - 2016$.

We passed the questionnaire once, (March 2012) at different locations such as a multi-located private gym also a private franchised gym and a public beach. Then, with the result obtained from the survey, we adjust the Spanish women population into three subpopulation using data from [15].

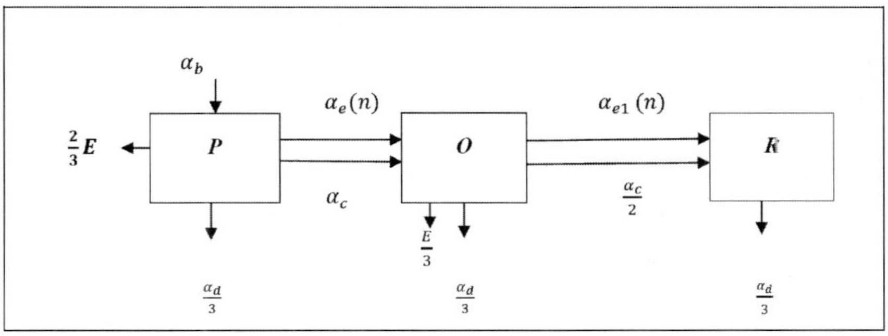

Figure 1. Dynamics of the population.

2.2.1. Hypothesis of the Model

a. The influence of the economy affects differently the subpopulations causing transits between them. Depending upon the economic situation, there is a transit from $P \longrightarrow O$ and from $O \longrightarrow R$.

b. The propagation effect caused by the personal relationship between P and O subpopulations producing the transit from $P \longrightarrow O$ and also, from $O \longrightarrow R$.

c. Therefore, the economic worsening produces that only the rich women transit from $O \longrightarrow P$.

The dynamics of the model is shown in Figure 1.

Thus, the dynamic model of the HIPS consumer's propagation can be modeled by the following equations:

$$H(n) = P(n) + O(n) + R(n). \tag{1}$$

$$\left.\begin{array}{rcl}
P(n+1) & = & \left(1+\alpha_b - \frac{\alpha_d}{3}\right)P(n) - \alpha_c P(n) \\
 & - & \alpha_e(n)P(n) - \frac{2}{3}E, \\
O(n+1) & = & \left(1-\frac{\alpha_d}{3}\right)O(n) + \alpha_e(n)P(n) + \alpha_c P(n) \\
 & - & \alpha_{e1}(n)O(n) - \frac{\alpha_c}{2}O(n) - \frac{E}{3}, \\
R(n+1) & = & \left(1-\frac{\alpha_d}{3}\right)R(n) + \frac{\alpha_c}{2}O(n) + \alpha_{e1}(n)O(n).
\end{array}\right\} \tag{2}$$

2.2.2. Parameters of the Model

The values of all parameters were computed from different sources of information and hypotheses as follows:

- $\alpha_c = 0.027$ is the annual contagion rate. The contagion effect is based on low levels of self-esteem combined with the mimetic behavior [6] that incentives the transit from $P \longrightarrow O$ and from $O \longrightarrow R$ due to contact between women. However the contagion effect of HIPS practice is also related to the economic situation. As consequence, we estimate that from the period of time $2009 - 2011$, there has been an average annual increase of the unemployment rate of 2.5% following, while the HIPS practice has increased annually a 2.7% [16, 17]. Assuming a conservative economic scenario in which the unemployment rate remains stable or even starts decreasing, the contagion rate is considered constant for the period of study. We estimate O subpopulation is less impacted by the contagion effect due to their previous HIPS experiences.

- $\frac{2}{3}\times E : \frac{2}{3}\times 150\,000 = 100\,000$ is the constant approximated value of Spanish women that leave Spain looking for a job abroad due to the economic crisis. Furthermore, $E = 150\,000 \div 3 = 50\,000$ are the occasional-consumers that leave Spain due to the crisis. We assume these values remain constant for the period of study. Also, we assume R subpopulation are high-income Spanish women who do not leave Spain since they are minimally affected by the crisis.

- $\alpha_e(n)$ and $\alpha_{e1}(n)$ are the economic effects on women who undergo plastic surgery. We assume than an economic improvement (decrease of unemployment rate) produces an increase of the HIPS and vice versa. When the economy is worse only rich women practices plastic surgery. The economic effects came from two situations:

  Firstly, when the unemployment rate decreases 1%, there is an increase of 0.027% HIPS practice of the total Spanish women. If the rate of unemployment increases of 1% there is no transit. Following these conditions for occasional-consumers:

$$\alpha_e(n) = \begin{cases} -0.027 \times (\rho(n+1) - \rho(n)) & \text{if,} \quad \rho(n+1) < \rho(n), \\ \\ 0 & \text{if,} \quad \rho(n+1) \geq \rho(n). \end{cases} \tag{3}$$

Where$\rho(n)$ is the Spanish unemployment rate at year n.

Secondly, when the economy deteriorates, just rich women practices plastic surgery. For every 1% increase in unemployment, the HIPS consumption increases 0.001%. We assume 0.001% remains constant for the period of study. However, when the unemployment rate decreases 1% the consumption of HIPS increase 0.026%.

$$\alpha_{e1}(n) = \begin{cases} 0.001 \times (\rho(n+1) - \rho(n)) & \text{if,} \quad \rho(n+1) \geq \rho(n), \\ \\ -0.026 \times (\rho(n+1) - \rho(n)) & \text{if,} \quad \rho(n+1) < \rho(n). \end{cases} \tag{4}$$

Where $\rho(n)$ is the Spanish unemployment rate at year n.

3. Results and Simulations

The mathematical model is helping us to predict the subpopulations $P(n)$, $O(n)$ and $R(n)$ at any year n.

We assumed the economic forecast from [17, 18] and [19] from 2011 until 2016. We introduced two economics scenarios one optimistic and one pessimistic for the total period of study; thus, any possible economic situation is enclosed in the range of variation of our scenarios, see Table 1.

Table 1. Economic forecast of the Spanish unemployment rate

	Pessimistic	Optimistic
2011	21.6	21.7
2012	25.0	25.1
2013	26.9	27.0
2014	28.1	26.0
2015	30.0	24.7
2016	29.0	23.2

Table 2 shows the collected results performed by the computation of the system expressed in volume of HIPS consumers. Therefore, the percentages of O and R increase over time what justifies the robustness of our model.

As Figure 2 shows the trend lines of subpopulations P, O and R overlap at both economic scenarios. The O subpopulation on pessimistic scenario evolves from 8.40% in 2011 to 17.45% in 2016, however, on optimistic scenario evolves $[8.40\%, 17.50\%]$; while the R subpopulation for pessimistic scenario increases from 1.76% in 2011 to 2.64% in 2016 but, for optimistic scenario evolves $[1.76\%, 2.65\%]$.

Table 2. Subpopulations forecasts in volume of HIPS according to the simulated scenarios

		Pessimistic	Optimistic
2011	**P**	11,254,967	11,254,967
	O	1,053,420	1,053,420
	R	221,772	221,772
2012	P	10,950,593	10,950,593
	O	1,292,682	1,292,682
	R	235,978	235,978
2013	P	10,653,255	10,653,255
	O	1,520,436	1,520,435
	R	253,391	253,391
2014	P	10,359,874	10,356,899
	O	1,737,022	1,739,626
	R	273,863	274,235
2015	P	10,075,168	10,068,484
	O	1,942,669	1,948,374
	R	297,277	298,226
2016	P	9,796,280	9,788,309
	O	2,140,141	2,146,733
	R	323,917	325,201

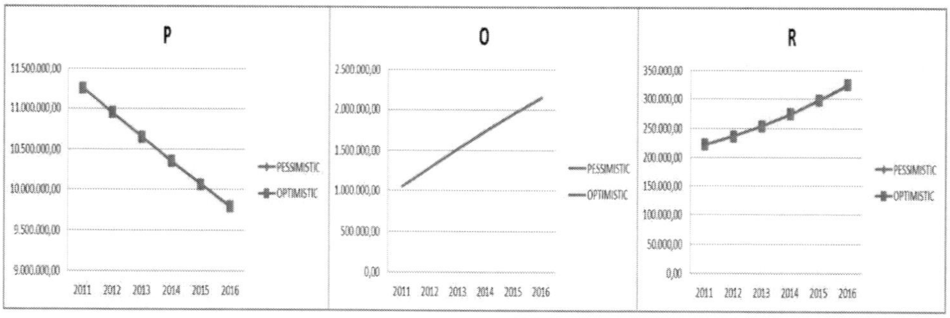

Figure 2. Simulations of HIPS consumption in Spain.

Conclusion

HIPS consumption increases for the period of time analyzed. The increase is performed by high-income level women, mainly R-consumers but, also by middle class women O-consumers.

The HIPS procedures such as breast augmentation, breast reduction, liposuction an others become a popular good of consumption between rich and medium class Spanish women even with uncertainty about the improvement of the Spanish economy. Between the causes that incentive the HIPS consumption are: professional success, body care, emotional stability.

As a result of our study we can conclude that the practice of these procedures among occasionally and regular-consumers (richer) may lead to develop body dysmorphic disorders (BDD). Public authorities should control the advertising/marketing through television and internet due to the relevance of the consumption of these goods from the medical point of view.

References

[1] Sarwer, D. B.; Crerand, C. E. Body image and cosmetic medical treatments *Body image*. 2004, 1(1), 99-111.

[2] Swami, V.; Taylor, R.; Carvalho, C. Acceptance of cosmetic surgery and celebrity worship: Evidence of associations among female undergraduates *Journal of Personality and Individual Differences*. 2009, 47(8), 869-872.

[3] Grant, R. T. (2012). Cosmetic surgery and the modern man: A Simple Plan for Every Man. Available at: http://www.pearlplasticsurgery.com/pdf/articles/Cosmetic%20Surgery%20and%20the%20Modern%20Man%2010-2-2012.pdf [last access: octuber 2013].

[4] Nassab, R.; Harris, P. Cosmetic surgery growth and correlations with financial indices: A comparative study of United Kingdom and United States from 2002-2011 *Aesthetic Surgery Journal*. 2013, 33(4), 604-608.

[5] Sarwer, D. B.; Wadden, T. A.; Petrschuk, M. J.; Whitaker, L. A. The psychology of cosmetic surgery: A review and reconceptualization *Clin Psychol Rev*. 1998, 18 (1), 1-22.

[6] Raafat, R. M.; Chater, N.; Frith, C. Herding in humans *Trends Cogn Sci*. 2009, 13(10), 420-428.

[7] Paik, A. M.; Hoppe, I. C.; Pastor, C. J. An analysis of leading, lagging, and coincident economic indicators in the United States and its relationship to volume of plastic surgery procedures performed : An update 2012 *Ann Plast Surg*. 2013, 71(3), 316-319.

[8] Sarwer, D. B.; Pruzinsky, T.; Cash, T. F.; Goldwyn, R. M.; Persing, J. A; Whitaker, L. A. *Psychological aspects of reconstructive and cosmetic plastic surgery: Clinical, empirical, and ethical perspectives* Lippincott, Williams & Wilkins: Philadelphia, USA, 2006.

[9] Duncan, C. O.; Ho-Asjoe, M.; Hittinger, R.; Nishikawa, H.; Waterhouse, N.; Coghlan, B.; Jones, B. Demographics and macroeconomic effects in aesthetic surgery in the UK *Br J Plast Surg*. 2004, 57(6), 561-566.

[10] De la Poza, E.; Alkasadi, M.; Jódar, L. Mathematical modeling of the consumption of low invasive plastic surgery practices: The case of Spain *Abstract and Applied Analysis*. 2013 available at: http://dx.doi.org/10.1155/2013/169253.

[11] Swami, V.; Arteche, A.; Chamorr-Premuzic, T.; Furnham, A.; Stieger, S.; Haubner, T.; Voracek, M. Looking good: Factors affecting the likelihood of having cosmetic surgery *Eur J Plast Surg*. 2008, 30, 211-218.

[12] Coughlin, J. W.; Schreyer, C. C.; Sarwer, D. B.; Heinberg, L. J.; Redgrave, G. W; Guarda, A. S. Cosmetic surgery in inpatients with eating disorders: Attitudes and experience *Body Image*. 2012, 9(1), 180-183.

[13] Leone, J. E.; Sedory, E. J.; Gray, K. A. Recognition and treatment of muscle dysmorphia and related body image disorders *J Athl Train*. 2005, 40(4), 352-359.

[14] Henderson-King, D.; Brooks, K. D. Materialism, sociocultural appearance messages, and parental attitudes predict college women's attitudes about cosmetic surgery *Psychol Women Quart*. 2009, 33(1), 133-142.

[15] Spanish Statistical Institute(INE).(2013). Available at: http://www.ine.es/ [last access: octuber 2013].

[16] International Society of Asthetic Plastic Surgery (ISAPS).(2013). Available at: http://www.isaps.org/isaps-global-statistics-2012.html [last access: September 2013].

[17] The Organization for Economic Cooperation and Development (OECD). (2013). Available at: http://www.oecd.org/eco/outlook/spaineconomicforecastsummary.htm[last access: October 2013].

[18] International Monetary Fund (IMF). (2013). Available at: http://www.imf.org/external/pubs/ft/scr/2013/cr1354.pdf [last access: October 2013].

[19] Cross Asset Research. Socit Gnrale (SG). (2013). Available at: https://publication.sgresearch.com/en/3/0/172963/125179.html?sid=5b4256d8671034005116a674000337f9 [last access: October 2013].

In: Mathematical Modeling in Social Sciences ...
Editors: J. C. Cortés López et al. ISBN: 978-1-63117-335-6

Chapter 18

AN OPTIMAL SCHEME FOR SOLVING THE NONLINEAR GLOBAL POSITIONING SYSTEM PROBLEM

Manuel Abad*, Alicia Cordero† and Juan R. Torregrosa‡
Instituto de Mathemática Multidisciplinar,
Universitat Politècnica de València,
Valencia, Spain

Abstract

A new eighth-order family of iterative method for solving the nonlinear system obtained in the Global Positioning System problem is presented. We extend the seventh-order scheme for solving nonlinear equations, designed by Soleymani et al in [12], to nonlinear systems improving its order of convergence. To generate our class of methods we use the weight functions procedure, with matricial functions. Numerical comparisons are made to confirm the theoretical results.

Keywords: Global Positioning System, nonlinear system, iterative method, efficiency, order of convergence

1. Introduction

The search of solutions of nonlinear systems of equations $F(x) = 0$, where $F : D \subseteq \mathbb{R}^n \to \mathbb{R}^n$, is an old and difficult problem with wide applications in sciences and engineering. The best known method, for being very simple and effective, is Newton's method. Its generalization to a nonlinear system of equations was proposed by Ostrowski [10] and is given by

$$x^{(k+1)} = x^{(k)} - F'(x^{(k)})^{-1}F(x^{(k)}), \quad k = 0, 1, \ldots,$$

*E-mail address: maabrod@mat.upv.es
†E-mail address: acordero@mat.upv.es
‡E-mail address: jrtorre@mat.upv.es

where $F'(x^{(k)})$ is the Jacobian matrix of the function F evaluated in the kth iteration $x^{(k)}$. In the literature, several modifications have been made on classical methods in order to accelerate the convergence or to reduce the number of operations and functional evaluations in each step of the iterative process. Recently, many researchers have designed Newton-type iterative methods with these goals, see for example [11, 4, 5, 8, 6, 7, 2] among others. In particular, the authors shown in [1] fourth and fifth-order methods for solving nonlinear systems of equations, applying them to solve the equations of the Global Positioning System (GPS).

The main goal of this chapter is to improve the results obtained with the software used mostly in the GPS. For that we are going to design eighth-order family of iterative schemes by using the weight functions procedure, with matricial functions.

The rest of this paper is organized as follows: in Section 2 we make an introduction to the Global Positioning System, focusing on the way that the receiver calculates the user position using the ephemeris data of the artificial satellites. In Section 3 we describe our new family of iterative methods, analyzing its convergence order. In Section 4 we show an application of these schemes solving the nonlinear system of the GPS. A comparison among the results of the new methods and Newton's one is shown.

2. Basics on Global Positioning System

In this section we introduce the basic concepts for understanding how a GPS receiver determines the user position. From the satellite constellation, the equations required for solving the user position conform a nonlinear system of equations. In addition, some practical considerations, (i.e. the inaccuracy of the user clock) will be included in these equations. These equations are usually solved through a linearization and a fixed point iteration method. The obtained solution is in a Cartesian coordinate system and after that the result will be converted into a spherical coordinate system. However, the Earth is not a perfect sphere; therefore, once the user position is estimated, the shape of the Earth must be taken into consideration. The user position is then translated into the Earth-based coordinate system. In this chapter we are going to focus our attention in solving the nonlinear system of equations of the GPS giving the results in a Cartesian coordinate system. We can find further information about GPS in [3].

2.1. GPS Performance Requirements

Some of the performance requirements are listed below:

1. The user position root mean square (rms) error should be 10-30 m.

2. It should be applicable to real-time navigation for all users including the high-dynamics user, such as in high-speed aircraft with flexible maneuverability.

3. It should have worldwide coverage. Thus, in order to cover the polar regions the satellites must be in inclined orbits.

4. The transmitted signals should tolerate, to some degree, intentional and unintentional interference. For example the harmonics from some narrow-band signals should not

disturb its operation. Intentional jamming of GPS signals is a serious concern for military applications.

5. It cannot require that every GPS receiver utilize a highly accurate clock such as those based on atomic standards.

6. When the receiver is first turned on, it should take minutes rather than hours to find the user position.

7. The size of the receiver antenna should be small. The signal attenuation through space should be kept reasonably small.

These requirements combining with the availability of the frequency band allocation determines the carrier frequency of the GPS to be in the L band (1-2 GHz) of the microwave range.

2.2. Basic GPS Concepts

The position of a point in space can be found by using the distances measured from this point to some known position in space. We are going to use an example to illustrate this point.

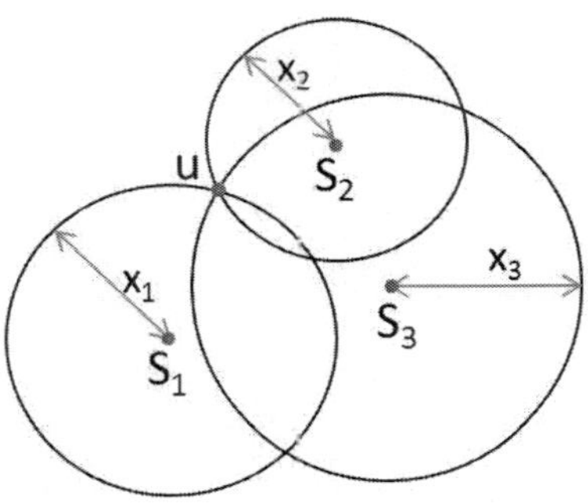

Figure 1. Two-dimensional user position.

Figure 1 shows a two-dimensional case. In order to determine the user position U, three satellites S_1, S_2 and S_3 and three distances are required. The trace of a point with constant distance to a fixed point is a circle in the two-dimensional case. Two satellites and two distances give two possible solutions because two circles intersect at two points. A third circle is needed to uniquely determine the user position. For similar reasons in a three-dimensional case four satellites and four distances are needed. The equal-distance trace to a fixed point is a sphere in a three-dimensional case. Two spheres intersect to make a circle. This circle intersects another sphere and this intersection produces two points. In order to determine which point is the user position, one more satellite should be needed. In GPS the position of the satellite is known from the ephemeris data transmitted by the satellite. By measuring the distance from the receiver to the satellite, the position of the receiver can be determined. In the above discussion, the distance measured from the user to the satellite is assumed to be very accurate and there is no bias error. However, the distance measured between the receiver and the satellite has a constant unknown bias, because the user clock usually is different from the GPS clock. In order to solve this bias error one more satellite is required. Therefore, in order to find the user position five satellites are needed. If one uses four satellites and the measured distance with bias error to measure a user position, two possible solutions can be obtained. Theoretically, one cannot determine the user position. However, one of the solutions is close to the Earth's surface and the other one is in the space. Since the user position is usually close to the surface of the Earth, it can be uniquely determined. Therefore, the general statement is that four satellites can be used to determine a user position, even though the distance measured has a bias error. The method of solving the user position discussed in the next subsections is through iteration. The initial position is often selected at the center of the Earth. In the following discussion four satellites are considered as the minimum number required for finding the user position.

2.3. Basic Equations for Finding User Position

In this section, the basic equations for determining the user position will be presented. Assume that the distance measured is accurate and under this condition three satellites should be sufficient. Let us suppose that there are three known points at locations r_1 or (x_1, y_1, z_1), r_2 or (x_2, y_2, z_2) and r_3 or (x_3, y_3, z_3), and an unknown point at r_u or (x_u, y_u, z_u). If the distances between the three known points to the unknown point can be measured as ρ_1, ρ_2, and ρ_3, these distances can be written as

$$\begin{aligned}
\rho_1 &= \sqrt{(x_1 - x_u)^2 + (y_1 - y_u)^2 + (z_1 - z_u)^2},\\
\rho_2 &= \sqrt{(x_2 - x_u)^2 + (y_2 - y_u)^2 + (z_2 - z_u)^2},\\
\rho_3 &= \sqrt{(x_3 - x_u)^2 + (y_3 - y_u)^2 + (z_3 - z_u)^2}.
\end{aligned} \tag{1}$$

Because there are three unknowns and three equations, the values of x_u, y_u and z_u can be determined from these equations. Theoretically, there should be two sets of solutions as they are second-order equations. These equations can be solved linearizing them and making an iterative approach. The solution of these equations will be discussed later in Section 2.4. In GPS operation, the positions of the satellites are given. This information can be obtained from the data transmitted from the satellites. The distances from the user (the unknown position) to the satellites must be measured simultaneously at a certain time instance. Each satellite transmits a signal with a time reference associated with it. By measuring the time of the signal traveling from the satellite to the user the distance between the user and the satellite can be found. The distance measurement is discussed in the next section.

2.4. Measurement of Pseudorange

Every satellite sends a signal at a certain time t_{si}. The receiver will receive the signal at a later time t_u. The distance between the user and the satellite i can be determined as

$$\rho_{iT} = c(t_u - t_{si}),$$

where c is the speed of light, ρ_{iT} is often referred to as the true value of pseudorange from user to satellite i, t_{si} is referred to as the true time of transmission from satellite i, and t_u is the true time of reception. From a practical point of view it is difficult, if not impossible, to obtain the correct time from the satellite or the user. The actual satellite clock time t'_{si} and actual user clock time t'_u are related to the true time as

$$t'_{si} = t_{si} + \Delta b_i, \qquad t'_u = t_u + b_{ut},$$

where Δb_i is the satellite clock error, and b_{ut} is the user clock bias error. Besides the clock error, there are other factors affecting the pseudorange measurement. The measured pseudorange ρ_i can be written as

$$\rho_i = \rho_{iT} + \Delta D_i - c(\Delta b_i - b_{ut}) + c(\Delta T_i + \Delta I_i + v_i + \Delta v_i),$$

where ΔD_i is the satellite position error effect on range, ΔT_i is the tropospheric delay error, ΔI_i is the ionospheric delay error, v_i is the receiver measurement noise error and Δv_i is the relativistic time correction. Some of these errors can be corrected; for example, the tropospheric delay can be modeled and the ionospheric error can be corrected in a two-frequency receiver. The errors will cause inaccuracy of the user position. However, the user clock error cannot be corrected through receiver information. Thus, it will remain as an unknown. So, the system of equations (1) must be modified as

$$\begin{aligned} \rho_1 &= \sqrt{(x_1 - x_u)^2 + (y_1 - y_u)^2 + (z_1 - z_u)^2} + b_u, \\ \rho_2 &= \sqrt{(x_2 - x_u)^2 + (y_2 - y_u)^2 + (z_2 - z_u)^2} + b_u, \\ \rho_3 &= \sqrt{(x_3 - x_u)^2 + (y_3 - y_u)^2 + (z_3 - z_u)^2} + b_u, \\ \rho_4 &= \sqrt{(x_4 - x_u)^2 + (y_4 - y_u)^2 + (z_4 - z_u)^2} + b_u, \end{aligned} \tag{2}$$

where b_u is the user clock bias error expressed in distance, which is related to the quantity b_{ut} by $b_u = cb_{ut}$. In system (2), four equations are needed to solve for four unknowns x_u, y_u, z_u and b_u. Thus, in a GPS receiver, a minimum of four satellites is required to solve the user position.

2.5. Solution of User Position from Pseudoranges

One common way to solve the system of equations (2) is to linearize them. The system can be written in a simplified form as

$$\rho_i = \sqrt{(x_i - x_u)^2 + (y_i - y_u)^2 + (z_i - z_u)^2} + b_u, \tag{3}$$

with $i = 1, 2, 3, 4$ and x_u, y_u, z_u and b_u are the unknowns. The pseudorange ρ_i and the positions of the satellites x_i, y_i, z_i are known. By differentiating (3),

$$\begin{aligned} \delta\rho_i &= \frac{(x_i - x_u)\delta x_u + (y_i - y_u)\delta y_u + (z_i - z_u)\delta z_u}{\sqrt{(x_i - x_u)^2 + (y_i - y_u)^2 + (z_i - z_u)^2}} + \delta b_u \\ &= \frac{(x_i - x_u)\delta x_u + (y_i - y_u)\delta y_u + (z_i - z_u)\delta z_u}{\rho_i - b_u} + \delta b_u. \end{aligned} \tag{4}$$

In (4), δx_u, δy_u, δz_u, and δb_u can be considered as the only unknowns. The quantities x_u, y_u, z_u and b_u are treated as known values because one can assume some initial values for these quantities. From these initial values a new set of δx_u, δy_u, δz_u, and δb_u can be calculated. These values are used to modify the original x_u, y_u, z_u and b_u to find another new set of solutions. This new set of x_u, y_u, z_u and b_u can be considered again as known quantities. This process continues until the absolute values of δx_u, δy_u, δz_u, and δb_u are very small and within a certain predetermined limit. The final values of x_u, y_u, z_u and b_u are the desired solution. This method is often referred to as an iteration method of fixed point. With δx_u, δy_u, δz_u and δb_u as unknowns, the above equation becomes a set of linear equations. This procedure is often referred to as linearization. The expression (4) can be written in matrix form as

$$\begin{bmatrix} \delta\rho_1 \\ \delta\rho_2 \\ \delta\rho_3 \\ \delta\rho_4 \end{bmatrix} = \begin{bmatrix} \alpha_{11} & \alpha_{12} & \alpha_{13} & 1 \\ \alpha_{21} & \alpha_{22} & \alpha_{23} & 1 \\ \alpha_{31} & \alpha_{32} & \alpha_{33} & 1 \\ \alpha_{41} & \alpha_{42} & \alpha_{43} & 1 \end{bmatrix} \begin{bmatrix} \delta x_u \\ \delta y_u \\ \delta z_u \\ \delta b_u \end{bmatrix}, \tag{5}$$

where

$$\alpha_{i1} = \frac{x_i - x_u}{\rho_i - b_u}, \qquad \alpha_{i2} = \frac{y_i - y_u}{\rho_i - b_u}, \qquad \alpha_{i3} = \frac{z_i - z_u}{\rho_i - b_u}, \qquad i = 1, 2, 3, 4.$$

The solution of (5) is

$$\begin{bmatrix} \delta x_u \\ \delta y_u \\ \delta z_u \\ \delta b_u \end{bmatrix} = \begin{bmatrix} \alpha_{11} & \alpha_{12} & \alpha_{13} & 1 \\ \alpha_{21} & \alpha_{22} & \alpha_{23} & 1 \\ \alpha_{31} & \alpha_{32} & \alpha_{33} & 1 \\ \alpha_{41} & \alpha_{42} & \alpha_{43} & 1 \end{bmatrix}^{-1} \begin{bmatrix} \delta\rho_1 \\ \delta\rho_2 \\ \delta\rho_3 \\ \delta\rho_4 \end{bmatrix}.$$

This process obviously does not provide the needed solutions directly. However, the desired solutions can be obtained from it. In order to find the desired position solution, this procedure must be used repetitively in an iterative way. A quantity is often used to determine whether the desired result is reached and this quantity can be defined as

$$\delta v = \sqrt{\delta x_u^2 + \delta y_u^2 + \delta z_u^2 + \delta b_u^2}. \tag{6}$$

When δv is lower than a certain predetermined threshold, the iteration will stop. Sometimes, the clock bias b_u is not included in (6). In this chapter we use as stopping criterion the quantity $||x^{(k+1)} - x^{(k)}|| + ||F(x^{(k+1)})||$ because it is more restrictive than (6). As we can verify in [13], the above iterative method used to calculate via software the receiver position in the GPS is the Newton's method, a well known method of second-order of convergence. Here, we improve the GPS software by means of a method of order eight, that converges to the solution with less number of iterations and more efficiency than Newton scheme.

3. Description of the Family and Convergence Analysis

In this section, we display a new class of eighth-order iterative methods for solving nonlinear systems, obtained by extending the seventh-order idea developed for solving nonlinear equations by Soleymani et al in [12] to nonlinear systems, by using the operator $[x, y; F]$, defined by Ortega and Rheinboldt in [9], and the weigh functions procedure, with matricial functions. The iterative expression is

$$\begin{aligned} y^{(k)} &= x^{(k)} - F'(x^{(k)})^{-1}F(x^{(k)}), \\ z^{(k)} &= y^{(k)} - G(t^{(k)})[x, y; F]^{-1}F(y^{(k)}), \\ x^{(k+1)} &= z^{(k)} - H(u^{(k)})[y, z; F]^{-1}F(z^{(k)}), \end{aligned} \tag{7}$$

where $y^{(k)}$ is the kth iteration of the Newton's method, G and H are weight matricial functions that should be chosen in order to obtain the eighth-order of convergence. These functions have as variables t and u, respectively, where

$$t = I - F'(x)^{-1}[x, y; F]$$

and

$$u = 2I - F'(x)^{-1}[x, y; F] - G(t)[x, y; F]^{-1}[y, z; F].$$

The next result establishes the conditions for obtaining the order of convergence eight.

Theorem 3.1. *Let $F : D \subseteq \mathbb{R}^n \to \mathbb{R}^n$, be sufficiently differentiable at each point of an open neighborhood D of $\overline{x} \in \mathbb{R}^n$, that is a solution of the nonlinear system $F(x) = 0$. Let us suppose that $F'(x)$ is continuous and nonsingular in $\overline{x}$. Then, for functions G and H enough differentiable, the sequence $\{x^{(k)}\}_{k\geq 0}$ obtained by using the iterative expression (7) converges to $\overline{x}$ with order eight when $G(0) = G'(0) = I$ and also $H(0) = I$, $H'(0) = 0$, $H''(0) = 4I$ and $H'''(0) = -12 + 12G''(0)$.*

The following functions

$$G_1(t) = I + t + \frac{1}{2}t^2,$$

$$H_1(u) = I + 2u^2,$$

satisfy the conditions of Theorem 3.1 and lead an element of family (7) denoted by $M8_1$. If we choose functions

$$G_2(t) = I + t,$$

$$H_2(u) = I + 2u^2 - 2u^3 - \frac{1}{2}u^4,$$

then the eighth-order is also achieved and the method is called $M8_2$.

4. Numerical Results

Variable precision arithmetics has been used, with 50 digits of mantissa, in order to make the numerical tests. The software used is MATLAB R2010b and we have checked that the iterate sequence converges to an approximation of the solution of the nonlinear system. For every method we calculate the error estimation made with values of $||x^{(k+1)} - x^{(k)}||$ and $||F(x^{(k+1)})||$ at the first three iterations.

In order to test the proposed scheme on the problem of a user position of a GPS device we have requested to the Cartographic Institute of Valencia to provide us data of known geocentric coordinates. Specifically, they provided us:

* An example of a fixed point located in Alcoy (Alicante, Spain) with geocentric coordinates: $x = 4984687.426, y = -41199.155$ and $z = 3966605.952$.

* Observations from that fixed point (file *.09o) for a day.

* Positions of the satellites for that day:*.09n and *.sp3 files.

* Description of RINEX format (*.09o file): http://www.igs.org/components/formats.html

* Description of the ephemeris file and satellite positions sp3: http://igscb.jpl.nasa.gov/igscb/data/format/sp3c.txt

* Links to other libraries for analysis calculations: http://www.ngs.noaa.gov/gps-toolbox/exist.htm

Table 1. Results for the GPS system with $x^{(0)} = (0, 0, 0, 0)^T$

Method	$\|x^{(1)} - x^{(0)}\|$	$\|F(x^{(1)})\|$	$\|x^{(2)} - x^{(1)}\|$	$\|F(x^{(2)})\|$	$\|x^{(3)} - x^{(2)}\|$	$\|F(x^{(3)})\|$
N	2.3889e6	1.8258e5	7.6772e6	1.5085e6	2.3646e6	1.9473e5
$M8_1$	6.6484e6	4.0014e5	2.8601e5	0.0436	0.1735	1.5934e-34
$M8_2$	6.7584e6	5.0083e5	4.0308e5	0.5165	2.4149	1.4908e-28

By using these data we have calculated the positions of the visible satellites at the instant associated to the provided data. Then, we calculate the approximated pseudoranges for every satellite and then we define the GPS nonlinear system (3) using four of the satellites. Then, we can compare the performance of Newton and two eighth-order methods, elements of the class (7), denoted by $M8_1$ and $M8_2$. The notation AeB that appears in the numerical results correspond to Ae^B.

In Tables 1 to 3 we can find a comparative among the iterative methods of Newton (N) and our eighth-order methods $M8_1$ and $M8_2$ for the GPS nonlinear system, using different initial estimations. We recall that the coordinates of the center of the Earth and $b_u = 0$, this is, $x^{(0)} = (0, 0, 0, 0)^T$, are usually used as initial estimation and the solution reached by all the methods is $x^* \approx (4984687.426,\ -41199.155,\ 3966605.952,\ 0.116e-8)^T$. Despite this, we have also tested the methods with some other initial conditions.

Table 2. Results for the GPS system with $x^{(0)} = 10^4(1, 1, 1, 1)^T$

Method	$\|x^{(1)} - x^{(0)}\|$	$\|F(x^{(1)})\|$	$\|x^{(2)} - x^{(1)}\|$	$\|F(x^{(2)})\|$	$\|x^{(3)} - x^{(2)}\|$	$\|F(x^{(3)})\|$
N	2.3454e6	1.7611e5	7.8559e6	1.5789e6	2.4388e6	2.0826e5
$M8_1$	6.5884e6	3.8444e5	2.3941e5	0.0233	0.1023	1.7178e-35
$M8_2$	6.6644e6	4.6607e5	3.2207e5	0.2918	1.5830	2.0896e-29

We can observe, from the data in Tables 1 to 3, that the proposed schemes reach the position of the user, with precision enough, when Newton's method is still far from the the user position for all the initial estimations.

Table 3. Results for the GPS system with $x^{(0)} = -10^4(1,1,1,1)^T$

Method	$\|x^{(1)} - x^{(0)}\|$	$\|F(x^{(1)})\|$	$\|x^{(2)} - x^{(1)}\|$	$\|F(x^{(2)})\|$	$\|x^{(3)} - x^{(2)}\|$	$\|F(x^{(3)})\|$
N	2.4355e6	1.8935e5	7.5006e6	1.4407e6	2.2918e6	1.819e5
$M8_1$	6.7056e6	4.1541e5	3.3063e5	0.0723	0.2661	1.2029e-33
$M8_2$	6.8510e6	5.3556e5	4.8335e5	0.8546	3.5821	9.2809e-28

Conclusion

In this chapter we have made an step in the GPS receivers software improvement. Actually, software-based GPS receivers use Newton's method to solve the associated nonlinear system in order to obtain the position of the user with the information obtained from signals received from the GPS constellation of satellites. We propose an eighth-order family of methods for solving systems of equations using weight functions and vectorial divided differences. The tests made have shown that this class is efficient and competitive in terms of the error estimation.

Acknowledgments

This research was supported by Ministerio de Ciencia y Tecnología MTM2011-28636-C02-02 and FONDOCYT 2011-1-B1-33 República Dominicana.

References

[1] Abad, M.F.; Cordero, A.; Torregrosa, J.R. Fourth-and fifth-order methods for solving nonlinear systems of equations: An application to the Global Positioning System. *Abstract and Applied Analysis*, 2013, vol. 2013, article ID. 586708.

[2] Babajee, D.K.R.; Dauhoo, M.Z.; Darvishi, M.T.; Karami, A.; Barati, A. Analysis of two Chebyshev-like third order methods free from second derivatives for solving systems of nonlinear equations, *J. Comput. Appl. Math.*, 2010, 233(8), 2002-2012.

[3] Bao-Yen Tsui, J. *Fundamentals of Global Positioning System Receivers, a Software Approach*, Wiley Interscience, Hoboken, New Jersey, 2005.

[4] Cordero, A.; Hueso, J.L.; Martínez, E.; Torregrosa, J.R. A modified Newton Jarratt's composition, *Numerical Algorithms*, 2010, 55, 87-99.

[5] Cordero, A.; Torregrosa, J.R. Variants of Newton's Method using fifth-order quadrature formulas, *Appl. Math. Comp.*, 2007, 190, 686-698.

[6] Cordero, A.; Torregrosa, J.R. On interpolation variants of Newton's method for functions of several variables, *J. Comp. Appl. Math.* 2010, 234, 34-43.

[7] Darvishi, M.T. Some three-step iterative methods free from second order derivative for finding solutions of systems of nonlinear equations, *Intern. J. Pure Appl. Math*, 2009, 57(4), 557-573.

[8] Frontini, M.; Sormani, E. Third-order methods from quadrature formulae for solving systems of nonlinear equations, *Appl. Math. Comput.*, 2004, 149, 771-782.

[9] Ortega, J.M.; Rheinboldt, W.G. *Iterative solutions of nonlinear equations in several variables*, Academic Press, New York, 1970.

[10] Ostrowski, A.M. *Solution of equations and systems of equations*, Prentice-Hall, Englewood Cliffs, New Jersey, 1964.

[11] Sharma, J.R.; Kumar, R.; Sharma, R. An efficient fourth order weighted-Newton method for systems of nonlinear equations, *Numerical Algorithms* 2013, 62, 307-323.

[12] Soleymani, F.; Mousavi, B.S. On Novel Classes of Iterative Methods for Solving Nonlinear Equations, *Computational Mathematics and Mathematical Physics* 2012, 52, 203-210.

[13] Sun, X.; Ji, Y.; Shi, H.; Li, Y. Evaluation of Two Methods for Three Satellites Position of GPS With Altimeter Aiding, *5th International Conference on Information Technology and Applications (ICITA 2008)*, 667-670.

In: Mathematical Modeling in Social Sciences ... ISBN: 978-1-63117-335-6
Editors: J. C. Cortés López et al.

Chapter 19

How to Make a Comparison Matrix in AHP without All the Facts

J. Benítez*, L. Carrión, J. Izquierdo and R. Pérez-García
Instituto Universitario de Matemática Multidisciplinar
I. M. M. Fluing Universitat Politècnica de València, Valencia, Spain

Abstract

AHP (analytic hierarchy process) is a leading multi-attribute decision-aiding model designed to help make better choices when faced with complex decisions. AHP is a multiple criteria decision analysis that uses hierarchical structured pairwise comparisons. One of the drawbacks of AHP is that a pairwise comparison cannot be completed by an actor or stakeholder not fully familiar with all the aspects of the problem. Here, we characterize when an incomplete, positive, and reciprocal matrix can be completed to become a consistent matrix. We show that this characterization reduces the problem to the solution of a linear system of equations. Some properties of such a completion are also developed using graph theory, including explicit calculation formulas. In real decision-making processes, facilitators conducting the study could use these characterizations to accept an incomplete comparison body given by an actor or to encourage the actor to further develop the comparison for the sake of consistency.

Keywords: AHP; Decision making; Consistent matrices; Graph theory

1. Introduction

The so-called analytic hierarchy process (AHP) [3, 4], has been accepted as a leading multi-attribute decision-aiding model both by practitioners and academicians, since it is designed to make better choices when faced with complex decisions. As a multiple criteria decision analysis (MCDA) technique AHP solves optimization decision problems, which involve choosing one of several possible alternatives. The AHP approach, which enables qualitative analysis using a combination of subjective and objective information/data, is a MCDA approach that uses hierarchical structured pairwise comparisons.

*E-mail address: jbenitez@mat.upv.es

However, some actors may not be completely familiar with one or more of the elements about which they have to issue their judgement or opinion. As a result, it is difficult to gather complete information about the preferences of such a stakeholder at a given moment. It seems reasonable to enable such an actor to express their preferences several times at his or her own convenience. Meanwhile, partial results based on partial preference data may be generated from data collected at various times – and this data may eventually be consolidated when the information is complete. Based on a process of linearization [1] that minimizes a matrix distance defined in terms of the Frobenius norm, in [2] the authors have initiated a line towards a dynamic model of AHP.

In addition, uncertainty coming from lack of comprehensive knowledge of any of the stakeholders must be handled suitably. In this regard, facilitators conducting the processes need robust tools enabling them discernment when collecting opinions from the various stakeholders.

We will provide a solution to this issue by solving the following problem: to characterize when an incomplete, positive, reciprocal matrix can be completed to be a consistent matrix. We show that this characterization reduces the consistent completion of an incomplete, positive, reciprocal matrix to the solution of a linear system of equations –a straightforward procedure. Finally, by using graph theory the uniqueness of the completion is studied and we give several ways to find such completion when it exists. In a real DM process the facilitator in charge of conducting the study could use these characterizations to accept an incomplete comparison body given by an actor or, on the contrary, to encourage the actor to work the comparison further for the sake of consistency.

2. Prerequisites and Formal Statement of the Problem

2.1. A Brief Review of AHP

As a result of the comparison performed, an $n \times n$ matrix $A = (a_{ij})$ is formed, n being the number of the decision elements and a_{ij} measuring the relative importance of element i over element j. To extract priority vectors from the comparison matrices, the eigenvector method, which was first proposed by Saaty in his seminal paper [3], is one of the most used methods.

A comparison matrix, A, exhibits a basic property, namely *reciprocity*:

$$a_{ij} = \frac{1}{a_{ji}}, \qquad 1 \leq i, j \leq n. \tag{1}$$

Besides the reciprocity property, another property, consistency, should theoretically be desirable for a comparison matrix. A positive $n \times n$ matrix is *consistent* if

$$a_{ij}a_{jk} = a_{ik}, \qquad 1 \leq i, j, k \leq n. \tag{2}$$

Consistency expresses the coherence that may exist between judgements about the elements of a set. Since preferences are expressed in a subjective manner it is reasonable for some kind of incoherence to exist. When dealing with intangibles, judgements are rarely consistent unless they are forced in some artificial manner.

In addition, a comparison matrix is not generally consistent because it contains comparison values obtained through numerical judgement using a fixed scale. For most problems, estimates of these values by an expert are assumed to be small perturbations of the 'right' values. For a consistent matrix A, the leading eigenvalue and the principal (Perron) eigenvector of A provide information to deal with complex decisions, the normalized Perron eigenvector giving the sought priority vector. In the general case, however, as said, A is not consistent. For non-consistent matrices, the problem to solve is the eigenvalue problem $A\mathbf{w} = \lambda_{\max}\mathbf{w}$, where $\lambda_{\max}$ is the unique largest eigenvalue of A that gives the Perron eigenvector as an estimate of the priority vector.

2.2. Notations and Basic Facts

The set of $n \times m$ real matrices is denoted by $\mathbb{R}_{n,m}$. We write $\mathbb{R}^+_{n,m} = \{A = (a_{ij}) \in \mathbb{R}_{n,m} : a_{ij} > 0 \text{ for all } i, j\}$. If A is a matrix, then $\mathrm{tr}(A)$ and A^T will denote the trace and the transpose of A, respectively. The standard basis of $\mathbb{R}^n$ is denoted by $\{\mathbf{e}_1, \ldots, \mathbf{e}_n\}$. The vector $(1, \ldots, 1)^T \in \mathbb{R}^n$ will be denoted by $\mathbf{1}_n$. As it can be seen from (1) and (2), any consistent matrix is reciprocal.

We will use the mappings $L : \mathbb{R}^+_{n,m} \to \mathbb{R}_{n,m}$ and $E : \mathbb{R}_{n,m} \to \mathbb{R}^+_{n,m}$ given by $L(A) = (\log(a_{ij}))$ and $E(A) = (\exp(a_{ij}))$, respectively, where $A = (a_{ij})$. Evidently one has that for $A \in \mathbb{R}^+_{n,n}$,

$$A \text{ is reciprocal} \iff L(A) \text{ is skew Hermitian.}$$

The image by L of the set of consistent matrices will play an important role in the sequel. Precisely, we define

$$\mathcal{L}_n = \{L(A) : A \in \mathbb{R}^+_{n,n} \text{is consistent}\}.$$

A basic property of $\mathcal{L}_n$ is established in the next restult.

Theorem 1. (Theorem 2.2 of [1]) *If we define*

$$\phi_n : \mathbb{R}^n \to \mathbb{R}_{n,n}, \qquad \phi_n(\mathbf{v}) = \mathbf{v}\mathbf{1}_n^T - \mathbf{1}_n\mathbf{v}^T, \tag{3}$$

then ϕ_n *is linear,* $\ker \phi_n = \mathrm{span}\{\mathbf{1}_n\}$, $\mathrm{Im}\phi_n = \mathcal{L}_n$, *and* $\dim \mathcal{L}_n = n - 1$.

2.3. Problem Definition

Our purpose is to characterize when a reciprocal and incomplete matrix can be completed to be a consistent matrix. Although the following result can be dealt by means of the general characterization given in Theorem 3., it can be proved by using Theorem 1..

Theorem 2. *Let* $A \in \mathbb{R}^+_{n,n}$. *The following affirmations are equivalent:*

(i) *There exist* $A_1 \in \mathbb{R}^+_{n,m}$, $A_2 \in \mathbb{R}^+_{m,n}$, *and* $A_3 \in \mathbb{R}^+_{m,m}$ *such that* $B = \begin{bmatrix} A & A_1 \\ A_2 & A_3 \end{bmatrix}$ *is consistent.*

(ii) *A is consistent.*

If we want to find all consistent completions of the matrix

$$B = \begin{bmatrix} A & \star & \cdots & \star \\ \star & 1 & \cdots & \star \\ \vdots & \vdots & \ddots & \vdots \\ \star & \star & \cdots & 1 \end{bmatrix} \in \mathbb{R}_{n+m,n+m},$$

where $A \in \mathbb{R}_{n,n}$ is consistent, then we apply the following procedure: Since A is consistent, there exists $\mathbf{u} \in \mathbb{R}^n$ such that $L(A) = \phi_n(\mathbf{u})$. Now, it is enough to pick any $\mathbf{w} \in \mathbb{R}^m$ and define B as

$$A_1 = E(\mathbf{u}\mathbf{1}_m^T - \mathbf{1}_n\mathbf{w}^T), \quad A_2 = E(\mathbf{w}\mathbf{1}_n^T - \mathbf{1}_m\mathbf{u}^T), \quad A_3 = E(\mathbf{w}\mathbf{1}_m^T - \mathbf{1}_m\mathbf{w}^T).$$

To motivate the notation and the precise establishment of the problem considered here, let us consider the following example. Let

$$A = \begin{bmatrix} 1 & 2 & 3 & \star \\ 1/2 & 1 & 3 & 4 \\ 1/3 & 1/3 & 1 & \star \\ \star & 1/4 & \star & 1 \end{bmatrix}. \tag{4}$$

By taking logarithms of the entries of the matrix, the aforementioned completion problem can be managed. Since the image by L of any consistent matrix is skew-Hermitian, in order to find a consistent completion of an incomplete reciprocal matrix, it is enough to restrict ourselves to the subset of reciprocal matrices of order n. From (4) we obtain

$$L(A) = \begin{bmatrix} 0 & \log 2 & \log 3 & \star \\ -\log 2 & 0 & \log 3 & \log 4 \\ -\log 3 & -\log 3 & 0 & \star \\ \star & -\log 4 & \star & 0 \end{bmatrix}, \tag{5}$$

then any skew-Hermitiam completion of $L(A)$ is of the form

$$\begin{bmatrix} 0 & \log 2 & \log 3 & 0 \\ -\log 2 & 0 & \log 3 & \log 4 \\ -\log 3 & -\log 3 & 0 & 0 \\ 0 & -\log 4 & 0 & 0 \end{bmatrix} + \lambda \begin{bmatrix} 0 & 0 & 0 & 1 \\ 0 & 0 & 0 & 0 \\ 0 & 0 & 0 & 0 \\ -1 & 0 & 0 & 0 \end{bmatrix} + \mu \begin{bmatrix} 0 & 0 & 0 & 0 \\ 0 & 0 & 0 & 0 \\ 0 & 0 & 0 & 1 \\ 0 & 0 & -1 & 0 \end{bmatrix}, \tag{6}$$

where $\lambda, \mu \in \mathbb{R}$.

From now on, we define for $1 \le i < j \le n$ the following skew-Hermitian matrices

$$B_{ij} = \mathbf{e}_i\mathbf{e}_j^T - \mathbf{e}_j^T\mathbf{e}_i. \tag{7}$$

Thus, with this notation, the skew-Hermitian completion considered in equalities (4), (5), and (6) takes the simpler form

$$C(\lambda, \mu) = C_0 + \lambda B_{14} + \mu B_{34}. \tag{8}$$

Furthermore, observe that matrix C_0 appearing in (8) can be written as

$$C_0 = \sum_{(i,j)\in N_4\setminus\{(1,4),(3,4)\}} \rho_{ij} B_{ij},$$

where $N_n = \{(i,j) : 1 \le i < j \le n\}$ and ρ_{ij} are real numbers that can be easily determined from the incomplete matrix A given in (4). In an informal way, we can think of C_0 as the incomplete skew-Hermitian matrix to be completed, and $(1,4)$, $(3,4)$ – and their symmetric positions with respect to the principal diagonal – as the void positions that must be filled.

3. Characterization of the Completion of a Reciprocal Matrix

Now we are ready to establish the first main result.

Theorem 3. *Let $1 \le i_1, j_1, \dots, i_k, j_k \le n$ be indices such that $i_r < j_r$ for $r = 1, \dots, k$. Denote $I = \{(i_1, j_1), \dots, (i_k, j_k)\}$ and $J = N_n \setminus I$. Let $C_0 = \sum_{(i,j)\in J} \rho_{ij} B_{ij}$. The following statements are equivalent*

(i) *There exist $\lambda_1, \dots, \lambda_k \in \mathbb{R}$ such that $C_0 + \sum_{r=1}^{k} \lambda_r B_{i_r j_r} \in \mathcal{L}_n$.*

(ii) *There exists $\mathbf{w} = (w_1, \dots, w_n)^T \in \mathbb{R}^n$ such that $\rho_{pq} = w_p - w_q$ for any $(p,q) \in J$.*

Furthermore, in the case that the statements hold, then

$$\lambda_r = w_{i_r} - w_{j_r}, \qquad \forall\, r \in \{1, \dots, k\}. \tag{9}$$

Example 1. We will apply Theorem 3. in order to show that matrix A in (4) cannot be completed to be consistent. If this completion were feasible, then by Theorem 3., there would exist $\mathbf{w} = (w_1, w_2, w_3, w_4)^T \in \mathbb{R}^4$ such that

$$\log 2 = w_1 - w_2, \qquad \log 3 = w_1 - w_3, \qquad \log 3 = w_2 - w_3, \qquad \log 4 = w_2 - w_4. \tag{10}$$

It can be quickly shown that this linear system has no solution.

Example 2. We will see if

$$A = \begin{bmatrix} 1 & \star & 1/3 \\ \star & 1 & 2/3 \\ 3 & 3/2 & 1 \end{bmatrix} \tag{11}$$

has a consistent completion. If there is a consistent completion, by item (iii) of Theorem 3., then there will exist $\mathbf{w} = (w_1, w_2, w_3)^T \in \mathbb{R}^3$ such that

$$-\log 3 = w_1 - w_3, \qquad \log 2 - \log 3 = w_2 - w_3. \tag{12}$$

This system, clearly, is solvable. Hence, the completion is possible. We will see how Theorem 3. enables us to find such completion(s). The general solution of (12) is

$$w_1 = -\log 3 + \alpha, \qquad w_2 = \log 2 - \log 3 + \alpha, \qquad w_3 = \alpha, \qquad \alpha \in \mathbb{R}.$$

If X is any consistent completion of A, then item (i) of Theorem 3. guarantees that exists $\lambda \in \mathbb{R}$ such that $L(X) = C_0 + \lambda B_{12}$, and such λ can be obtained from (9) obtaining $\lambda = w_1 - w_2 = -\log 2$. Thus, $L(X) = C_0 - \log 2 \cdot B_{12}$. We conclude that there is a unique consistent completion of A and is given by $X = \begin{bmatrix} 1 & 1/2 & 1/3 \\ 2 & 1 & 2/3 \\ 3 & 3/2 & 1 \end{bmatrix}$.

4. Completion of Reciprocal Matrices and Graph Theory

In this section we develop several useful results that enable us to study the uniqueness of the consistent completion and to compute in a straightforward manner all possible completions.

For an arbitrary $n \times n$ incomplete reciprocal matrix $A = (a_{ij})$, we use the following procedure to construct a directed graph, denoted by G_A : If $i \geq j$, then there is no arrow from i to j. If $i < j$ and we do not know the entry a_{ij}, then there is no arrow from i to j. If $i < j$ and we know the entry a_{ij}, then there is an arrow from i to j. Now, we easily construct the incidence matrix of G_A, denoted in the sequel by M_A.

To describe the linear system that appears in item (ii) of Theorem 3., we define, for an incomplete reciprocal matrix $A \in \mathbb{R}_{n,n}$, the vector $\mathbf{b}_A = (b_1, \ldots, b_m)^T \in \mathbb{R}^m$ [being m the number of columns of M_A] by the next procedure: For $r = 1, \ldots, m$, let us pay attention to the r-th column of M_A and let i, j be the unique indices such that the entry (i, r) of M_A is 1 and the entry (j, r) of M_A is -1. We set $b_r = \log(a_{ij})$.

Theorem 3. can be rephrased as follows: If A is an incomplete reciprocal matrix, then A can be completed to be a consistent matrix if and only if the system $M_A^T \mathbf{w} = \mathbf{b}_A$ is consistent.

Theorem 4. *Let $A \in \mathbb{R}_{n,n}$ be a reciprocal incomplete matrix and $2k$ be the number of void entries (located up and down the main diagonal of A). If G_A has p connected components and $2k \geq n^2 - 3n + 2p$, then A can be completed to be consistent.*

Table 1. Notation

	Incomplete matrix A	Directed graph G_A	Incidence matrix M_A
n	Size of A	Points of G_A	Rows of M_A
m		Arrows of G_A	Columns of M_A
$2k$	Entries of A to be filled		
p		Connected components of G_A	

Example 3. This example shows that the graph G_A can be disconnected. Let $a > 0$ and

$$A = \begin{bmatrix} 1 & \star & \star \\ \star & 1 & a \\ \star & 1/a & 1 \end{bmatrix}. \tag{13}$$

Obviously, G_A has two connected components. To find all possible consistent completions of A, we consider the system $M_A^T\mathbf{w} = \mathbf{b}_A$:

$$\begin{bmatrix} 0 & 1 & -1 \end{bmatrix} \begin{bmatrix} w_1 \\ w_2 \\ w_3 \end{bmatrix} = \log a.$$

Its solution is $w_1, w_2 \in \mathbb{R}$, $w_3 = w_2 - \log a$. If X is any consistent completion of A, then Theorem 3. assures that $\lambda_1 = w_1 - w_2$ and $\lambda_2 = w_1 - w_3 = w_1 - w_2 + \log a$ are such that

$$L(X) = \begin{bmatrix} 0 & 0 & 0 \\ 0 & 0 & \log a \\ 0 & -\log a & 0 \end{bmatrix} + \lambda_1 B_{12} + \lambda_2 B_{13}.$$

By denoting $b = \exp(w_1 - w_2)$ we obtain

$$X = \begin{bmatrix} 1 & b & ab \\ 1/b & 1 & a \\ 1/(ab) & 1/a & 1 \end{bmatrix}.$$

Observe that the consistent completion of A is not unique since $b \in \mathbb{R}^+$ is arbitrary.

Theorem 5. *Let $A \in \mathbb{R}_{n,n}$ be a reciprocal incomplete matrix and $2k$ be the number of void entries (located up and down the main diagonal of A). If $2k < n(n-1)$, G_A is connected, and there exists a consistent completion of A, then this completion is unique.*

Observe that if exists a consistent completion of A, then the general solution of $M_A^T\mathbf{w} = \mathbf{b}_A$ is given by $\mathbf{w}_0 + \mathcal{N}(M_A^T)$, where $\mathbf{w}_0$ is a particular solution of $M_A^T\mathbf{w} = \mathbf{b}_A$. It is simple to prove that if N is a $\{1\}$-inverse of M_A^T, then $N\mathbf{b}_A$ verifies the system $M_A^T\mathbf{w} = \mathbf{b}_A$. Hence the general solution of this latter system is

$$\{N\mathbf{b}_A + \mathbf{x} : \mathbf{x} \in \mathcal{N}(M_A^T)\}. \tag{14}$$

We can choose $N = (M_A^T)^\dagger$, where the superindex $\dagger$ means the Moore-Penrose inverse of a matrix.

Another result that can be useful is the following: "Let A be a $m \times n$ matrix and $\mathbf{b} \in \mathbb{R}^m$ such that the system $A\mathbf{x} = \mathbf{b}$ is consistent. If N is any matrix satisfying $ANA = A$, then the general solution of the $A\mathbf{x} = \mathbf{b}$ is given by $N\mathbf{b} + (I - NA)\mathbf{y}$ for arbitrary $\mathbf{y} \in \mathbb{R}^n$".

Finally, let us notice that to find the consistent completion of A when the corresponding graph G_A is connected, we can discard the vector $\mathbf{x}$ in $\mathcal{N}(M_A^T)$ appearing in (14).

Example 4. (This is the revisited previous example 2).

Let A be the incomplete matrix given in (11). Following the notation of Table 1 we have $k = 1, n = 3, m = 2, p = 1$. By Theorem 4. we obtain that there is a consistent completion. A solution of $M_A^T\mathbf{w} = \mathbf{b}_A$ is (by employing N as the Moore-Penrose inverse of M_A^T)

$$\mathbf{w} = (M_A^T)^\dagger \mathbf{b}_A = \frac{1}{3}\begin{bmatrix} 2 & -1 \\ -1 & 2 \\ -1 & -1 \end{bmatrix}\begin{bmatrix} -\log 3 \\ \log 2 - \log 3 \end{bmatrix} = \frac{1}{3}\begin{bmatrix} -\log 2 - \log 3 \\ 2\log 2 - \log 3 \\ -\log 2 + 2\log 3 \end{bmatrix}$$

and $\lambda = w_1 - w_2 = -\log 2$. This example finishes as before.

Example 5. (This is the revisited example 3) Let A be the incomplete matrix given in (13). Following the notation of Table 1 we have $n = 3, m = 1, k = 2, p = 2$. Any solution of $M_A^T \mathbf{w} = \mathbf{b}_A$ is given by

$$\mathbf{w} = (M_A^T)^\dagger \mathbf{b}_A + \mathbf{x} = \begin{bmatrix} 0 \\ 1/2 \\ -1/2 \end{bmatrix} \log a + \mathbf{x},$$

being $\mathbf{x} \in \mathcal{N}(M_A^T)$. But any vector of $\mathcal{N}(M_A^T)$ is of the form $(x, y, y)^T$. Thus

$$\mathbf{w} = \begin{bmatrix} x \\ y + \log a/2 \\ y - \log a/2 \end{bmatrix}. \tag{15}$$

Theorem 2 assures that $\lambda_1 = w_1 - w_2 = x - y - \log a/2$ and $\lambda_2 = w_1 - w_3 = x - y + \log a/2$ satisfy that $Y = \log a \cdot B_{23} + \lambda_1 \cdot B_{12} + \lambda_2 \cdot B_{13}$ is a matrix such that $E(Y)$ is any consistent completion of A. By denoting $b = \exp(x - y)/\sqrt{a}$ we obtain the same solution of the example 3 obtained before.

Another way of obtaining the same solution is by means of

$$\mathbf{w} = (M_A^T)^\dagger \mathbf{b}_A + (I - (M_A^T)^\dagger M_A^T)\mathbf{y} = \begin{bmatrix} 0 \\ 1/2 \\ -1/2 \end{bmatrix} \log a + \begin{bmatrix} 1 & 0 & 0 \\ 0 & 1/2 & 1/2 \\ 0 & 1/2 & 1/2 \end{bmatrix}\begin{bmatrix} y_1 \\ y_2 \\ y_3 \end{bmatrix}.$$

Obviously, one obtains the same solution as in (15) by doing $y_1 \to x$ and $y_1 + y_2 \to y$.

Let us observe that the linear system $M_A^T \mathbf{w} = \mathbf{b}_A$ is consistent if and only if $\mathbf{b}_A \in \mathcal{R}(M_A^T)$. But by standard linear algebra one has $\mathcal{R}(M_A^T) = \mathcal{N}(M_A)^\perp$. Hence we have that the linear system $M_A^T \mathbf{w} = \mathbf{b}_A$ is consistent if and only $\mathbf{b}_A^T \mathbf{x} = 0$ for any $\mathbf{x} \in \mathcal{N}(M_A)$.

In next result we find the null space of $\mathcal{N}(M_A)$ for some kind of graphs. To this end, we recall the concept of *cycle* in a graph. A cycle is a chain starting at a point and finishing at the same point.

Some further properties of the consistent completion of A can be deduced if the associated graph G_A is planar. Let us remark that it is possible that G_A is not planar as the following example shows: Let

$$B = \begin{bmatrix} 1 & \star & \star \\ \star & 1 & \star \\ \star & \star & 1 \end{bmatrix}, \quad C = \begin{bmatrix} a & b & c \\ d & e & f \\ g & h & i \end{bmatrix}, \quad D = \begin{bmatrix} a^{-1} & d^{-1} & g^{-1} \\ b^{-1} & e^{-1} & h^{-1} \\ c^{-1} & f^{-1} & i^{-1} \end{bmatrix},$$

$a, b, \ldots, i$ being positive numbers. The matrix $A = \begin{bmatrix} B & C \\ D & B \end{bmatrix}$ leads to a non planar graph, namely, the complete $(3, 3)$ bipartite graph.

Theorem 6. *Let G be a planar oriented graph and M its incidence matrix. If $\mathbf{x}_1, \ldots, \mathbf{x}_f$ correspond to the bounded faces of the graph, then $\{\mathbf{x}_1, \ldots, \mathbf{x}_f\}$ is a basis of $\mathcal{N}(M)$.*

Corollary 1. *Let A be an incomplete reciprocal matrix. If G_A is planar and has no bounded faces, then there exists a consistent completion of A.*

Let us now consider an incomplete reciprocal matrix that cannot be completed to be consistent. How can the known entries be modified to complete the matrix to be consistent? The answer will be clear if we recall the following summary: For an incomplete reciprocal matrix A, the following affirmations are equivalent:

(i) There is a consistent completion of A.

(ii) The linear system $M_A^T \mathbf{w} = \mathbf{b}_A$ is consistent.

(iii) $\mathbf{b}_A^T \mathbf{x} = 0$ for any $\mathbf{x} \in \mathcal{N}(M_A)$.

Example 6. If we want to modify some entries of matrix A given in (4) in order to have a consistent completion, let us start by writing

$$A = \begin{bmatrix} 1 & a_1 & a_2 & \star \\ a_1^{-1} & 1 & a_3 & a_4 \\ a_2^{-1} & a_3^{-1} & 1 & \star \\ \star & a_4^{-1} & \star & 1 \end{bmatrix}, \qquad \mathbf{b}_A = \begin{bmatrix} \log a_1 \\ \log a_2 \\ \log a_3 \\ \log a_4 \end{bmatrix}.$$

Now we can choose the entries $a_1, \ldots, a_4$ by using one of the above items. But as we know the null space of $\mathcal{N}(M_A)$, we choose item (iii):

$$\text{There is a consistent completion of } A \iff \mathbf{b}_A^T \mathbf{x} = 0 \iff a_1 a_3 = a_2.$$

It is noteworthy that the value of a_4 is arbitrary.

5. Assessing Consistency in Front of Incomplete Judgment

This paragraph is intended to provide a protocol that can be easily implemented in a decision support tool and/or followed by a facilitator in charge of a decision problem when assessing the consistency of an incomplete judgment given by a specific stakeholder.

Given the reciprocal incomplete matrix A, build matrix M_A and vector $\mathbf{b}_A$, and determine the solvability of the linear system $M_A^T \mathbf{w} = \mathbf{b}_A$ (Theorem 3.):

(a) Solvable: A can be consistently completed, and (9) gives the possible completions.

(b) Unsolvable: A cannot be consistently completed. In this case, by using least squares theory, the optimal solution of $M_A^T \mathbf{w} = \mathbf{b}_A$ can be used to find a completion of A that is close to be consistent.

Build the graph G_A and determine the numbers n, m, k, p of Table 1.

(a) If $2k \geq n^2 - 3n + 2p$, then the completion is possible (Theorem 4.).

(b) If $m > 0$ and $p = 1$, then the completion is unique (G_A is connected).

Regarding the calculations: calculate the Moore-Penrose inverse of matrix M_A, $M_A^\dagger$, and consider that the completion is possible if and only if $M_A^T (M_A^\dagger)^T \mathbf{b}_A = \mathbf{b}_A$. Then:

1. If the completion is unique, find the completion from $\mathbf{w} = (M_A^\dagger)^T \mathbf{b}_A$ and (9).
2. Otherwise, find the completions from $\mathbf{w} = (M_A^\dagger)^T \mathbf{b}_A + [I - (M_A^\dagger)^T M_A^T]\mathbf{y}$ for arbitrary $y \in \mathbb{R}^n$ and (9).

If G_A is planar and there are no bounded faces, then there exists a consistent completion of A. Otherwise, find the cycles of G_A (i.e., find a basis of the null space of M_A). Then

1. There is a consistent completion of A if and only if $\mathbf{b}_A^T \mathbf{x} = 0$ for any $\mathbf{x}$ belonging to the null space of M_A.
2. If there is not a consistent completion of A, then by forcing $\mathbf{b}_A^T \mathbf{x} = 0$ for any $\mathbf{x} \in \mathcal{N}(M_A)$, we can modify some entries of A to obtain a possible completion.

Conclusion

AHP has emerged as a decision support tool to integrate various technical information and stakeholder values. Frequently, decisions are based on multiple, conflicting criteria that, in addition, are subject to various types and levels of subjectivity and uncertainty. Also, performance of alternative decision options against criteria typically measured in different units, or even intangible, is sought.

To add more ambiguity and/or incompleteness, stakeholders –not always completely familiar with all the aspects involved– are gaining increased access importance in many decision making processes. In some cases, public participation is enforced explicitly. AHP provides a systematic approach to combine information inputs with benefit/cost information and decision-maker or stakeholder views to rank alternatives.

In participatory processes, specifically, the person (or team) in charge of conducting the decision process, the facilitator, needs powerful tools to help stakeholders to consistently complete their judgment, thus helping guarantee an optimal decision.

In this chapter we have characterized incomplete comparison matrices that can be consistently completed. And in Section 5 we have provided a set of simple and clear rules for the facilitator to apply in the case of opinion bodies issued in an incomplete way by any of the stakeholders. These rules are straightforward and can easily be implemented in any decision support tool based on AHP.

References

[1] Benítez J.; Delgado-Galván X.; Izquierdo; J.; Pérez-García R. Achieving matrix consistency in AHP through linearization. *Appl Math Model.* 2011, 35, 4449-4457.

[2] Benítez J.; Delgado-Galván X.; Izquierdo; J.; Pérez-García R. An approach to AHP decision in a dynamic context. *Decis Support Syst.* 2012, 53, 499-506.

[3] Saaty T.L. A scaling method for priorities in hierarchical structures *J Math Psychol.* 1977, 15, 234-281.

[4] Saaty T.L. *Theory and Applications of the Analytic Network Process*; RWS Publications: Pittsburgh, PA, 2009.

In: Mathematical Modeling in Social Sciences ...
Editors: J. C. Cortés López et al.
ISBN: 978-1-63117-335-6

Chapter 20

On Optimal Gaussian Preliminary Orbit Determination by Using a Generalized Class of Iterative Methods

Alicia Cordero*, Juan R. Torregrosa† and María P. Vassileva‡
Instituto de Matemáticas Multidisciplinar,
Universitat Politècnica de València, Valencia, Spain
Instituto Tecnológico de Santo Domingo (INTEC),
Santo Domingo, República Dominicana

Abstract

A class of optimal methods for solving nonlinear equations is extended up to sixteenth-order of convergence. Some numerical test are made to solve the orbit determination problem of artificial satellites in order to confirm the theoretical results and to compare the new methods with other known ones.

Keywords: Preliminary orbit determination, artificial satellites, nonlinear equation, Potra-Pták's method, multipoint scheme, optimal order efficiency

1. Introduction

In this chapter we present a technique to derive multi-point methods with optimal and arbitrary order of convergence. The algorithms are based on Traub's scheme (see [15]) and further developed by using weight functions procedure.

A variety of problems in different fields of science and technology require finding the solution of a nonlinear equation. Iterative methods for approximating solutions are the most used technique. The interest of the multipoint iterative methods has been renewed in the first decade of the 21st century as they are of great practical importance because they

*E-mail address: acordero@mat.upv.es
†E-mail address: jrtorre@mat.upv.es
‡E-mail address: maria.vassilev@gmail.com

exceed the theoretical limits of the methods point-to-point on the order of convergence and computational efficiency.

The existence of an extensive literature on higher order methods reveals that they are only limited by the nature of the problem to be solved: in particular, the numerical solution of nonlinear equations and systems are needed in the study of dynamical models of chemical reactors [1], or in radioactive transfer [6]. Moreover, many of numerical applications use high precision in their computations; in [16], high-precision calculations are used to solve interpolation problems in Astronomy; in [7] the authors describe the use of arbitrary precision computations to improve the results obtained in climate simulations; the results of these numerical experiments show that the high order methods associated with a multiprecision arithmetic floating point are very useful, because it yields a clear reduction in iterations. A motivation for an arbitrary precision in interval methods can be found in [11], in particular for the calculation of zeros of nonlinear functions.

We are going to design multipoint iterative methods to find a simple root ξ of a nonlinear equation $f(x) = 0$, where $f : I \subseteq \mathbb{R} \to \mathbb{R}$ for an open interval I. Many modified schemes of Newton's method, probably the most widely used iterative method, have been proposed to improve the local order of convergence and the efficiency index over the last years. The efficiency index, introduced by Ostrowski in [10] as $I = p^{1/d}$, where p is the order of convergence and d the number of functional evaluations per step, establishes the effectiveness of the iterative method. In this sense, Kung and Traub conjectured in [9] that a multipoint iterative scheme without memory, requiring $d + 1$ functional evaluations per iteration, has order of convergence at most 2^d. The schemes which achieve this bound are called optimal methods.

The outline of this chapter is as follows. In Section 2, the problem of preliminary orbit determination of artificial satellites is studied by using the classical fixed point method. In Section 3 the different families of methods are constructed and the convergence analysis is discussed. Finally, in Section 4 numerical experiments on the modified Gaussian preliminary orbit determination are performed and the proposed methods are compared with recent optimal known schemes.

2. Preliminary Orbit Determination

A classical reference in preliminary orbit determination is F. Gauss (1777-1855), who deduced the orbit of the minor planet Ceres, discovered in 1801 and afterwards lost. The so-called Gauss' method is based on the rate y between the triangle and the ellipse sector defined by two position vectors, from astronomical observations. This proportion is related with the geometry of the orbit and the observed position by

$$y = 1 + X(l + x), \tag{1}$$

where $l = \frac{r_1 + r_2}{4\sqrt{r_1 r_2}\cos\left(\frac{\nu_2 - \nu_1}{2}\right)} - \frac{1}{2}$, $x = \sin^2\left(\frac{E_2 - E_1}{4}\right)$ and $X = \frac{E_2 - E_1 - \sin(E_2 - E_1)}{\sin^3\left(\frac{E_2 - E_1}{2}\right)}$. The angles E_i, ν_i, $i = 1, 2$, are the eccentric and true anomalies, respectively, associated to the observed positions $\vec{r_1}$ and $\vec{r_2}$ (let us denote by r_i the modulus of vector $\vec{r_i}$, $i = 1, 2$).

Equation (1) is, actually, the composition of the First and Second Gauss Equation

$$y^2 = \frac{m}{l + x} \qquad \text{and} \qquad y^2(y - 1) = mX,$$

where $m = \frac{\mu\tau^2}{[2\sqrt{r_1 r_2}\cos(\frac{\nu_2-\nu_1}{2})]^3}$, μ is the gravitational parameter of the motion and τ is a modified time variable.

The original iterative procedure used to solve the nonlinear Gauss equation (1) is the Fixed Point method (FP) (see, for example, [5]) and is described in the following scheme:

(i) From the initial estimation $y_0 = 1$, $x_0 = \frac{m}{y_0^2} - l$ is obtained (it is possible to calculate m and l from the observed positions $\overrightarrow{r_1}$ and $\overrightarrow{r_2}$ and the time τ).

(ii) From x_0 and $\cos(\frac{E_2 - E_1}{2}) = 1 - 2x_C$, $\sin(\frac{E_2-E_1}{2}) = +\sqrt{4x_0(1-x_0)}$, we calculate $E_2 - E_1$. Then, we obtain $X_0 = \frac{E_2 - E_1 - \sin(E_2 - E_1)}{\sin^3(\frac{E_2-E_1}{2})}$.

(iii) By using the combined Gauss equation (1) a new iteration y_1 is calculated and the process start again.

The iterative process follows as described above, getting new estimations of the ratio, until it does not vary within a given tolerance. Once the method has converged, the semi-major axis a, can be calculated by means of equation

$$y = \frac{\sqrt{\mu p}\cdot\tau}{r_2 r_1 \sin(\nu_2 - \nu_1)} = \frac{\sqrt{\mu}\cdot\tau}{2\sqrt{a}\sqrt{r_2 r_1}\sin(\frac{E_2-E_1}{2})\cos(\frac{\nu_2-\nu_1}{2})},$$

from the last estimations of ratio and difference of eccentric anomalies, and the last phase is then initiated, to determine velocity and orbital elements.

Let us note that the original Gauss' scheme has a restriction when the angle formed by the two position vectors is greater than $\pi/4$, since in this case the areas of the triangle and the ellipse sector are not similar. In this chapter, we are going to design a family of high-order iterative methods in order to improve the results of the original Gauss' scheme and reducing considerably the number of iterations and the error in the calculations.

3. Description of the Optimal Multipoint Methods

Our starting point is the Traub's scheme (see [15], also known as Potra-Pták's method) whose iterative expression is

$$x_{k+1} = y_k - \frac{f(y_k)}{f'(x_k)} = x_k - \frac{f(x_k) + f(y_k)}{f'(x_k)},$$

where y_k is the Newton's step. This method has order three but it requires three functional evaluations, so it is not optimal according to Kung-Traub conjecture and our purpose is to design optimal methods.

So, we begin the process from the iterative scheme (see [2])

$$\begin{aligned} y_k &= x_k - \beta\frac{f(x_k)}{f'(x_k)}, \\ x_{k+1} &= y_k - H(u(x_k))\frac{f(y_k)}{f'(x_k)}, \end{aligned} \tag{2}$$

where β is a real parameter and $H(u)$ is a real function with $u = \frac{f(y)}{f(x)}$.

Theorem 3.1. *Let $\xi \in I$ be a simple zero of a sufficiently differentiable function $f : I \subset \mathbb{R} \to \mathbb{R}$ in an open interval I and x_0 an initial guess close to ξ. The method defined by (2) has order four if $\beta = 1$ and a function H is chosen so that the conditions $H(0) = 1$, $H'(0) = 2$, $|H''(0)| < \infty$ are fulfilled. The error equation is*

$$e_{k+1} = \left[\left(5 - \frac{H''(0)}{2}\right) c_2^3 - c_2 c_3\right] e_k^4 + \mathcal{O}(e_k^5),$$

where $c_k = \dfrac{1}{k!}\dfrac{f^{(k)}(\xi)}{f'(\xi)}$, $k = 2, 3, \ldots$ *and* $e_k = x_k - \xi$.

Recently, taking as the first two steps the method (2) and adding a new step, Džunić et al. in [4] designed the following three-step method

$$\begin{aligned} z_k &= y_k - H(u(x_k))\frac{f(y_k)}{f'(x_k)}, \\ x_{k+1} &= z_k - G(u(x_k), v(x_k))\frac{f(z_k)}{f'(x_k)}, \end{aligned} \tag{3}$$

where y_k is the Newton's step and $G(u, v)$ is a function of two variables: $u = \frac{f(y)}{f(x)}$ and $v = \frac{f(z)}{f(y)}$.

They proved in [4] that the method defined by (3) has order of convergence 8, under some conditions on functions H and G.

Theorem 3.2. *Let $\xi \in I$ be a simple zero of a sufficiently differentiable function $f : I \subset \mathbb{R} \to \mathbb{R}$ in an open interval I and x_0 an initial guess close to ξ. The method defined by (3) has optimal eighth-order convergence if sufficiently differentiable functions H and G are chosen so that the following conditions are satisfied:*

$$\begin{array}{llll} H(0) = 1, & H'(0) = 2, & G(0,0) = 1, & G_u(0,0) = 2, \\ G_v(0,0) = 1, & G_{uu}(0,0) = 2 + H''(0), & G_{uv}(0,0) = 4, & \end{array}$$

and $G_{uuu}(0,0) = -24 + 6H''(0) + H'''(0)$. *The error equation of the method is*

$$\begin{aligned} e_{k+1} &= -\frac{1}{2}c_2(3c_2^2 - c_3)(9(-6 + G_{vv}(0,0))c_2^4 + 2(17 - 3G_{vv}(0,0))c_2^2 c_3 \\ &\quad + (-2 + G_{vv}(0,0))c_3^2 - 2c_2c_4)e_k^8 + \mathcal{O}(e_k^9), \end{aligned}$$

where $c_k = \dfrac{1}{k!}\dfrac{f^{(k)}(\xi)}{f'(\xi)}$, $k = 2, 3, \ldots$ *and* $e_k = x_k - \xi$.

Weight functions H and G should be chosen as simple as possible. One of the simplest forms is that obtained by using the Taylor polynomials of these functions according to conditions of Theorem 3.2, that is

$$H(u) = 1 + 2u, \qquad G(u, v) = 1 + 2u + v + u^2 + 4uv - 4u^3.$$

The iterative method resulting by using these functions is denoted by M8.

Now, we wonder if it is possible to find a sixteenth-order iterative method by adding a new step with the same settings accompanied with a weight function T that depends on three variables u, v and $w = \dfrac{f(s)}{f(z)}$, where s is the last step of the eighth-order method (3). The iterative expression of the new scheme is

$$\begin{aligned} s_k &= z_k - G(u(x_k), v(x_k))\frac{f(z_k)}{f'(x_k)}, \\ x_{k+1} &= s_k - T(u(x_k), v(x_k), w(x_k))\frac{f(s_k)}{f'(x_k)}, \end{aligned} \tag{4}$$

where y_k and z_k are the same steps as in method (3). It can be proved the following result that establish the sixteenth-order of family (4).

Theorem 3.3. *Let $\xi \in I$ be a simple zero of a sufficiently differentiable function $f : I \subset \mathbb{R} \to \mathbb{R}$ in an open interval I and x_0 an initial guest close to ξ. The method defined by (4) has optimal sixteenth-order convergence if sufficiently differentiable functions H, G and T are chosen so that the conditions of Theorem 3.2 and the following requirements are satisfied:*

$$\begin{array}{llll} H''(0) = 0, & H^{(3)} = 24, & H^{(4)}(0) = -72, & G_{uu}(0,0) = 2, \\ G_{uuu}(0,0) = 0, & G_{uuuv}(0,0) = 24, & G_{uuvv}(0,0) = -16, & G_{uuv}(0,0) = 6, \\ G_{uuuu}(0,0) = 0, & T(0,0,0) = 1, & T_u(0,0,0) = 2, & T_v(0,0,0) = 1, \\ T_w(0,0,0) = 1, & T_{uu}(0,0,0) = 2, & T_{uv}(0,0,0) = 4, & T_{vv}(0,0,0) = G_{vv}(0,0), \\ T_{vw}(0,0,0) = 2, & T_{uuv}(0,0,0) = 8, & T_{uuu}(0,0,0) = 0, & T_{uvv}(0,0,0) = 4 + G_{uvv}(0,0), \\ T_{uvw}(0,0,0) = 8, & T_{uw}(0,0,0) = 2, & T_{uuw}(0,0,0) = 2, & \end{array}$$

$G_{uvv}(0,0) = 8 - \frac{1}{3}(G_{uvvv}(0,0) + 6G_{vv}(0,0))$ and $T_{vvv}(0,0,0) = -3 + 3G_{vv}(0,0) + G_{vvv}(0,0)$. The error equation of the method is

$$\begin{aligned} e_{k+1} = & -\frac{1}{48}c_2(5c_2^2 - c_3)(5(-14 + 5G_{vv}(0,0))c_2^4 + 2(16 - 5G_{vv}(0,0))c_2^2c_3 + \\ & (-2 + G_{vv}(0,0))c_3^2 - 2c_2c_4)(25N_1c_2^8 - 20N_2c_2^6c_3 - N_3c_3^4 + 60N_4c_2^5c_4 + \\ & 24N_5c_2^3c_3c_4 + 12N_6c_2c_3^2c_4 + 6c_2^4(N_7c_3^2 + 20c_5) - \\ & 4c_2^2(N_8c_3^3 + 3(2 - T_{ww}(0,0,0))c_4^2 + 6c_3c_5))e_k^{16} + \mathcal{O}(e_k^9), \end{aligned}$$

where $c_k = \dfrac{1}{k!}\dfrac{f^{(k)}(\xi)}{f'(\xi)}$, $k = 2, 3, \ldots$, $e_k = x_k - \xi$ and N_i, $i = 1, 2, \ldots, 8$ depend on the partial derivatives of order one, two and three of the weight functions G and T at zero.

A particular element of this family is obtained by choosing

$$\begin{aligned} H(u) &= 1 + 2u + 4u^3 - 3u^4, \\ G(u,v) &= 1 + 2u + v + u^2 + 4uv + 3u^2v + 4uv^2 + 4u^3v - 4u^2v^2, \\ T(u,v,w) &= 1 + 2u + v + w + u^2 + 4uv + 2uw + 4u^2v + u^2w + 6uv^2 \\ & \quad + 8uvw - v^3 + 2vw, \end{aligned}$$

which is denoted by M16, that we will use in the following sections.

In the following section, we are going to compare schemes M8 and M16 with other known ones of order 8 and 16, respectively. In particular, we analyze the behavior of these methods to obtain the preliminary orbit of an artificial satellite.

4. Numerical Results

All the iterative schemes introduced in the following are optimal in the sense of Kung-Traub's conjecture and have been designed with the weight-function technique, so they are fully comparable with the new ones designed in this paper. Let us refer now to the procedure that Kim present in [8]: a three-step eighth-order method, whose iterative expression is

$$\begin{aligned} z_k &= y_k - \frac{1+u_k+2/3u_k^2}{1-u_k-2u_k^2}\frac{f(y_k)}{f'(x_k)}, \\ x_{k+1} &= z_k - \frac{1-2u_k+v_k}{1-3u_k-2v_k}\frac{f(z_k)}{f'(x_k)+f[y_k,x_k,z_k](z_k-x_k)}, \end{aligned}$$

where y_k is Newton's step, $u_k = \frac{f(y_k)}{f(x_k)}$, $v_k = \frac{f(z_k)}{f(x_k)}$ and $f[\cdot,\cdot,\cdot]$ denotes the divided difference of order two. We will denote this scheme by K8.

We will also compare our new schemes with the method designed by Soleymani et al. in [13] (denoted by S8), initialized with Ostrowski's procedure,

$$\begin{aligned} z_k &= y_k - \frac{f(x_k)}{f(x_k)-2f(y_k)}\frac{f(y_k)}{f'(x_k)}, \\ x_{k+1} &= z_k - \frac{f(z_k)}{2f[y_k,x_k]-f'(x_k)+f[z_k,x_k,x_k](z_k-y_k)} \\ &\quad \left(1+w_k+2v_k-2u_k^3+\frac{2}{5}\frac{f(z_k)}{f'(x_k)}\right), \end{aligned}$$

where y_k is Newton's step, $u_k = \frac{f(y_k)}{f(x_k)}$, $v_k = \frac{f(z_k)}{f(x_k)}$ and $w_k = \frac{f(z_k)}{f(y_k)}$.

The proposed iterative scheme M16 will be compared with some known methods existing in the literature. In particular, the iterative scheme of the sixteenth-order scheme designed by Thukral in [14] is

$$\begin{aligned} z_k &= y_k - \frac{f[w_k,x_k]}{f[w_k,y_k]}\frac{f(y_k)}{f[x_k,y_k]}, \\ a_k &= z_k - \frac{1}{(1+2u_3u_4^2)(1-u_2)}\frac{f(z_k)}{f[y_k,z_k]-f[x_k,y_k]+f[x_k,z_k]}, \\ x_{k+1} &= a_k - Tf(a_k)\frac{f[y_k,z_k]}{f[y_k,a_k]f[z_k,a_k]} \end{aligned}$$

where y_k is Steffensen's step, $w_k = x_k + f(x_k)$, $u_1 = \dfrac{f(z_k)}{f(x_k)}$, $u_2 = \dfrac{f(z_k)}{f(w_k)}$, $u_3 = \dfrac{f(y_k)}{f(x_k)}$, $u_4 = \dfrac{f(y_k)}{f(w_k)}$, $u_5 = \dfrac{f(a_k)}{f(x_k)}$, $u_6 = \dfrac{f(a_k)}{f(w_k)}$ and $T = 1+u_1u_2-u_1u_3u_4^2+u_5+u_6+u_1^2u_4+u_2^2u_3+3u_1u_4^2\dfrac{u_3^2-u_4^2}{f[x_k,y_k]}$. We will denote this scheme by T16.

We will also use the sixteenth-order procedure designed by Sharma et al. in [12], that will be denoted by S16, whose iterative expression is

$$\begin{aligned} z_k &= y_k - \frac{f(x_k)}{f(x_k)-2f(y_k)}\frac{f(y_k)}{f'(x_k)}, \\ t_k &= x_k - \frac{f(x_k)(p+q+r)}{pf[z_k,x_k]+qf'(x_k)+rf[y_k,x_k]}, \\ x_{k+1} &= x_k - \frac{p_1 f[z_k,y_k]+q_1 f[y_k,x_k]+rf[t_k,y_k]}{p_1 l+q_1 m+rn}f(x_k), \end{aligned}$$

where y_k is Newton's step, $p = (x_k - y_k)f(x_k)f(y_k)$, $q = (y_k - z_k)f(z_k)f(y_k)$, $r = (z_k - x_k)f(z_k)f(x_k)$, $p_1 = (x_k - t_k)f(x_k)f(t_k)$, $q_1 = (t_k - z_k)f(t_k)f(z_k)$, $l = \dfrac{f(y_k)f[z_k,x_k] - f(z_k)f[y_k,x_k]}{y_k - z_k}$, $m = \dfrac{f(y_k)f'(x_k) - f(x_k)f[y_k,x_k]}{y_k - x_k}$ and $n = \dfrac{f(y_k)f[x_k,t_k] - f(t_k)f[y_k,x_k]}{y_k - t_k}$.

In the numerical test made, variable precision arithmetics has been used, with 4000 digits of mantissa in Matlab R2011b. Some reference orbits have been used in the test, that can be found in [5]. As orbital elements of each one of the test orbits are known, the vector position in the instants t_1 and t_2 have been re-calculated with 3998 exact digits. Then, our aim is to solve the unified Gauss' equation from these positions, with the highest possible precision. In this terms, the orbital elements can be calculated with the best accuracy.

- Test Orbit I has the position vectors

  $$\vec{r_1} \approx [2.46080928705339, 2.04052290636432, 0.14381905768815],$$

  $$\vec{r_2} \approx [1.98804155574820, 2.50333354505224, 0.31455350605251],$$

  measured in Earth radius (e.r.) at the julian days (J.D.) from the perigee $t_1 = 0$ and $t_2 = 0.01044412000000$. The orbital elements corresponding to the geometry of the orbit are the semimajor axis $a = 4$ e.r., the eccentricity $e = 0.2$, the epoch of the perigee $T_0 = 0h0m0s$, and the Euler angles which fit the orbit in space are the Right Ascension of the ascending node, $\Omega = 30^o$, the argument of the perigee $\omega = 10^o$ and the inclination of the orbit $i = 15^o$.

- Test Orbit II. Position vectors and times:

  $$\vec{r_1} \approx [-1.75981065999937, 1.68112802634201, 1.16913429510899] \text{ e.r.}, \; t_1 = 0 \text{ J.D.},$$

  $$\vec{r_2} \approx [-2.23077219993536, 0.77453561301361, 1.34602197883025] \text{ e.r.},$$

  $$t_2 = 0.01527809 \text{ J.D.},$$

  Orbital elements: $\Omega = 80^o, \omega = 60^o, i = 30^o, a = 3$ e.r., $e = 0.1, T_0 = 0h0m0s$.

- Test Orbit III. Position vectors and times:

  $$\vec{r_1} \approx [0.41136206679761, -1.66250000000000, 0.82272413359522] \text{ e.r.}, \; t_1 = 0 \text{ J.D.},$$

  $$\vec{r_2} \approx [0.97756752977209, -1.64428006097667, -0.04236299091612] \text{ e.r.},$$

  $$t_2 = 0.01316924 \text{ J.D.},$$

  Orbital elements: $\Omega = 120^o, \omega = 150^o, i = 60^o, a = 2$ e.r., $e = 0.05, T_0 = 0h0m0s$.

We will compare the different error estimations at the first three iterations of the proposed eighth-order method M8 and the known schemes K8, and S8 and the sixteenth-order method M16 and the schemes T16 and S16. We also include, in Tables 1 to 3, the approximated computational order of convergence (ACOC) (see [3]), in order to check the computational efficiency of the schemes related to their theoretical rate of convergence. This index is evaluated by the formula

$$p \approx ACOC = \frac{\log |(x_{k+1} - x_k)/(x_k - x_{k-1})|}{\log |(x_k - x_{k-1})/(x_{k-1} - x_{k-2})|}.$$

The different test orbits have been chosen with increasing angle $\nu_2 - \nu_1$. It measures the

Table 1. Comparison of modified-Gauss schemes for Orbit I

	$\|x_1 - x_0\|$	$\|F(x_1)\|$	$\|x_2 - x_1\|$	$\|F(x_2)\|$	$\|x_3 - x_2\|$	$\|F(x_3)\|$	ACOC
FP	0.6450e-2	-	0.8288e-4	-	0.1055e-5	-	1.002
K8	0.6368e-2	0.2059e-21	0.2033e-21	0.6553e-158	0.647e-158	0.2164e-1113	7.001
S8	0.6368e-2	0.1377e-23	0.1359e-23	0.5565e-197	0.5495e-197	0.3967e-1584	8.001
M8	0.6368e-2	0.1382e-23	0.1365e-23	0.5791e-197	0.5718e-197	0.5488e-1584	8.000
T16	0.6368e-2	0.2662e-63	0.2628e-63	0.1642e-1045	NaN	NaN	-
S16	0.6368e-2	0.7454e-48	0.7361e-48	0.6647e-783	0.6563e-783	0.	16.000
M16	0.6368e-2	0.6998e-47	0.6910e-47	0.2286e-766	0.2258e-766	0.	16.000

spread in the observations and, by the design of Gauss' procedure, it induces instability in the system when it gets higher. The difference between the true anomalies of the observations is, for the test orbits I to III, 12.23^o, 22.06^o and 31.46^o, respectively. It can be observed in Tables 1 to 3 that, when the spread of the observations increases, the precision obtained in the calculations per step reduces in the same rate for any method of the same order. It is clear that the application of high-order schemes to the problem of preliminary

Table 2. Comparison of modified-Gauss schemes for Orbit II

	$\|x_1 - x_0\|$	$\|F(x_1)\|$	$\|x_2 - x_1\|$	$\|F(x_2)\|$	$\|x_3 - x_2\|$	$\|F(x_3)\|$	ACOC
FP	0.2397e-1	-	0.1132e-2	-	0.5163e-4	-	1.011
K8	0.2289e-1	0.2830e-15	0.2707e-15	0.7343e-113	0.7023e-113	0.5810e-796	7.007
S8	0.2289e-1	0.6075e-17	0.5809e-17	0.8328e-142	0.7964e-142	0.1039e-1140	8.006
M8	0.2289e-1	0.3696e-17	0.3534e-17	0.9933e-144	0.9500e-144	0.2705e-1156	8.005
T16	0.2289e-1	0.2913e-45	0.2786e-45	0.4103e-748	0.3924e-748	0.	16.000
S16	0.2289e-1	0.1482e-34	0.1417e-34	0.4368e-556	0.4195e-556	0.	16.010
M16	0.2289e-1	0.4590e-34	0.4389e-34	0.1062e-557	0.1016e-557	0.	16.000

orbit calculation by Gauss procedure gets an important success, as the gain in speed and the precision obtained in the calculations are increased. Let us note that the precision of the orbital elements calculated with the third estimation provided by any sixteenth-order method is total, as all the 4000 decimal digits of the solution considered as exact are reached with only three iterations.

Table 3. Comparison of modified-Gauss schemes for Orbit III

	$\|x_1 - x_0\|$	$\|F(x_1)\|$	$\|x_2 - x_1\|$	$F(x_2)\|$	$\|x_3 - x_2\|$	$\|F(x_3)\|$	ACOC
FP	0.5499e-1	-	0.5830e-2	-	0.5723e-3	-	1.034
K8	0.4968e-1	0.1579e-11	0.1437e-11	0.1661e-85	0.1512e-85	0.2376e-603	7.02
S8	0.4968e-1	0.5842e-13	0.5317e-13	0.6265e-109	0.5701e-109	0.1095e-876	8.017
M8	0.4968e-1	0.1092e-13	0.9941e-14	0.2294e-115	0.2087e-115	0.8667e-929	8.007
T16	0.4968e-1	0.2742e-34	0.2495e-34	0.1560e-567	0.1419e-567	0.7e-3998	16.010
S16	0.4968e-1	0.1550e-26	0.1411e-26	0.1066e-435	0.9702e-436	0.1e-3998	16.020
M16	0.4968e-1	0.3967e-27	0.3610e-27	0.1512e-445	0.1376e-445	0.1e-3998	16.010

Conclusion

The Gaussian procedure for determining preliminary orbits has been modified in order to use modern and efficient iterative schemes of any optimal order of convergence and achieve high-level accuracy. From the obtained results, it can be deduced that the proposed schemes are, at least, as competitive as recently published methods of the same order of convergence, being better in some cases. It has also shown to be robust enough to hold the theoretical order of convergence when an exigent precision is demanded.

Acknowledgments

This research was supported by Ministerio de Ciencia y Tecnología MTM2011-28636-C02-02 and FONDOCYT 2011-1-B1-33 República Dominicana.

References

[1] Bruns, D.D.; Bailey, J.E. Nonlinear feedback control for operating a nonisothermal CSTR near an unstable steady state, *Chem. Eng. Sc.*, 1977, 32, 257–264.

[2] Chun, C. Some fourth-order iterative methods for solving nonlinear equations, *Appl. Mathematics. Comput.*, 2008, 195, 454-459.

[3] Cordero, A., Torregrosa, J.R. Variants of Newton's method using fifth-order quadrature formulas, *Appl. Mathematics. Comput.*, 2007, 190, 686-698.

[4] Džunić, J.; Petković, M.S.; Petković, L.D. A family of optimal three-point methods for solving nonlinear equations using two parametric functions, *Appl. Mathematics. Comput.*, 2011, 217, 7612-7619.

[5] Escobal, P.R. *Methods of orbit determination*, Robert E. Krieger Publishing Company, 1965.

[6] Ezquerro, J.A.; Gutiérrez, J.M.; Hernández, M.A.; Salanova, M.A. Chebyshev-like methods and quadratic equations, *Revue d'Anal. Num. et de Th. de l'Approximation*, 1999, 28, 23–35.

[7] He, Y.; Ding, C. Using accurate arithmetics to improve numerical reproducibility and stability in parallel applications, *J. Supercomput.*, 2001, 18, 259–277.

[8] Kim, Y.I. A triparametric family of theree-step optimal eighth-order multipoint iterative methods for solving nonlinear equations, *Intern. J. Comp. Math.*, 2012, 89(8), 1051-1059.

[9] Kung H.T.; Traub, J.F. Optimal order of one-point and multi-point iterations, *J. Assoc. Comput. Math.*, 1974, 21, 643-651.

[10] Ostrowski, A.M. *Solution of equations and systems of equations*, Prentice-Hall, Englewood Cliffs, New Jersey, USA, 1964.

[11] Revol N.; Rouillier, F. Motivation for an arbitrary precision interval arithmetic and the MPFI Library, *Reliable Comput.*, 2005, 11, 275–290.

[12] Sharma, J.R.; Guha, R.K.; Gupta, P. Improved King's method with optimal order of convergence based on rational approximations, *Appl. Math. Lett.*, 2013, 26, 473-480.

[13] Soleymani, F.; Sharifi, M.; Mousavi, B.S. An Improvement of Ostrowski's and King's Techniques with Optimal Convergence Order Eight, *J. Optim. Theory Appl.*, 2012, 153, 225-236.

[14] Thukral, R. New Sixteenth-Order Derivative-Free Methods for Solving Nonlinear Equations, *Amer. J. Comput. Appl. Math.*, 2012, 2(3), 112-118.

[15] Traub, J.F. *Iterative Methods for the Solution of Equations*, Prentice Hall, New York, 1964.

[16] Zhang, Y.; Huang, P.; High-precision Time-interval Measurement Techniques and Methods, *Progr. Astronomy*, 2006, 24(1), 1–15.

In: Mathematical Modeling in Social Sciences ...
Editors: J. C. Cortés López et al.
ISBN: 978-1-63117-335-6

Chapter 21

SOLVING ENGINEERING MODELS WHICH USE MATRIX HYPERBOLIC SINE AND COSINE FUNCTIONS

Emilio Defez [†], ***Jorge Sastre*** [♯], ***Javier J. Ibáñez*** [◇], ***Jesús Peinado*** [◇]
*† Instituto Universitario de Matemática Multidisciplinar,
♯ Instituto de Telecomunicaciones y Aplicaciones Multimedia,
◇ Instituto de Instrumentación para Imagen Molecular,
Universitat Politècnica de València, Valencia, Spain

Abstract

Matrix functions have multiple applications in different areas of applied mathematics. Methods to calculate the matrix exponential and the sine and cosine matrix functions in terms of orthogonal matrix polynomials are introduced recently. In this chapter a method for computing hyperbolic matrix cosine and sine, based on Hermite matrix polynomial series expansions, is presented. This approach allows us to approximate both functions simultaneously. An error bound analysis is given. Based on the ideas above, an efficient and highly-accurate Hermite algorithm is presented. A MATLAB implementation of this algorithm has also been developed. This implementation has been compared with MATLAB function `funm` on a large class of matrices for different dimensions, obtaining higher accuracy and lower computational costs in the majority of cases.

Keywords: Hermite matrix polynomial, Hyperbolic matrix sine and hyperbolic matrix cosine, computation, error bound

1. Introduction

Coupled partial differential systems are frequent in many different fields of science and technology: magnetohydrodynamic flows Sezgin [1987], biochemistry King and Chou [1976], elastic and inelastic contact problems of solids Jódar et al. [2000], cardiology Winfree [1987], diffusion problems Morimoto [1962]. Coupled hyperbolic systems appear in

*E-mail address: edefez@imm.upv.es, jorsasma@iteam.upv.es, {jjibanez, jpeinado}@dsic.upv.es

microwave heating processes Pozar [1991] and optics Das [1991] for instance. The exact solution of a class of this problems, see Jódar et al. [2003], is given in terms of matrix functions, in particular, of hyperbolic sine and cosine of a matrix, $\sinh(A), \cosh(A)$, defined respectively by

$$\cosh(Ay) = \frac{e^{Ay} + e^{-Ay}}{2}, \ \sinh(Ay) = \frac{e^{Ay} - e^{-Ay}}{2}. \tag{1}$$

For the numerical solution of these problems, analytic-numerical approximations are most suitably obtained by using the hyperbolic matrix functions $\sinh(A)$ and $\cosh(A)$, see Jódar et al. [2003]. It is well known that the computation of both functions can be reduced to the cosine of a matrix, because

$$\cosh(A) = \cos(iA), \sinh(A) = i\cos(A - \frac{i\pi}{2}I).$$

Thus, the matrix cosine can be effectively calculated, Defez et al. [2009, 2013], with the disadvantage, however, to require complex arithmetic even though the matrix A is real, which contributes substantially to the computational overhead. Direct calculation through the exponential matrix using (1) is costly. In Defez et al. [2012], a method to approximate $\cosh(A)$ using Hermite matrix polynomials series expansion was given. The proposed method use the bound for Hermite matrix polynomials given in Defez et al. [2013] by

$$\left\| H_{2n}\left(x, \frac{1}{2}A^2\right)\right\|_2 \leq (2n)!\, e \cosh\left(x \left\|A^2\right\|^{\frac{1}{2}}\right). \tag{2}$$

Lamentably, a similar bound for polynomials of odd degree has not been yet obtained, and the computation of $\sinh(A)$ by the relation

$$\sinh(A) = -i\cosh\left(-A - \frac{\pi}{2}iI\right). \tag{3}$$

requires also complex arithmetic even though the matrix A is real. In this chapter we propose a method to evaluate both matrix functions, $\sinh(A)$ and $\cosh(A)$, simultaneously and avoiding complex arithmetic when not needed.

This chapter is organized as follows. Section 2 summarizes previous results of Hermite matrix polynomials and includes a new Hermite series expansion of the matrix hyperbolic sine and cosine. Section 3 deals with the Hermite matrix polynomial series expansion of $\cosh(At)$ and $\sinh(At)$ for an arbitrary matrix as well as with its finite series truncation with a prefixed accuracy in a bounded domain, and an algorithm of the method is given. Section 4 deals with a selection of examples in order to investigate the accuracy of the new method proposed here. Finally, conclusions are presented in section 5.

Throughout this chapter, $[x]$ denotes the integer part of x and $\lfloor x \rfloor$ is the standard floor function which maps a real number x to its next smallest integer. The matrices I_r and $\theta_{r\times r}$ in $\mathbb{C}^{r\times r}$ denote the matrix identity and the null matrix of order r, respectively. Following Golub and Loan [1989], for a matrix A in $\mathbb{C}^{r\times r}$, its infinite-norm will be denoted by $\|A\|_\infty$ and its 2-norm will be denoted by $\| A \|_2$.

2. Some Results on Hermite Matrix Polynomials

For the sake of clarity in the presentation of the following results we recall some properties of Hermite matrix polynomials which have been established in Defez and Jódar [1998] and Jódar and Company [1996]. From (3.4) of [Jódar and Company, 1996, p. 25] the nth Hermite matrix polynomial satisfies

$$H_n\left(x, \frac{1}{2}A^2\right) = n! \sum_{k=0}^{\left[\frac{n}{2}\right]} \frac{(-1)^k (xA)^{n-2k}}{k!(n-2k)!}, \tag{4}$$

for an arbitrary matrix A in $\mathbb{C}^{r\times r}$. Observe that the nth scalar Hermite polynomial coincides with the $n-$th matrix Hermite polynomial when $r = 1$ and $A = 2$. Taking into account the three-term recurrence relationship (3.12) of [Jódar and Company, 1996, p. 26], it follows that

$$\left.\begin{array}{c} H_n\left(x, \frac{1}{2}A^2\right) = xAH_{n-1}\left(x, \frac{1}{2}A^2\right) - 2(n-1)H_{n-2}\left(x, \frac{1}{2}A^2\right) \quad , \quad n \geq 1 \\ H_{-1}(x, \frac{1}{2}A^2) = \theta_{r\times r} \quad H_0(x, \frac{1}{2}A^2) = I_r \end{array}\right], \tag{5}$$

and from its generating function in (3.1) and (3.2) [Jódar and Company, 1996, p. 24] one gets

$$e^{xtA-t^2I} = \sum_{n\geq 0} H_n\left(x, \frac{1}{2}A^2\right) t^n/n!, \quad |t| < \infty, \tag{6}$$

where $x, t \in \mathbb{C}$. Taking $y = tx$ and $\lambda = 1/t$ in (6) it follows that

$$e^{Ay} = e^{\frac{1}{\lambda^2}} \sum_{n\geq 0} \frac{1}{\lambda^n n!} H_n\left(\lambda y, \frac{1}{2}A^2\right), \; \lambda \in \mathbb{C}, \; y \in \mathbb{C}, \; A \in \mathbb{C}^{r\times r}. \tag{7}$$

Now, we look for the Hermite matrix polynomials series expansion of the matrix hyperbolic cosine $\cosh(Ay)$. Given an arbitrary matrix $A \in \mathbb{C}^{r\times r}$, with (1) and using (7) in combination with [Jódar and Company, 1996, p. 25], it follows that $H_n(-x, A) = (-1)^n H_n(x, A)$. Thus, one gets the looked for expression:

$$\cosh(Ay) = e^{\frac{1}{\lambda^2}} \sum_{n\geq 0} \frac{1}{\lambda^{2n}(2n)!} H_{2n}\left(y\lambda, \frac{1}{2}A^2\right). \tag{8}$$

Denoting by $CH_N(A, \lambda)$ the Nth partial sum of series (8) for $y = 1$, one gets

$$CH_N(\lambda, A) = e^{\frac{1}{\lambda^2}} \sum_{n=0}^{N} \frac{1}{\lambda^{2n}(2n)!} H_{2n}\left(\lambda, \frac{1}{2}A^2\right) \approx \cosh(A), \; \lambda \in \mathbb{C}, \; A \in \mathbb{C}^{r\times r}. \tag{9}$$

Similarly, one gets the looked for expression for :

$$\sinh(Ay) = e^{\frac{1}{\lambda^2}} \sum_{n\geq 0} \frac{1}{\lambda^{2n+1}(2n+1)!} H_{2n+1}\left(y\lambda, \frac{1}{2}A^2\right). \tag{10}$$

Denoting by $SH_N(A, \lambda)$ the Nth partial sum of series (10) for $y = 1$, one gets

$$SH_N(\lambda, A) = e^{\frac{1}{\lambda^2}} \sum_{n=0}^{N} \frac{1}{\lambda^{2n+1}(2n+1)!} H_{2n+1}\left(\lambda, \frac{1}{2}A^2\right) \approx \sinh(A), \; \lambda \in \mathbb{C}, \; A \in \mathbb{C}^{r\times r}. \tag{11}$$

3. Accurate and Error Bounds

From reference Defez et al. [2011] we have the bound $\left\|H_n\left(x, \frac{1}{2}A^2\right)\right\|_2 \leq n! e^{\left(|x|\|A\|_2+1\right)}$, and thus

$$\left\|H_{2n}\left(\lambda, \frac{1}{2}A^2\right)\right\|_2 \leq (2n)! e^{\left(|\lambda|\|A\|_2+1\right)}. \tag{12}$$

Taking the approximate value $CH_N(\lambda, A)$ given by (9) and taking into account (12) and $\lambda > 1$, it follows that

$$\begin{aligned} \|\cosh(A) - CH_N(\lambda, A)\|_2 &\leq e^{\frac{1}{\lambda^2}} \sum_{n \geq N+1} \frac{1}{\lambda^{2n}(2n)!} \left\|H_{2n}\left(\lambda, \frac{1}{2}A^2\right)\right\|_2 \\ &\leq e^{\left(\frac{1}{\lambda^2}+\lambda\|A\|_2+1\right)} \sum_{n \geq N+1} \frac{1}{\lambda^{2n}} \\ &= \frac{e^{\left(\frac{1}{\lambda^2}+\lambda\|A\|_2+1\right)}}{\left(\lambda^2-1\right)\lambda^{2N-1}}. \end{aligned} \tag{13}$$

Now, let $\varepsilon > 0$ be an *a priori* error bound. Using (13), if N is the first positive integer so that

$$N \geq \frac{\log\left(\frac{e^{\left(\frac{1}{\lambda^2}+\lambda\|A\|_2+1\right)}}{\varepsilon(\lambda^2-1)}\right)}{2\log(\lambda)} + \frac{1}{2} \tag{14}$$

from (13) and (13) one gets $\|\cosh(A) - CH_N(\lambda, A)\|_2 \leq \varepsilon$. By other hand, from reference Defez et al. [2011] we have the bound:

$$\left\|H_{2n+1}\left(\lambda, \frac{1}{2}A^2\right)\right\|_2 \leq (2n+1)! e^{\left(|\lambda|\|A\|_2+1\right)}. \tag{15}$$

Taking into account (10), $\lambda > 1$ and proceeding as above, we obtain the error bound for approximation (11):

$$\left\|\sinh(A) - SH_N(\lambda, A^2)\right\|_2 \leq \frac{e^{\left(\frac{1}{\lambda^2}+\lambda\|A\|_2+1\right)}}{\left(\lambda^2-1\right)\lambda^{2N}}. \tag{16}$$

Now, let $\varepsilon > 0$ be an *a priori* error bound. Using (16), if N is the first positive integer so that

$$N \geq \frac{\log\left(\frac{e^{\left(\frac{1}{\lambda^2}+\lambda\|A\|_2+1\right)}}{\varepsilon(\lambda^2-1)}\right)}{2\log(\lambda)}, \tag{17}$$

from (16) one gets $\|\sinh(A) - SH_N(\lambda, A)\|_2 \leq \varepsilon$.

3.0.1. EXAMPLE

Let A be a matrix defined by $A = \begin{pmatrix} 3 & -1 & 1 \\ 2 & 0 & 1 \\ 1 & -1 & 2 \end{pmatrix}$ with $\sigma(A) = \{1, 2\}$. Matrix A is non-diagonalizable. Using the minimal theorem [Dunford and Schwartz, 1957, p. 571], see also Defez and Jódar [1998], the exact value of $\cosh(A)$ is

$$\cosh(A) = \begin{pmatrix} 7.38905609893065 & -3.62686040784702 & 3.62686040784702 \\ 5.84597546411541 & -2.08377977303177 & 3.62686040784702 \\ 2.21911505626839 & -2.21911505626839 & 3.76219569108363 \end{pmatrix}.$$

It is easy to check that $\|A\| = 4.41302$. Taking $\lambda = 12$ and $\varepsilon = 10^{-5}$, one gets that we have to take $N = 13$ because by (14) one gets

$$\frac{\log\left(\frac{e^{\left(\frac{1}{\lambda^2}+\lambda\|A\|+1\right)}}{\varepsilon(\lambda^2-1)}\right)}{2\log(\lambda)} + \frac{1}{2} \approx 12.6762.$$

Thus, we have to take $N = 13$ to obtain:

$$CH_{13}(12, A) = \begin{pmatrix} 7.38905609893065 & -3.62686040784702 & 3.62686040784702 \\ 5.84597546411541 & -2.08377977303177 & 3.62686040784702 \\ 2.21911505626839 & -2.21911505626839 & 3.76219569108363 \end{pmatrix},$$

$$\|\cosh(A) - CH_{13}(12, A)\|_2 = 8.12612599689556 \times 10^{-21}.$$

The number of terms required to obtain a prefixed accuracy uses to be smaller than the one provided by (14). For instance, taking $N = 6$ one gets

$$CH_6(12, A) = \begin{pmatrix} 7.38905494469012 & -3.62685939161559 & 3.62685939161559 \\ 5.84597430987726 & -2.08377875680273 & 3.62685939161559 \\ 2.21911491826167 & -2.21911491826167 & 3.76219555307453 \end{pmatrix},$$

and

$$\|\cosh(A) - CH_6(12, A)\|_2 = 2.617702 \times 10^{-6}.$$

The choice of parameter λ can still be refined. For example, for the same $N = 6$, taking $\lambda = 4.95$ one gets

$$\|\cosh(A) - C_6(4.95, A)\|_2 = 3.66321 \times 10^{-7}.$$

This illustrates how the error norm depends on the varying parameter λ and it becomes evident that an adequate choice of λ may provide results with higher accuracy. By other hand

$$\sinh(A) = \begin{pmatrix} 7.38905609893065 & -3.76219569108363 & 3.76219569108363 \\ 6.21385490528685 & -2.58699449743983 & 3.76219569108363 \\ 2.45165921420322 & -2.45165921420322 & 3.62686040784702 \end{pmatrix},$$

and taking $\varepsilon = 10^{-5}$ and $N = 10$, we obtain:

$$SH_{10}(4, A) = \begin{pmatrix} 7.38905609893065 & -3.76219569108363 & 3.76219569108363 \\ 6.21385490528685 & -2.58699449743983 & 3.76219569108363 \\ 2.45165921420322 & -2.45165921420322 & 3.62686040784702 \end{pmatrix},$$

$$\|\sinh(A) - SH_{10}(4, A)\|_2 = 1. \times 10^{-17}.$$

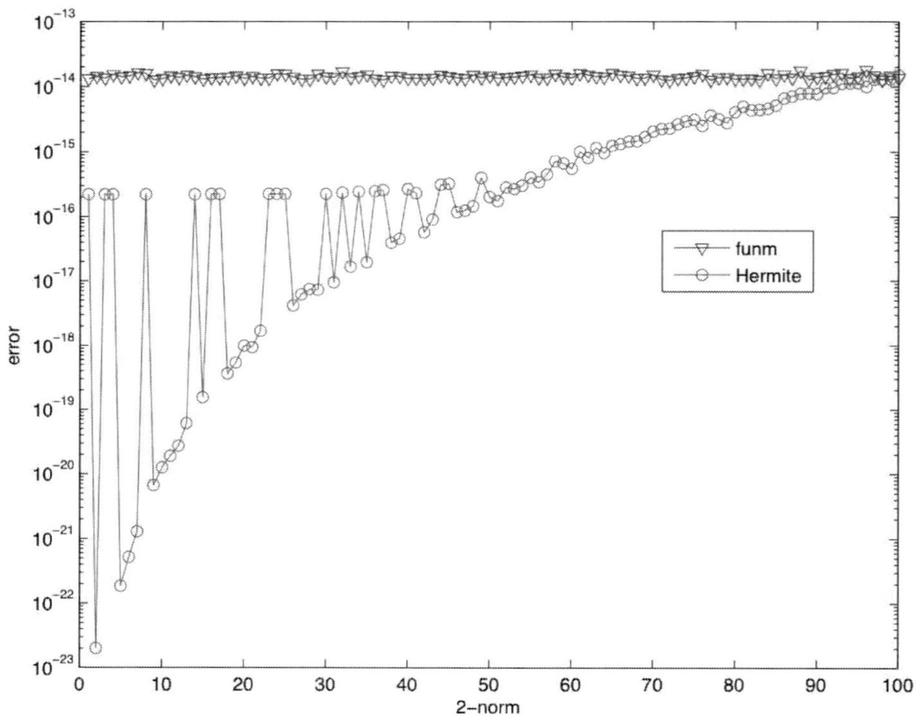

Figure 1. Error comparing MATLAB `funm` with Hermite hyperbolic matrix cosine approximation for $r = 512$, $\lambda = 2$.

4. Algorithm and Test

Starting with expressions (9) and (11), it is possible to compute simultaneously the hyperbolic matrix cosine and sine using the algorithm 1. We have determined for each value of N and $M \in \mathbb{N}$, $1 \leq M \leq 100$, the optimal value of λ, i. e. the minimal of

$$\frac{e^{\left(\frac{1}{\lambda^2}+\lambda M+1\right)}}{\left(\lambda^2-1\right)\lambda^{2N-1}}. \tag{18}$$

Since the minimal value of Expression (18) when $M \to \infty$ is obtained for $\lambda \to 1$, we have selected $\lambda = 1$, for $M \geq 100$. A MATLAB implementation of this algorithm has been compared with the built-in Matlab function `funm`. In tests, 100 diagonalizable matrices of dimension r equal to 512 were used. These matrices were generated as $A = QDQ$, where $Q = H/\sqrt{512}$, with H a Hadamard matrix of dimension 512. Diagonal matrices D were randomly generated, with 2-norm varying between 1 and 100. The hyperbolic cosine of A were computed as $\cosh(A) = Q\cosh(D)Q$, using 32 digits of precision. We used an Apple Macintosh iMac (mid 2011) with a quadcore i5-2400S 2.5 Ghz processor and 12Gb of RAM. All the tests were carried out using MATLAB R2012a and OS X 10.6.8. In 100 test (varying) the 2-norm, our Hermite algorithm has a better error behaviour in 98 times

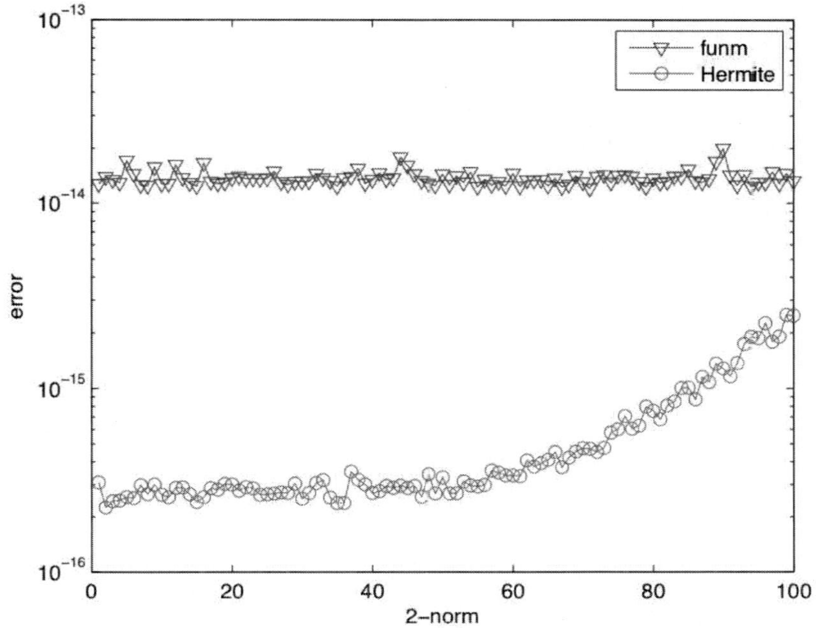

Figure 2. Error comparing MATLAB `funm` with Hermite hyperbolic matrix sine approximation for $r = 512$, $\lambda = 2$.

and it is worse in 2 times (see Figure 1). The total time media Te for all the 100 executions for our algorithm is $Te = 9.407$ seconds and for MATLAB `funm` is $Te = 11.371$. Similar results are obtained with the matrix hyperbolic sine, see Figure 2.

Conclusion

In this chapter a modification of the algorithm proposed in Defez and Jódar [1998] for computing matrix cosine and sine based on Hermite matrix polynomial expansion is presented. The numerical experiments show that the MATLAB implementation of the new algorithm has lower execution times and higher accuracy than the MATLAB function `funm`. Also, the new algorithm allows the simultaneous evaluation of the hyperbolic matrix sine and cosine. The algorithm depends on the parameter λ, whose impact on the numerical efficiency is currently studied. Furthermore, pending work focuses on the optimal scaling of the matrix and the study of the evaluation Paterson and Stockmeyer [1973] of the approximations (9) and (11). To do parallel implementation of the algorithms presented in this work in a distributed memory platform, using the message passing paradigm, MPI and BLACS for communications, and PBLAS and ScaLAPACK Blackford et al. [1997] for computations.

Algorithm 1 computes hyperbolic sine and cosine of a matrix.

Function $[C, S] = sinhcoshher(A, N)$
Inputs: Matrix $A \in \mathbb{R}^{r \times r}$; $2N + 1$ is the order of the Hermite approximation ($N \in \mathbb{N}$) of hyperbolic sine/cosine function; parameter $\lambda \in \mathbb{R}$
Output: Matrices $C = \cosh(A) \in \mathbb{R}^{r \times r}$ and $S = \sinh(A) \in \mathbb{R}^{r \times r}$

$M = \lfloor \|A\|_2 \rfloor$
Select the optimal value of λ depending on N and M
$H_0 = I_r$
$H_1 = \lambda A$
$C = H_0$
$S = H_1/\lambda$
$\alpha = 1/\lambda$
for $n = 2 : 2N + 1$ **do**
  $H = \lambda A H_1 - 2(n-1)H_0$
  $H_0 = H_1;$
  $H_1 = H$
  $\alpha = \alpha/(\lambda n)$
  if $\mod(n, 2) == 0$ **then**
    $C = C + \alpha H$
  else
    $S = S + \alpha H$
  end if
end for
$C = e^{1/\lambda^2} C$
$S = e^{1/\lambda^2} S$

References

L. S. Blackford, J. Choi, A. Cleary, E. D'Azevedo, J. Demmel, and I. Dhillon. *ScaLAPACK Users' Guide*. SIAM, 1997.

P. Das. *Optical Signal Processing*. Springer, New York, 1991.

E. Defez and L. Jódar. Some applications of Hermite matrix polynomials series expansions. *Journal of Computational and Applied Mathematics*, 99:105–117, 1998.

E. Defez, J. Sastre, Javier J. Ibáñez, and Pedro A. Ruiz. Computing matrix functions solving coupled differential models. *Mathematical and Computer Modelling*, 50(5-6):831–839, 2009.

E. Defez, Michael M. Tung, and Jorge Sastre. Improvement on the bound of hermite matrix polynomials. *Linear Algebra and its Applications*, 434:1910–1919, 2011.

E. Defez, J. Sastre, Javier J. Ibáñez, and Pedro A. Ruiz. Computing hyperbolic matrix functions using orthogonal matrix polynomials. In *The 17th European Conference on Mathematics for Industry 2012*, 2012. In Press.

E. Defez, J. Sastre, Javier J. Ibáñez, and Pedro A. Ruiz. Computing matrix functions arising in engineering models with orthogonal matrix polynomials. *Mathematical and Computer Modelling*, 57(7-8):1738–1743, 2013.

N. Dunford and J. Schwartz. *Linear Operators, Part I.* New York, 1957.

G. H. Golub and C. F. Van Loan. *Matrix computations.* The Johns Hopkins University Press, Baltimore, MD, USA, second edition, 1989.

L. Jódar and R. Company. Hermite matrix polynomials and second order matrix differential equations. *Journal Approximation Theory Application*, 12(2):20–30, 1996.

L. Jódar, E. Navarro, and J.A. Martín. Exact and analytic-numerical solutions of strongly coupled mixed diffusion problems. *Proceedings of the Edinburgh Mathematical Society*, 43:269–293, 2000.

L. Jódar, E. Navarro, A.E. Posso, and M.C. Casabán. Constructive solution of strongly coupled continuous hyperbolic mixed problems. *Applied Numerical Mathematics*, 47 (34):477–492, 2003.

A. King and C. Chou. Mathematical modeling simulation and experimental testing of biochemical systems crash response. *Journal of Biomechanics*, 9:301–317, 1976.

H. Morimoto. Stability in the wave equation coupled with heat flows. *Numerische Mathematik*, 4:136–145, 1962.

M. S. Paterson and L.J. Stockmeyer. On the number of nonscalar multiplications necessary to evaluate polynomials. *SIAM Journal on Computing*, 2(1):60–66, 1973.

D. Pozar. *Microwave Engineering*. Addison-Wesley, New York, 1991.

M. Sezgin. Magnetohydrodynamics flows in a rectangular duct. *International Journal for Numerical Methods in Fluids*, 7(7):697–718, 1987.

A. Winfree. *When Times Breaks Down*. Princeton University Press, Princeton, New Jersey, 1987.

In: Mathematical Modeling in Social Sciences ... ISBN: 978-1-63117-335-6
Editors: J. C. Cortés López et al.

Chapter 22

RSV Modeling Using Genetic Algorithms in a Distributed Computing Environment Based on Cloud File Sharing

***J. Gabriel García Caro*[1,*], *Javier Villanueva-Oller*[2,†] *and J. Ignacio Hidalgo*[1,‡]**
[1]Universidad Complutense de Madrid, Madrid, Spain,
[2]CES Felipe II, Universidad Complutense de Madrid, Madrid, Spain

Abstract

Usually, when dealing with random network models, we find that the search for the best parameters is a difficult computing task. This is because, the usual way of tackling this problem is through an exhaustive evaluation of all solutions. This finding leads to an optimization of this process to reduce the cost in time and resources needed. In this chapter it is presented an alternative which combines evolutionary algorithms, distributed computation and cloud storage which has allowed us to work with elements created independently (computation system, networks model and genetic generator) and for different platforms without any additional modification.

Keywords: Parallel Genetics Algorithms, Respiratory Syncytial Virus, Network Model

1. Introduction and Motivation

Network models have become paramount in the analysis of complex systems. These systems range from evolutionary biology [1] to neural networks [4] via social networks [2], transport or economic [3]. For example, these models can be used to study infectious diseases [6, 7, 8]. In recent times random networks have become popular for simulating patterns of disease dissemination in large networks [9, 10, 11]. Furthermore, these networks provide an alternative to traditional schemes based on differential equations whose origin

*E-mail address: jggarciacaro@gmail.com
†E-mail address: jvillanueva@pdi.ucm.es
‡E-mail address: hidalgo@dacya.ucm.es

may be found in the Kermack and McKendrick research [12, 13, 14]. It is true that differential equations are a well known and powerful mechanism [14, 15] which allows studying the dynamics of many systems. Their main drawback is that, when they are used in the environment of an epidemiological model, they have many limitations because they cannot distinguish among specific individuals. For this reason the introduction of new elements in the model (such as age, sex, previous illnesses, etc) is very complex. Alternatively, networks models are managed as a graph in which each node is an individual, with a number of specific attributes, such as age, sex, disease status (susceptible, infected, recovered in latency , etc.), and with a number of edges that connect them to other nodes representing the relationships (social or others) that define the structure of the network itself and how the disease spreads. There are different ways to implement this type of networks. The most traditional is the Erdos and Renyi [17], but other alternatives have emerged as the scale-free networks [18] or those of Watts and Strogatz [19].

Until now, the studies carried out using this type of networks has been restricted to a relatively small number of individuals, usually not more than 10000 [5], but in many cases (for example a pandemic), the number of involved subjects are millions. This has a high price in terms of computational cost, since, except for specific cases such as networking with potential probability distribution (allowing an adjustment similar to differential equations), the adjustment involves an exhaustive search, making it infeasible using traditional methods or forcing major limitations, such as reduced network size (in nodes and/or relationships) or restrictions in the exploration of the parameters.

Models of this type have been analysed and solved with good results [25, 26, 27], even using distributed computing such as Respiratory Syncytial Virus (RSV) [25, 26] or brain impulses [28]. Nevertheless the execution times are extremely high. Therefore, in this chapter we present a system capable of adjust these models with a much lower computational cost. This exhaustive search is replaced by a Genetic Algorithm (GA), which is combined with a parallel computing system. In addition, in order to maintaining a loose coupling among them, the two systems rely on a cloud file storage service like Dropbox.

This chapter is organized as follows: Section 2 details the principles of the neural random network model, Section 3 explains how to generate individuals through genetic logic for its later resolution by the model described above, Section 4 shows the experimental results and finally Section 5 details the conclusions reached after the calculations and the future lines.

2. RSV Model

RSV (Respiratory Syncytial Virus) is the main cause of respiratory diseases in infants and young children with annual epidemics of pneumonia and bronchiolitis, also it causes tracheobronchitis in older children and adults [20]. Its impact on health systems grows as the number of children hospitalized for bronchiolitis [21] (more than 15,000 visits / year to the primary care paediatrician in Spain), and has not been until recently that its effects are being studied in adults, being responsible for over 18% of hospitalizations for pneumonia in people higher than 65 years [22].

It is therefore of particular interest the simulation of this disease, and to study its own evolution, understanding the parameters that influence its spread, predict how it will spread in the population, and make decisions such as what type of vaccination strategy should be followed. In this simulation process is where models come into play, whose production (roughly) has associated the following steps:

1. To develop the model based on a set of parameters that characterize it (population, infection rate, mortality rate, duration of immunity, recovery time, etc.);
2. Adjust the model according to known results. If the model searches, for example, the number of fatalities over a period of time, seek the combination of parameters that makes the model outcome more closely resemble the mortality rates of years past;
3. Once found the values of the best fitted parameters, we use the model to predict some results in the future.

In our case, the model to be used has been proven in [25, 27]. We are going to use it in our approach in several steps of increasing complexity:

1. Particularised to adjust only the probability of transmission among the network nodes (infected and susceptible) linked by a relationship (i.e. sharing an edge). The average of the random network nodes is 48 and the number of nodes is 1000000. These specific values are chosen as a starting point because we know the best solution obtained by exhaustive search, and we want to check if we are able to converge to the same solution previously obtained and its cost;
2. Adapted to adjust the model using the longevity of the infection and RMS (explained in sections below) independently. This step, however, has a problem that needs to be solved. This will turn out to be quite complex because, as the results will show, some of the parameters are mutually dependant, something that will force us to try with the next step;
3. As said above the mutually dependent was a huge problem, then in this step is proposed a new way to fit the solutions rewarding the more long-lived, to do this, the solutions are adjusted using the relationship between longevity and the RMS.

The transitions among the states of the nodes follow a standard evolution SIRS (susceptible to infected to recovered to susceptible) (Figure 1).

3. GA and Its Implementation

To avoid the need for an exhaustive search, it is replaced this method by a genetic algorithm that can explore the solutions space and reduces the computation time. Specifically, for the model described in 2 has been developed an algorithm that, based on the parameters of network construction (explained above), is capable to find a solution at least as good as that obtained by exhaustive search with a much lower computational cost.

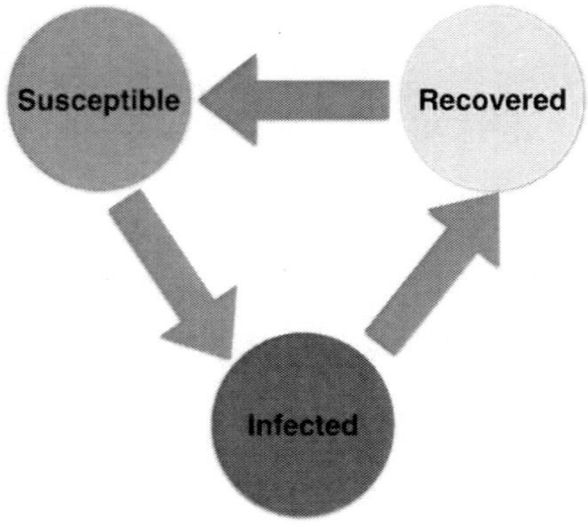

Figure 1. SIRS model.

3.1. Genetic Algorithm (GA)

The first step of the GA is the generation of a random initial population of N individuals. Do not confuse the individuals of this population (each individual is a complete network with its own parameters), with the network individuals (each node or neuron is an infected person or not). In this case each individual is a random neural network which parameters will provide an unique behavior to the RSV that will lead to a greater or lesser number of infected over time.

In our current researching process we have done three experiments that are explained below.

3.1.1. Model Adjustment Using One Parameter

The parameter to study is the constant component of the probability of transmission from an infected to a susceptible individual, called by the RSV model, `b0`. The initial value of `b0` is a random number ranging between 0 and 1 coded in binary which length is 12 bits.

The selection is done using binary tournament and crossover phase is based on uniform crossover [31].

The mutation operator negates the value of one bit chosen in randomly from each individual. In addition, an elitist scheme is used, this mean that the two best individuals in each generation are part of the next generation and are unaffected by the mutations.

When the RSV model calculates an individual (neural network) generates a file containing the number of infected people for each instant of time. These data are fitted with a root mean square (RMS) which is a set of values (or a continuous-time waveform) is the square root of the arithmetic mean (average) of the squares of the original values (or the square of the function that defines the continuous waveform).

In the case of a set of n values $\{x_1, x_2, ..., x_N\}$, the RMS value is given by this formula:

$$x_{RMS} = \sqrt{\frac{1}{N}\sum_{i=1}^{N} x_i^2}$$

This cycle is repeated until it reaches the termination condition, delimited by n iterations. The best individual has the lowest value in its RMS.

3.1.2. Model Adjustment Using Two Parameters

The parameters to study are the constant component of the probability of transmission from an infected to a susceptible individual ($b0$) as in previous subsection and the average connectivity degree of the random network (k). Doing so, the chromosome of each individual is composed by $b0$ and k coded in binary with a length of 20 bits.

The selection, crossover, mutation and evaluation are unmodified respect to the previous experiment.

3.1.3. Model Adjustment Using Two Parameters with Improved Fitness

The parameters to study, mutation, crossover and selection are the same than the section above.

The main difference with the other experiments is the fitness evaluation. This evaluation is quite more complex than the previously used because it establishes a relationship between the longevity, and the RMS of the solution. The longevity is a value incremented each day of the simulation if the number of the infected people is greater than 0. This value has been introduced in the fitness function because, at the beginning of the algorithm there are solutions whose RSV disappears quickly and it is impossible to fit them via RMS, thus, it is a way to identify the solutions at the beginning of the computing.

Then the improved fitness is given by the following formula:

$$Fitness = \frac{Longevity}{x_{RMS}}$$

3.2. Parallelization

To solve this RSV model by using exhaustive search, 20 years of CPU time on a personal computer would be needed [25]. This CPU time is very high due to the computational cost needed to calculate and simulate each neuronal network (about 90 minutes in this case).

Using a genetic algorithm as a "search and optimization" method, greatly improves the required computation time (originally it was necessary to solve hundreds of thousands of networks and now we have to solve only a few thousand), but still a single computer would need 525 days to process the models. While it is a huge improvement with respect to the original 20 years time, it is still impractical. To further improve this, we have chosen to calculate the networks in a parallel and distributed way. For this purpose, we will use the *Sísifo*[1] system, a distributed computing manager. We have already used it in the past, so have a good background acknowledge and can use that as starting point.

In this way, we move from a GA based architecture to a GA computationally distributed environment where all individuals are evaluated in parallel, letting us classify the combination of Sísifo plus GA as a "synchronous master-slave parallel genetic algorithm" (PGA) [23].

[1]For more detail see http://sisifo.imm.upv.es/

Sísifo is a client-server based system designed to allow a problem to be solved using distributed computations. Working in a conceptual way much like BOINC[2], Sísifo is able to assign tasks to a set of PCs, wait for the tasks to complete and collect the results for further analysis. However, and in contrast to BOINC (which requires a team of specialists working for weeks just to install and configure it), Sísifo is made with simplicity as main aim, giving as a result a system that requires almost no maintenance, needs very little configuration time, and can be deployed in just a couple of hours. Sísifo offers a limited access a limited access control and security, making it suitable only in a controlled environment such as an intranet.

Sísifo has the following characteristics:

- 10 PC XEON X3230 to 2.66 GHz with 6.6 GB of RAM
- 5 PC XEON X3430 a 2.4 GHz with 16 GB of RAM

3.3. Adaptation of the Genetic Generator to the Distributed Environment

In our first try, the **genetic generator** was included on the **Sísifo server**, and so was also the responsible for distributing the individuals to its clients for its computation. The problem of this architecture and work method is that, each time the genetic generator needs to be updated, it is necessary to include it again in the Sísifo server. This is something not quite simple, however, because Sísifo runs in **Linux O.S.** and the genetic generator is implemented on **Windows O.S.**.

Nowadays, it is of common use the so called *cloud*, whose basic principle is to allow sharing files between heterogeneous computers over the Internet. This feature fits perfectly to our needs, because in the calculation process we have to process individuals and solutions in the form of files, shared between Sísifo and the genetic generator. Therefore, we choose to use one of these services in the cloud, namely *Dropbox*[3].

Dropbox is a file hosting service operated by the company with the same name, that offers cloud storage, file synchronization, and client software. Dropbox allows users to create a special folder on each of their computers, which Dropbox then synchronizes so that it appears to be the same folder (with the same contents) regardless of which computer is used to view it. Files placed in this folder also are accessible through a website and mobile phone applications.

Linking the genetic generator and Sísifo via Dropbox, every element of the computation system works together and synchronized with the others despite of being in different computers with different OS in physically separated *networks*. As each component (i.e. software module) of the system communicates with the others via data files, the coupling is very low.

While this is something not new, and GAs have been used already tools like Dropbox to share files between computers for distributed computation [29, 30], the main difference with respect to our proposal is the architecture presented in Figure 2 composed by:

[2]For more info visit http://boinc.berkeley.edu/trac/wiki/BoincIntro

[3]For more detail visit www.dropbox.com

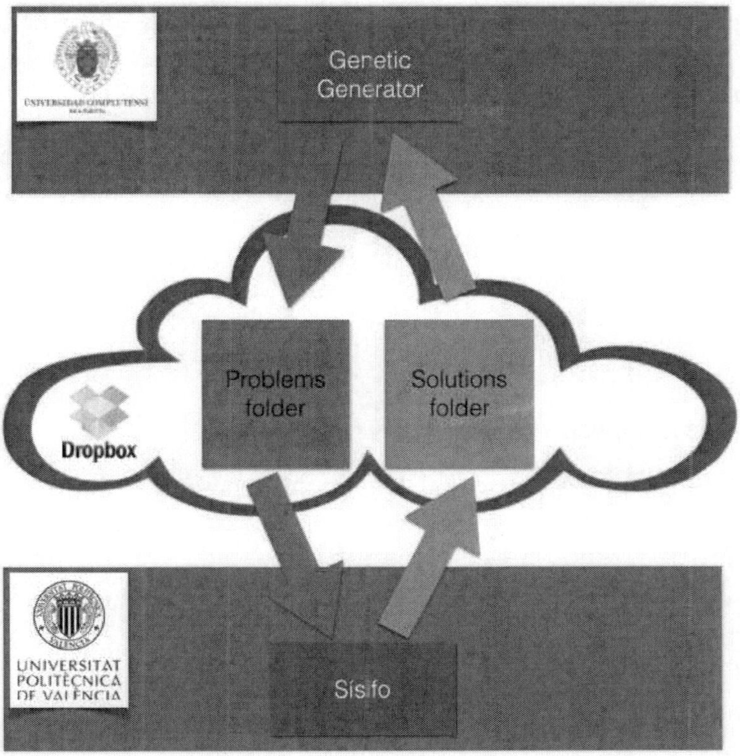

Figure 2. Distributed computing environment.

- **Genetic Generator:** Detailed in the section 3 It is based in the Universidad Complutense de Madrid and is in charge of the individuals generation on a file with a specific format that will be solved later in collaboration with Sísifo. Furthermore it operates according to genetic logic;
- **Sísifo:** Calculation server and clients based in the Universidad Politécnica de Valencia, it is in charge of collecting the individuals generated by the genetic generator to solve them;
- **Dropbox:** Is the medium whereby Sísifo and the GA share individuals and solution files in a synchronized way.

3.4. System Operation

This system (Figure 2) works as follows:

- In each iteration the **genetic generator** creates in a Dropbox share folder a number of N individuals and then begins the folder monitoring until all individuals have been solved and its solutions generated and shared in a different folder;
- **Sísifo** detects that there are new individuals to be computed, then it distributes the files among its clients;

- Once the individual has been processed, the client that has solved it, returns the simulation result to the **Sísifo server**, which validates its data integrity via a checksum and copies the solution to a specific shared folder in **Dropbox**;
- When **the genetic generator** detects that all the solutions of the current generation are available, it calculates via the genetic logic previously described in the section 3 for each solution the fitness and repeats all the process until the end condition is reached.

4. Experimental Results

4.1. Model Adjustment Using One Parameter

In this first experiment, it has been defined the parameters show in the Table 1, empirically adjusted.

Table 1. PGA Parameters

Number of individuals	100
Number of iterations	100
Mutation probability	2%
Crossover probability	80%

Figure 3 shows the best solution obtained using the PGA vs the best solution found via exhaustive search. The data in red color are the actual data of hospitalized people per week in the Valencian Community. Data in blue are the results calculated by the RSV according to the data provided by the PGA.

The network assessment is done during a simulated time of 7 years. With these features, each network takes about 90 minutes in the evaluation and the verifying the quality of the solution.

The evolution that we have calculated is composed of 60 generations for a total CPU time of 344 days. In practice, given that we have used Sísifo with 15 computers with four cores each one working in parallel, that means, we have the best solution in less than six days. Moreover, the best solution is already obtained at iteration number 19, around 2 days.

The solution we have obtained is at least as good as that found by means far more costly in [25], although we have limited experience in this initial search to a single parameter (b0). For this reason the adjust is not very good around the weeks number 100 and 150.

4.2. Model Adjustment Using Two Parameters

The paramters utilised in this operations are in the table 1.

Once we have proved the validity of our computation system for a single PGA pattern, we try to evolve the GA into a more complex one. To do that, we try to search not only for one but for two parameters in the solution space. Specifically, we want to study the

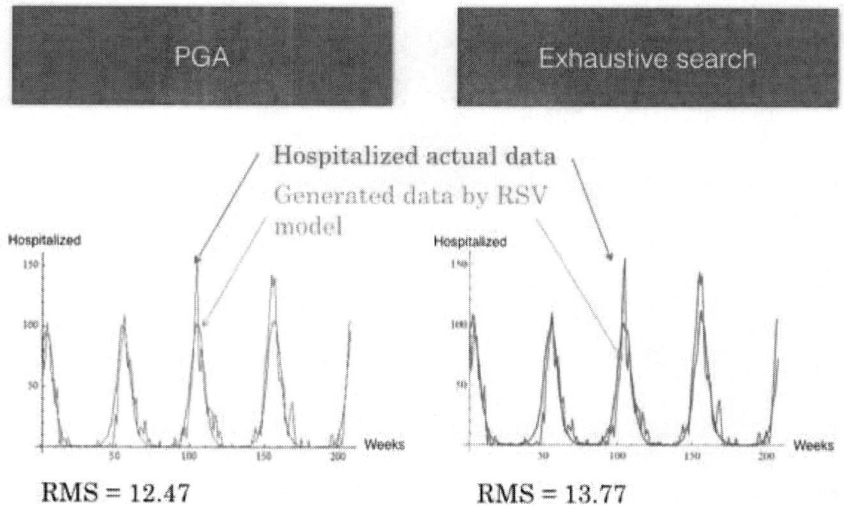

Figure 3. Comparation among the best solutions.

behaviour of the RSV model with respect to the parameter $b0$ (probability of transmission from an infected to a susceptible individual) and the average degree of the network (parameter k).

The evolution calculated is composed of 60 generations for a total CPU time of 223 days (with Sísifo less than 4 days). This CPU time is less than the previous experiment due to that the solutions have no activity (Figure 4), then the RVS model finishes each invalid execution.

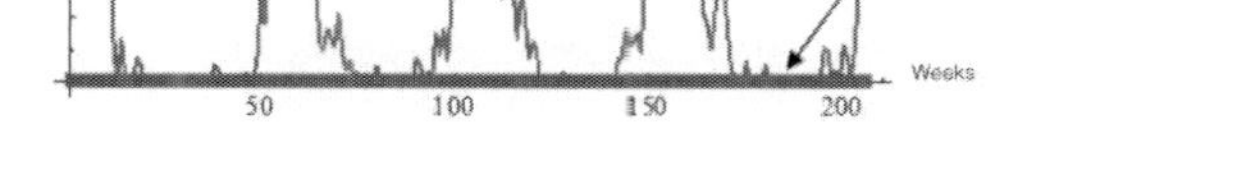

Figure 4. Solution without activity.

4.3. Model Adjustment Using Two Parameters with Improved Fitness

This experiment is needed because that was said in the section 2, the two parameters to fit are dependant among them, this mean that, it is not easy to find the suitable combination among them.With this approach, the best solution has a parameter k equal to 49. Note that on the first experiment the parameter k was a fixed value equal to 48 (found, as said before, using exhaustive search). We can see that both values are very similar. Furthermore, we found that the parameter $b0$ was slowly converging to the value found during the exhaustive search (0.000368) but stopped improvement in generation 68 with a value of 0.002686. From this point, the solution (Figure 5) ceased to improve because it fallen into a premature convergence, a typical problem when it deals with GAs.

The simulation was made of 68 generations with the parameters that were described in the Table 1 for a total CPU time of 435 days (with Sísifo around 7 days).

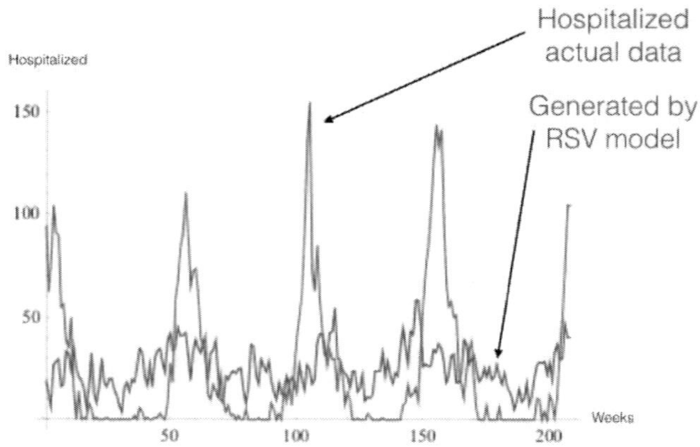

Figure 5. Solution with activity, RMS = 33.62.

5. Conclusion and Future Lines

5.1. Conclusion

According to our own experiences we can draw the following conclusions:

- GAs are able to find similar solutions to those found by exhaustive search at a much lower cost;
- For some very complex problems, including using GAs are necessary their combination with distributed computing in order to have acceptable time solution, this leads to PGA;
- The incorporate a cloud service like Dropbox, facilitates the monitoring and the integration of both systems, because everything is based on sharing folders and files;
- The RSV model, it difficult to fit thus the parameter that are being fitted are dependant among them.

5.2. Future Lines

Since the PGA has worked properly in finding a single parameter and that the results obtained until now in the searching of two parameters are converging to a similar solution obtained in the first experiment we want to adapt the PGA to an evolutionary strategy which aim is the comparing the performance of between algorithm, verifying if the RSV can be solved using other kind of algorithm and analyse the cost of that. In addition to this we want to study the RSV behaviour using multi-objective optimization through an NSGA-II. Finally when we have already finished all of our experiments, we will have a complete study of what algorithm is better to solve a RSV model.

Once it is checked the RSV model, we want to adapt the search using GAs to other virus model as meningococcus and HPV (Human papillomavirus).

Acknowledgments

J. Gabriel García Caro is supported by Spanish Government Iyelmo INNPACTO-IPT-2011-1198-430000 project. The work has also been supported by Spanish Government grants TIN 2008-00508 and MEC CONSOLIDER CSD00C-07-20811.

References

[1] S. R. Proulx, D. E. L. Promislow, P. C. Phillips, *Network thinking in ecology and evolution, TRENDS in Ecology and Evolution*, Vol. 20 N. 6(2005) 345-353.

[2] A. L. Traud, P. J. Mucha, M. A. Porter, Social structure of Facebook networks, *Physica A* 391 (2012) 4165-4180.

[3] M. J. van der Leij,The Economics of Networks: Theory and Empirics, Thesis dissertation, *Tinbergen Institute Research Series*, ISBN:905170973 0, Available Online at http://repub.eur.nl/res/pub/8212/

[4] Y. Bar-Yam, *Dynamics of Complex Systems*, Addison-Wesley, Reading, Massachusetts, 1997.

[5] N. A. Christakis, J. H. Fowler, The collective dynamics of smoking in a large social network, *New England Journal of Medicine* 358 (21) (2008) 2249-2258. doi:10.1056/NEJMsa0706154.

[6] E. Ahmed, H. N. Agiza, On modeling epidemics including latency, incubation and variable susceptibility, *Physica A: Statistical and Theoretical Physics* 253 (1-4) (1998) 347-352. doi:10.1016/S0378-4371(97)00665-1.

[7] R. M. Z. dos Santos, Immune responses: Getting close to experimental results with cellular automata models, in: D. Stauer (Ed.), *Annual Reviews of Computational Physics* VI, 1999.

[8] U. Hershberg, Y. Louzoun, H. Atlan, S. Solomon, HIV time hierarchy: winning the war while, losing all the battles, *Physica A: Statistical Mechanics and its Applications* 289 (1-2) (2001) 178-190. doi:10.1016/S0378-4371(00)00466-0.

[9] G. Witten, G. Poulter, Simulations of infectious diseases on networks, *Computers in Biology and Medicine* 37 (2) (2007) 195-205.

doi:10.1016/j.compbiomed.2005.12.002.

[10] L. Acedo, J.-A. Moraño, J. Díez-Domingo, Cost analysis of a vaccination strategy for respiratory syncytial virus (RSV) in a network model, *Mathematical and Computer Modelling* 52 (7-8) (2010) 1016-1022. doi:10.1016/j.mcm.2010.02.041.

[11] C. L. Barrett, K. R. Bisset, S. G. Eubank, X. Feng, M. V. Marathe, Episimdemics: ancient algorithm for simulating the spread of infectious disease over large realistic social networks, in: *Proceedings of the 2008 ACM/IEEE conference on Supercomputing*, SC'08, IEEE Press, Piscataway, NJ, USA, 2008, pp. 37:1-37:12. URL http://portal.acm.org/citation.cfm?id=1413370.1413408

[12] W. O. Kermack, A. G. McKendrick, Contributions to the mathematical theory of epidemics - Part I, *Proc. Roy. Soc.* 115 (1927) 33-55.

[13] L. Edelstein-Keshet, *Mathematical models in Biology*, SIAM, 2005.

[14] J. D. Murray, *Mathematical Biology: I. An Introduction*, Springer-Verlag, Berlin, 2002.

[15] H. W. Hethcote, The mathematics of infectious diseases, *SIAM Rev.* 42(2000) 599-653.

[16] A.Weber, M.Weber, P.Milligan, Modeling epidemics caused by respiratory syncytial virus (RSV), *Mathematical Biosciences* 172 (2) (2001) 95-113. doi:10.1016/S0025-5564(01)00066-9.

[17] B. Bollob as, *Random graphs*, Cambridge University Press, 2nd edition, 2001.

[18] A.-L. Barab asi, R. Albert, Emergence of scaling in random networks, *Science* 286 (5439) (1999) 509-512. doi:10.1126/science.286.5439.509.

[19] D. J. Watts, Small worlds: *The dynamics of networks between order and randomness*, Princeton University Press, 2003.

[20] Glezen, W.P., Taber, L.H., Frank, A.L., Kasel, J.A., Risk of primary infection and reinfection with respiratory syncytial virus, *Am. Jour. Dis. Ch.*, 140:441-456, 1986.

[21] Langley, J.M., Leblanc, J.C., Smith, B., Wang, E.E.L., Increasing incidence of hospitalization for bronchiolitis among canadian children 1980-2000, *J. Inf. Dis.*, 118:1764-1767, 2003.

[22] Han, L., Alexander, J., Anderson, L., Respiratory syncitial virus, pneumonia among the elderly: An assessment of disease burden, *J. Inf. Dis.*, 179:25-30, 2003.

[23] A. la Torre de la Fuente,*Algoritmos genticos paralelos*, 2005.

[24] M. Mitchell, *An introduction to genetic algorithms*. ISBN 0-262-13316-4(HB),0-262-63185-7(PB), 1996.

[25] L. Acedo, J.A. Moraño, R.J. Villanueva, J. Villanueva-Oller, Using random networks to study the dynamics of respiratory syncitial virus (RSV) in the Spanish region of Valencia, *Mathematical and Computer Modelling number* 54, ISSN 0895-7177, pp. 1650-1654 (year 2011) doi:10.1016/j.mcm.2010.11.068.

[26] I.C. Lombana, M. Rubio, E. Sánchez, F.J. Santonja, J. Villanueva-Oller, A network model for the short-term prediction of the evolution of cocaine consumption in Spain in the next few years, *Mathematical and Computer Modelling number* 52, ISSN 0895-7177, pp. 1023-1029 (year 2010) doi:10.1016/j.mcm.2010.02.032.

[27] J. Díez-Domingo, J. Villanueva-Oller, L. Acedo, R.J. Villanueva, J.A. Moraño, Seasonal Respiratory Syncytial Virus Epidemic in a Random Social Network, *Modelling for addictive behaviour, medicine and engineering* 2010, ISBN 978-84-693-9537-0, pp. 1-4 (Year 2010). Ed. Instituto Universitario de Matemática Multidisciplinar.

[28] L. Acedo, J.A. Moraño, R.J. Villanueva, J. Villanueva-Oller, The Neurona@Home project: Simulating a large-scale cellular automata brain, *12th Granada Seminar on Computational and Statistical Physics Universidad de Granada*, Instituto Carlos I, Granada, Spain, 17-21/09/2012 doi: 10.1063/1.4776528.

[29] Maribel García-Arenas, Juan Julián Merelo Guervós, Pedro Castillo,Juan Luis Jiménez Laredo, Gustavo Romero, y Antonio M. Mora, Using free cloud storage services for distributed evolutionary algorithms, *Proceedings of the 13th annual conference on Genetic and evolutionary computation (GECCO '11)*, Natalio Krasnogor (Ed.). ACM, New York, NY, USA, 1603-1610, 2011.

[30] K. Meri, M. G. Arenas, A. M. Mora, J. J. Merelo, P. A. Castillo, P. García-Sánchez, J. L. J. Laredo, Cloud-based evolutionary algorithms:An algorithmic study. *Natural Computing*, Volume 12, Issue 2, pp 135-147, 2013.

[31] J. Arranz de la Peña, A. Parra Truyol, *Algoritmos Genticos*, Universidad Carlos III, 2006.

In: Mathematical Modeling in Social Sciences ... ISBN: 978-1-63117-335-6
Editors: J. C. Cortés López et al.

Chapter 23

MULTI-AGENT AND CLUSTERING IN DATA ANALYSIS OF GPR IMAGES

D. Ayala-Cabrera*, E. P. Carreño-Alvarado, S. J. Ocaña-Levario, J. Izquierdo and R. Pérez-García
Instituto Universitario de Matemática Multidisciplinar, I.M.M. Fluing,
Universitat Politècnica de València, Valencia, Spain

Abstract

A combination of the multi-agent paradigm and a very well known clustering technique is used for unsupervised classification of subsoil characteristics working on a collection of ground penetrating radar (GPR) survey files. The main objective is to assess the feasibility of extracting features and patterns from radargrams. By optimizing both the field work and the interpretation of the raw images, our target is to obtain visualizations that are automatic, fast, and reliable so to suitably assess the characteristics of the prospected areas and extract relevant information. The system also helps characterize subsoil properties in a very natural and fast way, favors GPR files interpretation by non-highly qualified personnel, and does not require any assumptions about subsoil parameters.

Keywords: Ground penetrating radar, signal processing, image processing, multi-agent systems, pipe visualization, management of water supply systems

1. Introduction

Ground penetrating radar (GPR) has been extensively used as a nondestructive methodology to analyze components and anomalies in water supply systems (WSS). The components most frequently analyzed are pipes and, especially, metallic pipes. Only a few incipient attempts have been conducted regarding leaks. Information about components, undergone changes, and anomalies is completely necessary for productive control and management of WSS. This information is crucial to achieve the goals of WSS technical management.

*E-mail address: daaycab@upv.es

Recent studies, underline the use of non-destructive tools, as methodologies favoring technical management of WSS instead of other destructive testing. However, even though information retrieval by non-destructive methods is worthwhile, the huge volume of generated information and the interpretation of data usually require high levels of skill and experience.

Many GPR-based works have been developed in this regard trying to locate and detect components and anomalies in WSS. The success of the application of these methodologies hinges mainly on the cleanliness of the images obtained with some classification pre-processing. In most cases, the objective is the identification of the typical hyperbolae identifying the objects of interest in the image under study.

This work aims at generating a tool for analysis and simplification of GPR databases that could help decision-making in WSS management. The specific objective is to obtain a reduced number of clusters capturing the more relevant subsoil characteristics. The main idea behind the process boils down to gathering into clusters objects or anomalies within the inspected area, in a natural and fast way. As a result, the searching spectrum is enlarged and interpretation may be achieved in a fast way without requiring a high level of skill and experience.

The following section presents the proposed system. It also introduces a recently presented methodology to transform the wave signal space into a suitable framework for applying other pipe location processes on GPR images. Section 3 presents an experimental layout made out of a number of arrangements where the system has been tested. Also, a number of sensitivity analyses on various candidate metrics and linkage procedures has helped in the process of fine-tuning the proposed system architecture by selecting the most suitable combination of metrics and linkage procedures. A conclusions section closes the paper.

2. Proposed System Architecture

The architecture of the proposed system (Fig. 1) may be split into three interrelated processes: I) pre-processing, II) hierarchical agglomerative clustering, and III) information retrieval and visualization. The first process uses a methodology denoted *agent racing* [1]. It is a multi-agent process to develop a pre-analysis of the signals in GPR survey files. This technique builds two spaces, named warming-up and racing, using the agents' behavior into the world where the agents evolve. The output of this process is used as input for the clustering process. This process embodies an unsupervised technique to cluster the survey data in a natural way that, at the same time, allows easy and reliable interpretation. The methodology used in this paper is the so-called *Hierarchical Agglomerative Clustering* (HAC). To implement HAC we have evaluated some of the most common procedures and have chosen the one that best represents the various clusters showing the soil variability. Finally, the last process retrieves the data constituting the developed clusters and places everything back into the original space. The visualization herein obtained is the final objective since the sought results are easily obtained from the new images.

This architecture is run repeatedly by varying the metrics and the linkage methods used in process II). In this manuscript, process III) is applied to all the cases described in Section 3 The obtained images through these runs are crucial to determine the most suitable procedures to use in process II). These processes and a discussion about the performed selection

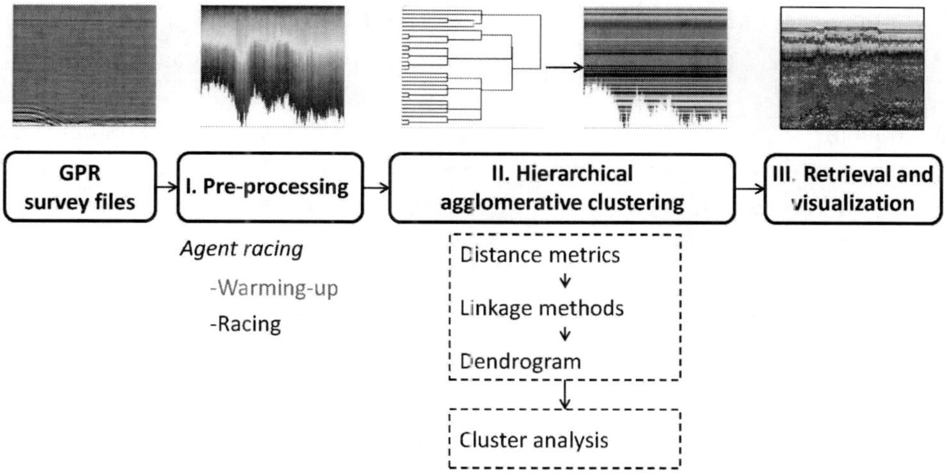

Figure 1. Architecture of the proposed system.

are presented in detail below.

2.1. Introduction to the Pre-Processing

The *agent racing* algorithm, based on Game Theory is used in process I). The agent racing will provide an interpretation and a grouping method for data from GPR radargrams. In this pre-process we try to reduce the number of data integrating the initial radargram, while preserving its initial properties and all the most relevant data, so that its ability to identify buried objects through suitable visualizations is preserved. The multi-agent approach enables significant reduction of the time needed for the analysis.

The input of the agent racing algorithm is the raw material obtained with the survey GPR radargram. The signals received in GPR prospects are stored in a matrix, A (radargram), that is made up of m-vectors, b_k, $k = 1, \ldots, n$, (traces), that represent the variation of the soil's electromagnetic properties in terms of depth. Let us represent this matrix by columns $A = [b_1, b_2, \ldots, b_{n-1}, b_n]$. The length, m, of vectors b_k, corresponds to the volume of signal data recorded in each trace.

The race is an endurance test for the agents, with a prize consisting in advancing one position (movement) depending on the effort made. Efforts are based on wave amplitude values in each column of A. The movement of agents during the racing will be conditioned by the changing trends of the traces they travel through. The racing will end once time t has elapsed. The winner(s) are the agent(s) that, according to the race conditions, have performed a bigger amount of movements. The output of this competition among the agents is a matrix, R, with columns $m1$-vectors X_s, $s = 1, \ldots, n$, n being the number of participating agents. X_s represents the race time variation, which collects the various movements performed by agent s. The lenght $m1$ is the number of movements performed by the winner(s). As a non-winner agent, s, have managed to fill a lower number of coordinates in X_s, zeros are used to complete its missing coordinates till the value $m1$ accomplished by the winner(s). The output produced in this racing phase is used as input for the hierarchical agglomerative clustering process (see Fig. 1 - process I).

The proposed game is a payoff function specified for each player. So, the game is a function $\pi : \prod_{s \in P} \Sigma^s \rightarrow \mathbb{R}^n$, [2], where P is the set of agents (players). It is a finite set, which we label $\{1, 2, \ldots, n\}$. Each agent s in P has a finite number of strategies making up a strategy profile set, Σ. The n traces generated by the GPR survey (columns of matrix A) are used as pseudo-parallel tracks for the n agents to compete.

During the race, each agent s in P builds its vector of strategies, whose i-th coordinate is the strategy taken by the agent at time i. A strategy for a player is a function that maps sequences of states to a natural number, corresponding to a move available to the player at the end of the sequence [3]. Here, the vector of strategies for an agent is determined by its respective column (vector b) in matrix A.

The agents' competition evolves in time from $i = 1$ till $i = m$. In the competition each agent s in P has four properties: a) interpretation, b) decision to move, c) movement time, and d) the race phases. The four properties of agents are explained next.

1. *Interpretation.* For each time during the race, an agent takes one value of the trace (b_i); then, this value is compared with two more signal values, the before-value b_{i-1} and the next-value b_{i+1}; as a result, a binary value ($bin \in \{0, 1\}$) is generated and stored for each agent at every time step.

2. *Decision to move.* An agent's decision to move is based on the binary value variation. According to this variation, a property called stamina varies positively (variable $StaIni$) or negatively (variable $StaEnd$). When the total stamina is zero, that is to say $StaIni$ equals $StaEnd$, the agent receives its payoff for the effort performed. This is accomplished by the variable $AgeMov$. As explained in *the race phases property*, this is applied during the 'official' race, just after the *warming-up*.

3. *Movement time.* Each effort developed by an agent happens between a start time and end time. These values, associated to the agent movement ($AgeMov$) are stored in two agent personal vectors, namely, $StaTiIni$ and $StaTiEnd$, respectively. Also, every agent movement ($AgeMov$), has one movement time associated $MovTi$ that we define as the average time between the stamina's time start ($StaTiIni$), and the stamina's time end ($StaTiEnd$). A component of $MovTi$ is defined every time the difference between these stamina values is 0.

4. *The race phases.* The race comprises two phases: a) *warming-up*, and b) *racing*. The phases are characterized by two times: a warming-up time (t_w), and a racing time (t_r), totaling a time $t = t_w + t_r$, where the t_w time corresponds to the time for the agent to overcome the end wave amplitude value ($AmplEnd$) in some percent of the average wave amplitude value for before values for the current time ($AmplProm$).

2.2. Hierachical Agglomerative Clustering

The hierarchical clustering process can be visualized in a dendrogram form, where each step in the clustering process is illustrated by a join in the tree (see Fig. 1 - process II). This process is divided into two parts: 1) dendrogram construction, and 2) cluster analysis. In the first phase, a dendrogram is built according to two aspects, which we want to evaluate: a) distance metrics, and b) linkage methods. These issues are related through the *cophenetic* correlation, which gives the goodness of the classification by comparing distances among input data and distances among output data. The second phase cuts the dendrogram according to a criterion previously determined. The criterion used in this document pursues a cut such the data are divided in a natural way. To this purpose, the so-called *inconsistency* coefficient is used. We explain now the elements involved in these two phases.

Distance Metrics. We use three distance metrics in our system, which are common in agglomerative hierarchical clustering. Given an $m1 \times n$ data matrix X, which is treated as $m1$ $(1 \times n)$ row vectors $x_1, x_2, \ldots, x_{m1}$, the different distances between vectors x_s and x_t are defined as follows:

1. *Euclidean distance:*

$$D^2_{Eu}(x_s, x_t) = (x_s - x_t)(x_s - x_t)'. \tag{1}$$

2. *Seuclidean distance.* Each coordinate difference between rows in X is scaled by dividing by the corresponding element of the standard deviation:

$$D^2_{Seu}(x_s, x_t) = (x_s - x_t) V^{-1} (x_s - x_t)', \tag{2}$$

where V is the $n \times n$ diagonal matrix whose j^{th} diagonal element is $S(j)^2$, where S is the vector of standard deviations.

3. *Cosine distance.* One minus the cosine of the included angle between points (treated as vectors):

$$D_{Co}(x_s, x_t) = 1 - \frac{x_s x_t'}{\sqrt{(x_s x_s')(x_t x_t')}}. \tag{3}$$

Linkage Methods. Once the proximity between objects in the data set has been computed, we can determine how objects in the data set should be grouped into clusters, using the linkage methods. The linkage methods take the distance information generated by metric measures and link pairs of objects that are close together into binary clusters (clusters made up of two objects). The linkage methods then link these newly formed clusters to each other and to other objects to create bigger clusters until all the objects in the original data set are linked together in a hierarchical tree. There are many possible choices in updating the similarity values. Among them, the most common linkage methods are 1) single, 2) average, and (3) complete. The following notation is used to describe the linkages used by the methods: r and s are two clusters; n_r and n_s are the number of object in clusters r and s, respectively; x_{ri} is the ith object in cluster r; x_{sj} is the jth object in cluster s. The linkage methods we use here are the following:

1. *Single Linkage:* also called *nearest neighbor clustering*, it is based on the minimum distance between clusters.

$$Z(r, s) = min\left(dist\left(x_{ri}, x_{sj}\right)\right), \tag{4}$$

where $i \in (1, \ldots, n_r), j \in (1, \ldots, n_s)$.

2. *Average Linkage:* also called *unweighted average distance* (UPGMA); it is calculated as the average distance between members of a pair of clusters. Average linkage tends to join clusters with small variances, and it is slightly biased towards producing clusters with the same variance.

$$Z(r,s) = \frac{1}{n_r n_s} \sum_{i=1}^{n_r} \sum_{j=1}^{n_s} dist(x_{ri}, x_{sj}). \tag{5}$$

3. *Complete Linkage*: also called *furthest neighbor clustering*; it distance is based on the points in all the clusters that are farthest apart.

$$Z(r,s) = max\left(dist\left(x_{ri}, x_{sj}\right)\right), \tag{6}$$

where $i \in (1, \ldots, n_r), j \in (1, \ldots, n_s)$.

Cophenetic Correlation. After linking the objects in a data set into a hierarchical cluster tree, we might want to verify that the distances (that is, heights) in the tree reflect the original distances accurately. The cophenetic correlation for a cluster tree is defined as the linear correlation coefficient between the cophenetic distances obtained from the tree, and the original distances (or dissimilarities) used to construct the tree. Thus, it is a measure of how faithfully the tree represents the dissimilarities among observations [4]. The output value, c, is the cophenetic correlation coefficient. The magnitude of this value should be very close to 1 for a high-quality solution. This measure can be used to compare alternative cluster solutions obtained using different algorithms. The cophenetic correlation is defined by

$$c = \frac{\sum_{i<j}\left(Y_{ij} - y\right)\left(Z_{ij} - z\right)}{\sqrt{\sum_{i<j}\left(Y_{ij} - y\right)^2 \sum_{i<j}\left(Z_{ij} - z\right)^2}}, \tag{7}$$

where

Y_{ij} is the distance between objects i and j given by Equations (1), (2) or (3); Z_{ij} is the cophenetic distance between objects i and j given by Equations (4), (5) or (6); y and z are the average of Y_{ij} and Z_{ij}, respectively.

Cluster Analysis. The hierarchical cluster tree may naturally divide the data into distinct, well-separated clusters. This can be particularly evident in a dendrogram diagram created from data where groups of objects are densely packed in certain areas and not in others. The inconsistency coefficient of the links in the cluster tree can identify these divisions where the similarities between objects change abruptly. We can use this value to determine where the cluster function creates cluster boundaries. In this way we define a label for each link in our dendrogram. This label shows to what extent two clusters are similar. With this measure, we can join clusters if the inconsistency value is less than certain specific threshold. The inconsistency coefficient characterizes each link in a cluster tree by comparing its length with the average length of other links at the same level of the dendrogram. The higher the value of this coefficient, the less similar the clusters connected by the link. For each link, k, the inconsistency coefficient is calculated as:

$$IC(k) = \left(L\left(k\right) - W1\left(k\right)\right)/W2\left(k\right), \tag{8}$$

where IC and L are $(m-1) \times 1$ vectors; IC: Inconsistency coefficient; $W1$: Mean of the lengths of all the links included in the calculation; $W2$: Standard deviation of all the links included in the calculation; and L: Vector of the lengths of the links.

For leaf nodes, nodes that have no further nodes under them, the inconsistency coefficient is set to 0. The threshold used in this paper is 1.

3. Experimental Study

In this section we evaluate the sensitivity of the proposed system to subsoil variations and determine the most suitable combinations of procedures. We have prepared a laboratory layout with 16 types of configurations. In a schematic way, these configurations are presented in Fig. 2.

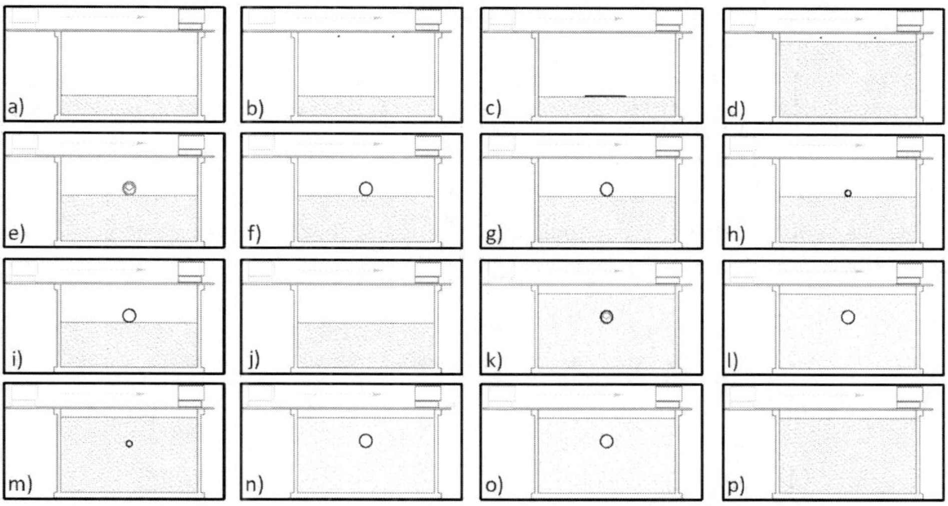

Figure 2. Layouts of the prepared configurations.

In Figure 2, configurations from e) to i) and from k) to o) correspond to the soil configurations j) and p), respectively. Those configurations differ from j) and p), respectively, due to the addition to the layout of one of the four different pipe materials commonly used in WSS. A description of the characteristics of the inserted pipes and the configuration of the ensemble is presented in Table 1. All the arrangements in Fig. 2 were tested with the GPR. In order to observe the responses of the various configurations, the same array of parameters was used in the equipment for all the surveys. The GPR equipment used in each prospection corresponds with a monostatic commercial antenna, with a control unit GSSI SIR 3000 and an antenna central frequency of 1.5 GHz.

In Fig. 3 we present the obtained raw results of the performed surveys by using a gray scale.

In the images corresponding to raw data shown in Figure 3, it can be easily observed that for PVC, PE1, and PE2 materials borderlines are not clear-cut and, as a result, poor visualization properties are obtained. This is attributed to the very low permittivity exhibited by these materials. This causes a low color intensity and, consequently demarcation

Table 1. Characteristics of the buried pipes used for testing

Material	Index	Index Figs. 2, 3, 5	Diam. (mm)
PVC	PVC	i), o)	100
Polietylene	PE1	h), m)	35
Polietylene	PE2	g), n)	76
Asbestos cement	Fib	e), k)	80
Cast Iron	Fund	f), l)	86

of these materials are very difficult, making them almost invisible. Even being more easily observed, configurations for the other materials (Fib and Fund) present some degree of difficulty. Clearly, these images in Fig. 3, which have undergone no processing at all, are really hard to understand and interpret, and even to discriminate. This fact underlines the need of methodologies enabling suitable interpretation.

Images in Fig. 3 have been treated by the system described in the previous section. We have used three different metrics and another three linkage methods. Table 2 presents the nine combinations among these metrics and linkage methods. The entries of the matrix symbolize the different combinations, and are used below.

Table 2. Combinations of metrics and linkage methods used within HAC

	Single	Average	Complete
Euclidean	SEu	AEu	CEu
Seuclidean	SSeu	ASeu	CSeu
Cosine	SCo	ACo	CCo

After evaluating these alternatives within the proposed system for the arrangements defined above, the *cophenet* correlation coefficient for each configuration was obtained. These coefficients are shown in Figure 4.

It can be easily observed that the use of *cosine* increases the *cophenet* correlation among the proposed subsoil configurations. We can also observe that the combinations ACo and CCo present the highest values. Finally, it is ACo the combination that provides the highest and most constant *cophenet* correlation coefficients. In view of these results, we have adopted the combination of *cosine*, as a distance metric, and *Average Linkage*, as the most suitable alternative for GPR images to obtain clusters through HAC. In figure 5 we show the clusters (taken back to the original space) obtained by the use of ACo.

Images shown in Fig. 5 present a lower amount of points than their corresponding images in Fig. 3, thus enabling more easy interpretation. It can also be observed when comparing Figures 2 and 5, that layouts with more similar configurations also produce more similar images. As a result, we claim that the application of HAC with the right combination of procedures improve the insight into the subsoil properties since these techniques manage

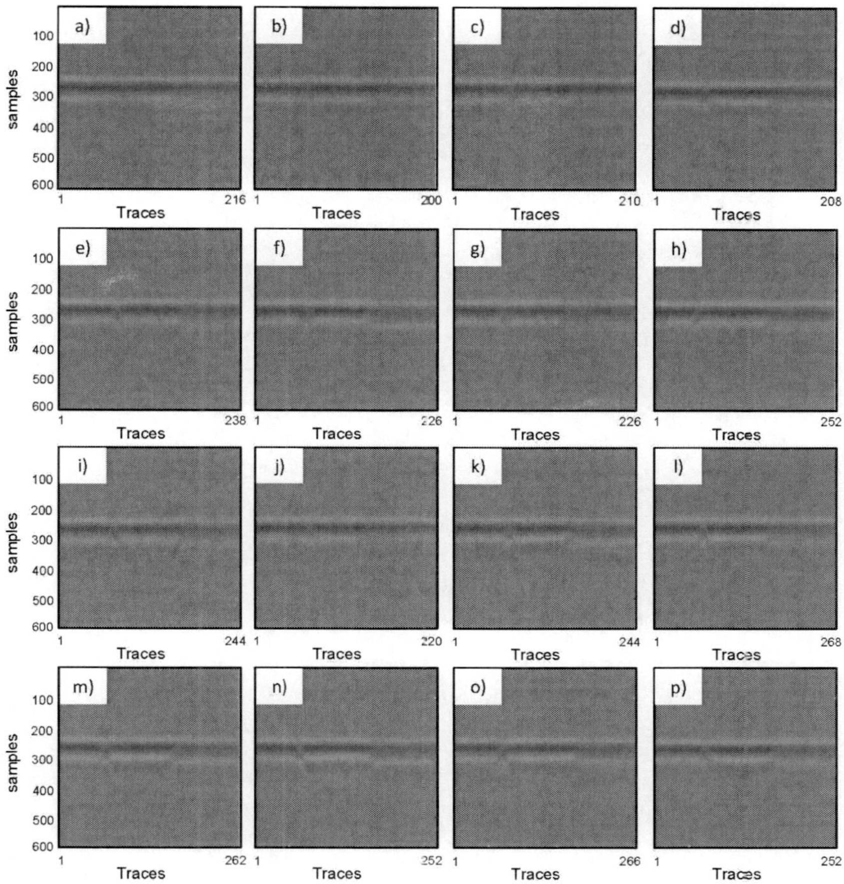

Figure 3. GPR survey files for the subsoil configurations in Fig. 2.

to cluster objects in a more natural and more identifiable way.

The proposed methodology was tested with 1140 images of pipes of various materials through suitable GPR prospections. By applying this methodology the identification was successful in 95% of the cases, since cleaner images with a significant reduction of points were rendered. Also, the decrease regarding the number of points constituting the processed image is worth mentioning: the processed images had on average less than 60% of information compared with the raw images. This represents an important step forward regarding posterior classification using intelligent methodologies, since a lower amount of information is considered, while simultaneously preserving the main characteristics of the image and obtaining more clear soil profiles. These profiles would ease 3D images interpolation and posterior production of maps of isoclines.

Conclusion

In this chapter, a tool for unsupervised classification of soil characteristics from GPR surveys, based on multi-agent and clustering approaches, has been presented. We have specifically focused on the identification of pipes of various materials in the prospected area. The

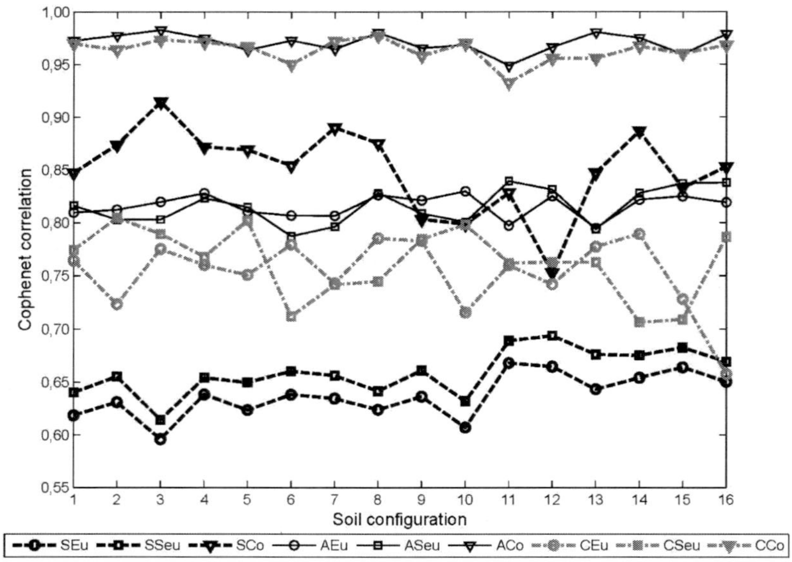

Figure 4. Obtained *cophenet* correlation coefficients.

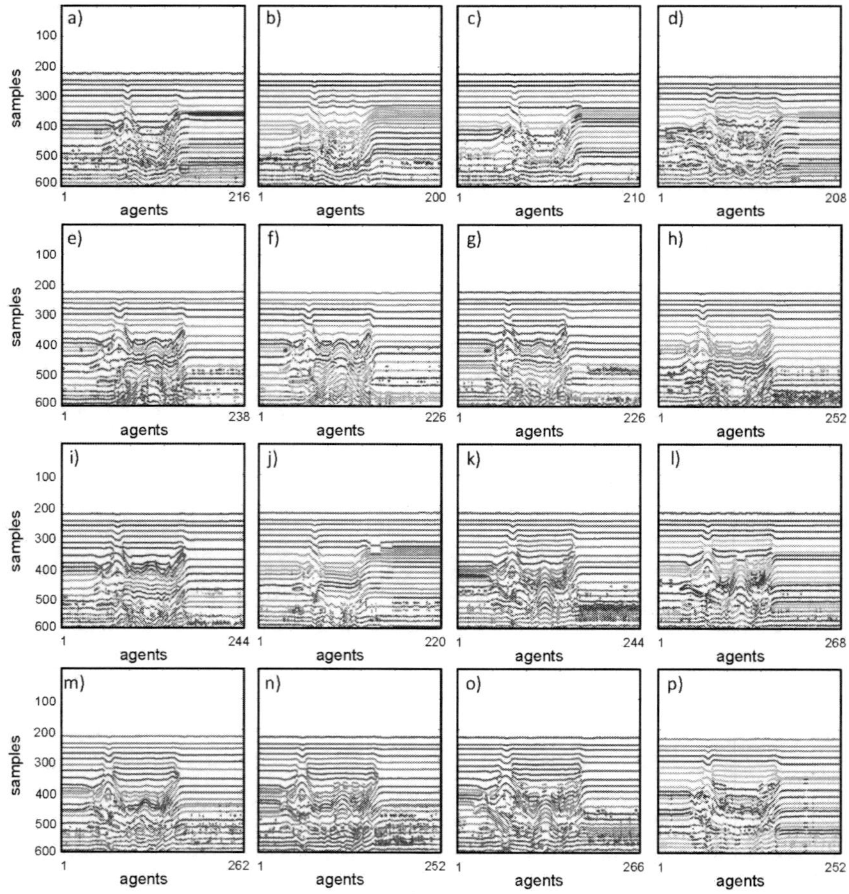

Figure 5. Clusters obtained for the configurations considered in Fig. 2 placed back in the original space.

proposed methodology is able to suitably group data into clusters, thus suitably classifying the information for better interpretation. As a result, soil is characterized in a natural and quick way, what favors the interpretation of GPR files performed by non-highly specialized personnel. Also, there is no need of any a priori parameter assumption. The results are promising, since they help reduce the amount of information needed to be dealt with, while preserving the essential data for materials identification. This produces more clear visualization of pipes in water distribution systems, thus favoring better identification of the system components. The tool developed, in conjunction with the non-destructive nature of a powerful technique as the GPR, can be used to create soil profiles with automatic procedures that will favor WSS technical management.

References

[1] Ayala-Cabrera, D.; Izquierdo, J.; Montalvo, I.; Pérez-García, R. Water supply system component evaluation from GPR radargrams using a multi-agent approach, *Math Comp Mod.* 2013, 57 (7-8), 1927-1932.

[2] Shoham, J.; Leyton-Brown, K. *Multiagent systems: algorithmic, game-theoretic and logical foundations*, Cambridge University Press, 2009.

[3] Lomuscio, A.; Raimondi, F. Model checking knowledge, strategies, and games in multi-agent systems, *in Proc. of the The fifth international joint conference on autonomous agents and multiagent systems*. 2006, 161-168.

[4] Wedel, M; Bijmolt, T.H.A. Mixed tree and spacial representations of dissimilarity judgments, *J Classif.* 2000, 17 (2), 243-271.

In: Mathematical Modeling in Social Sciences ...
Editors: J. C. Cortés López et al.
ISBN: 978-1-63117-335-6

Chapter 24

SEMI-AUTOMATIC SEGMENTATION OF IVUS IMAGES FOR THE DIAGNOSIS OF CARDIAC ALLOGRAFT VASCULOPATHY

Damián Ginestar[1]***, José L. Hueso***[1,*]***, Jaime Riera***[1]
and Ignacio Sánchez Lázaro[2]
[1]Instituto Universitario de Matemática Multidisciplinar
Universitat Politècnica de València, Valencia, Spain
[2]Hospital La Fe Universitari i Politècnic, Valencia, Spain

Abstract

Cardiac allograft vasculopathy is the leading cause of death after the first year post-heart transplantation. It consists of a proliferation of the intima layer in the coronary arteries of the transplanted heart, which progressively diminishes the supply of oxygen to the heart. Allograft vasculopathy is difficult to be detected by a standard coronariography and only using an intravascular ultrasound (IVUS) image can be accurately diagnosed. IVUS is a medical imaging technique that produces cross-sectional images as a catheter is pulled-back inside blood vessels and it provides quantitative assessment of the wall, information about the nature of atherosclerotic lesions as well as the plaque shape and size. In this work, a supervised technique for the IVUS images segmentation is proposed. This method consists of two steps. First the media-adventitia contour of the coronary artery is obtained and the second step consists of detecting the lumen contour using a classifying tool. To test the performance of the method, some IVUS images have been segmented and the obtained results have been compared with the segmentation performed manually by an expert.

Keywords: IVUS segmentation, cardiac allograft vasculopathy, Texture classification.

1. Introduction

Cardiac allograft vasculopathy (CAV) is the major cause of late death in patients undergoing heart transplantation (HT). It is manifested by a unique and unusually accelerated form of

*E-mail address: jlhueso@mat.upv.es (Corresponding author)

coronary disease affecting both intramyocardial and epicardial coronary arteries and veins [1]. Although the exact pathogenesis of CAV remains to be established, several lines of data suggest that it is primarily an immunemediated disease.

The most validated method for the diagnosis of CAV is intravascular ultrasound (IVUS), since there are no sufficiently reliable non-invasive methods. IVUS imaging is a relatively new medical tool that consists of placing a catheter, with a sensor inside the artery. This sensor rotates as it emits pulses of ultrasound. When it receives the echoes the tissues return, it generates an image like the one shown in Figure 1. In this image, the lumen-intima border and the media-adventitia border have to be detected and the zone between these two borders is known as the plaque, which is related with the CAV progression.

Automatic processing of large IVUS data sets represents an important challenge due to ultrasound speckle, catheter artefacts or calcification shadows. Moreover, a typical IVUS acquisition contains several hundred of images making nonautomatic analysis of the data long, fastidious and subject to intraobserver and interobserver variabilities. These could be serious constraints against the clinical usage of IVUS.

Usual techniques addressing segmentation of vessel borders rely on a single local image descriptors of edges. Energy minimization contour-based techniques either guide a snake towards the target structures [2], or minimize a cost function [3]. A common inconvenience of segmentation based on contour detection is that it requires some kind of image filtering to avoid fake responses. Recent approaches use either a probabilistic framework or classification strategies to better characterize coronary structures [4], [5]. Here a simple supervised method for the IVUS images segmentation is proposed that is mainly based on two steps, first the media-adventitia border is detected by preprocessing the image to enhance the contours information and a snake is used to determine the border. This determines a region of interest and, in this region, a classifier is used to distinguish between the lumen and the plaque. To use the method it is necessary the segmentation by an expert of the first IVUS image and the following images are segmented automatically.

2. Media-Adventitia Border Detection

IVUS images are quite noisy, so a noise reduction filtering has been considered. Particularly, a median filter is used. This filter runs through the pixels of the image substituting the pixel entry by the median of the pixels of a given window. The window used has been a square window of size 5×5 (see Figure 1).

To find the main edges present in the image a Canny's filter has been applied [6]. This method finds edges by looking for local maxima of the gradient of the image, which is previously smoothed with a Gaussian filter. To determine the edges from the magnitude of the gradient, the method uses two thresholds, to detect strong and weak edges. The parameters of this filter are the high threshold, the low threshold and the standard deviation of the Gaussian filter. The particular values used for these parameters are 0.1, 0.5 and 3, respectively. A typical output of the Canny's filter applied to an IVUS image is shown in Figure 2.a), where part of the media-adventitia border is recovered.

A snake model is used to close the contour of the vessel. The model used is a parametric active contours [7], which synthesizes parametric curves within an image domain and allows them to move toward the edges drawn by internal and external forces.

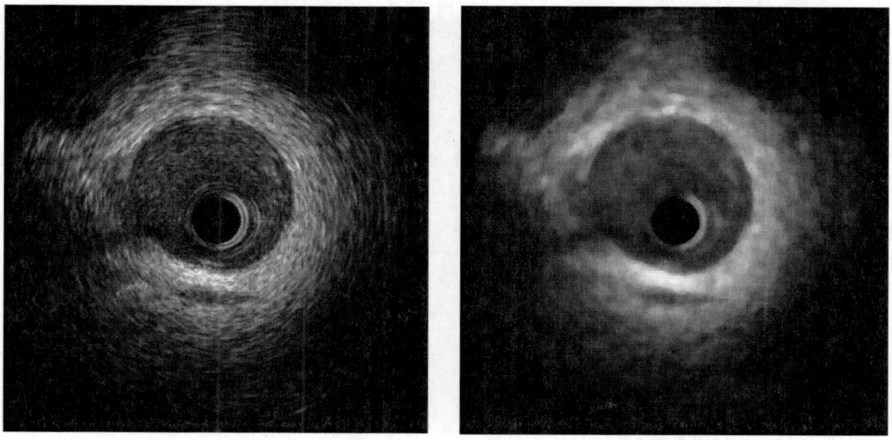

Figure 1. Original and filtered images of the first IVUS frame.

A parametric snake is a curve $\boldsymbol{x}(s) = (x(s), y(s))$, $s \in [0,1]$ that moves though the space defined by the image to minimize an energy functional

$$E = \int_0^1 \frac{1}{2}\left(\alpha\left|\mathbf{x}'(s)\right|^2 + \beta\left|\mathbf{x}''(s)\right|^2\right) + E_{\text{ext}}(\boldsymbol{x}(s))\, ds \;, \tag{1}$$

the parameters α and β control the snake tension and rigidity, respectively. The external energy takes its minimum values at the edges of the image. If the image $I(x,y)$ is understood as a derivable function, typical external forces are

$$E_{\text{line}}(x,y) = \pm I(x,y) \;, \quad E_{\text{edge}}(x,y) = -\left|\nabla I(x,y)\right|^2 \;, \tag{2}$$

and if $G_\sigma(x,y)$ is a Gaussian function with standard deviation σ, we define the termination energy

$$E_{\text{term}} = \frac{C_{yy}C_x^2 + C_{xx}C_y^2 - 2C_{xy}C_xC_y}{\left(C_x^2 + C_y^3\right)^{3/2}} \;, \quad C(x,y) = G_\sigma(x,y) * I(x,y) \;. \tag{3}$$

The external energy is $E_{\text{ext}} = w_{\text{line}}E_{\text{line}} + w_{\text{edge}}E_{\text{edge}} + w_{\text{term}}E_{\text{term}}$.

The snake that minimizes the energy E, satisfies the Euler equation

$$\alpha\boldsymbol{x}''(s) - \beta\boldsymbol{x}''''(s) - \nabla E_{\text{ext}} = 0 \;. \tag{4}$$

Equation (4) can be viewed as a forces balance $\boldsymbol{F}_{\text{in}} + \boldsymbol{F}_{\text{ext}} = 0$, where $\boldsymbol{F}_{\text{in}} = \alpha\boldsymbol{x}''(s) - \beta\boldsymbol{x}''''(s)$ and $\boldsymbol{F}_{\text{ext}} = -\nabla E_{\text{ext}}$.

The particular snake used in this work is based on the gradient vector flow (GVF) field [7]. To obtain this snake the forces balance is maintained changing the external force and the evolution of the snake is obtained by means of the dynamic equation

$$\frac{d\boldsymbol{x}}{dt} = \alpha\boldsymbol{x}''(s) - \beta\boldsymbol{x}''''(s) + \boldsymbol{F}_{\text{ext}} + \boldsymbol{v} \;. \tag{5}$$

The gradient vector flow field is the vector $\boldsymbol{v} = (u(x,y), v(x,y))$, that minimizes

$$\varepsilon = \iint \mu \left(u_x^2 + u_y^2 + v_x^2 + v_y^2\right) + |\nabla f|^2 \, |\boldsymbol{v} - \nabla f| \; dxdy \,, \tag{6}$$

where μ is a regularization parameter, which has to be set depending on the amount of noise present in the image, and f is an edge map of the image. A possible edge map is $f(x,y) = -E_{\text{ext}}(x,y)$.

The gradient vector flow field can be obtained solving the Euler equations

$$\mu\nabla^2 u - (u - f_x)\left(f_x^2 + f_y^2\right) = 0 \,, \quad \mu\nabla^2 v - (v - f_y)\left(f_x^2 + f_y^2\right) = 0 \,, \tag{7}$$

which are solved looking for the stationary solutions of the dynamic equations

$$\frac{du}{dt} = \mu\nabla^2 u - (u - f_x)\left(f_x^2 + f_y^2\right) \,, \quad \frac{dv}{dt} = \mu\nabla^2 v - (v - f_y)\left(f_x^2 + f_y^2\right) \,. \tag{8}$$

These equations are discretized using a finite differences approximation [7], to obtain first the gradient vector flow field and, with this field, the evolution of the snake is computed with a discrete version of equations (5).

To improve the performance of the snake dynamics, a balloon force [8] is added to the external forces $\boldsymbol{F}_{\text{ball}}(x(s), y(s)) = K_{\text{ball}}\boldsymbol{n}(x(s), y(s))$, where $\boldsymbol{n}(x(s), y(s))$ is the normal unitary vector to the curve at point $(x(s), y(s))$, and K_{ball} is the magnitude of this force.

Some of the parameters set in the algorithm are the following ones: time step, $\Delta t = 0.25$, the standard deviation of the Gaussian filter used to compute the image derivatives $\sigma_1 = 2$, standard deviation used to calculate the gradient of the edge energy, $\sigma_2 = 2$, and the external force weight $\kappa = 1$. The number of iterations of the snake has been set to 200. Figure 2.b) shows an example of snake that adjusts the vessel contour.

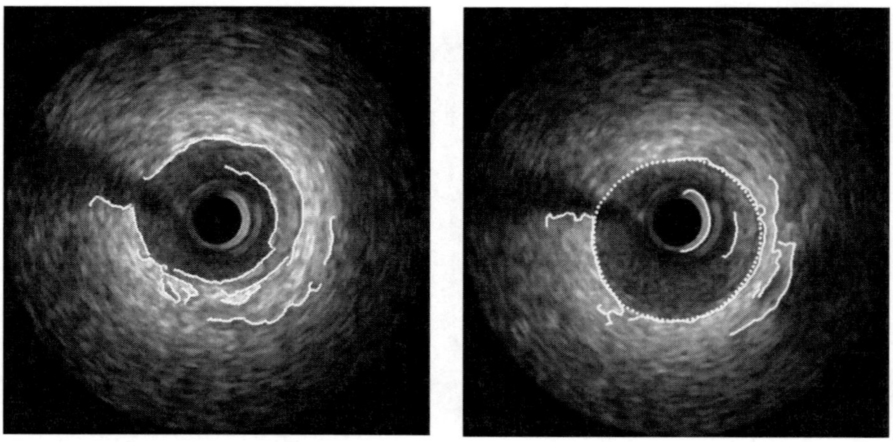

Figure 2. a) Borders obtained by Canny's method. b) Snake closing the contour.

3. Lumen Segmentation

Once the the media-adventitia border has been obtained the inner part of the vessel is considered as a region of interest. First, the catheter is removed using a circular mask. In the first photogram a region corresponding to the lumen and a region corresponding to the plaque are selected, as it is shown in Figure 3. The pixels of these regions are used to distinguish between the lumen and the plaque in the vessel using a classification system. The mean and the standard deviation of a square window of 11×11 pixels are used as characteristics for a linear discriminant analysis in two groups or categories.

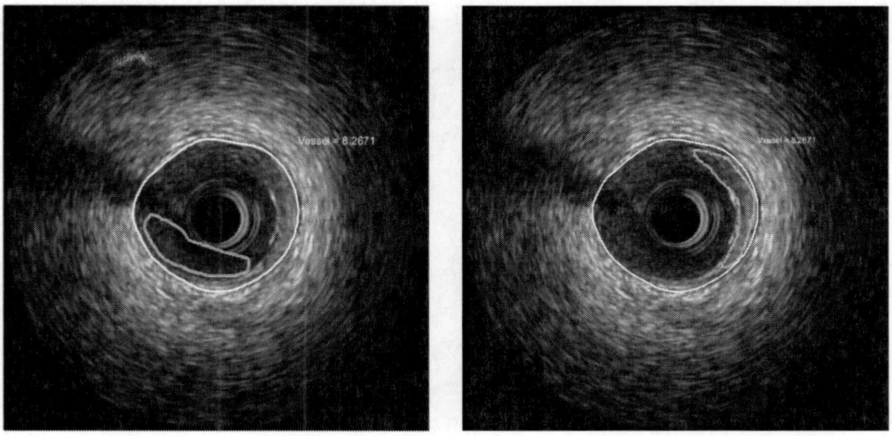

Figure 3. Lumen and plaque regions selection in the first frame.

The Fisher Linear discriminant analysis [9] mainly consists of projecting data from a d-dimensional space onto a line with a direction such that the projected data are as well separated as possible. Let us assume that we have a set of n d-dimensional samples $\boldsymbol{x}_1, \boldsymbol{x}_2, \ldots, \boldsymbol{x}_n$, n_1 in the subset W_1 and n_2 in the subset W_2. A linear combination of the components of $\boldsymbol{x}_i$ is of the form $y = w^T \boldsymbol{x}_i$, and this projection divides the corresponding n samples $y_1, y_2 \ldots, y_n$ in the subsets Y_1 and Y_2. If $\|w\| = 1$, y_i is the projection of the corresponding $\boldsymbol{x}_i$ onto a line in the direction of w. We want to find the best direction w that enables a good classification in two different categories.

Introducing the sample means and the mean of the projected points

$$m_i = \frac{1}{n_i} \sum_{x \in W_i} x \ , \quad \tilde{m}_i = \frac{1}{n_i} \sum_{y \in Y_i} y = \frac{1}{n_i} \sum_{x \in W_i} w^T x = w^T m_i \ , \tag{9}$$

the distance between the projected means is given by $|\tilde{m}_1 - \tilde{m}_2| = \left| w^T (m_1 - m_2) \right|$.

Defining

$$\tilde{s}_i^2 = \sum_{y \in Y_i} (y - \tilde{m}_i)^2 \ , S_i = \sum_{x \in W_i} (x - m_i)(x - m_i)^T \ , \quad S_W = S_1 + S_2 \ ,$$

we can write

$$\tilde{s}_i^2 = \sum_{x \in W_i} \left(w^T x - w^T m_i \right)^2 = \sum_{x \in W_i} w^T (x - m_i)(x - m_i)^T w = w^T S_i w \ ,$$

and $\tilde{s}_1^2 + \tilde{s}_2^2 = w^T S_w w$. Similarly $(\tilde{m}_1 - \tilde{m}_2)^2 == w^T S_B w$, with $S_B = (m_1 - m_2)$.

Fisher discriminant analysis looks for a vector w that maximizes the function

$$J(w) = \frac{w^T S_B w}{w^T S_W w} \,. \tag{10}$$

This vector is the dominant solution of the generalized eigenvalue problem $S_B w = \lambda S_W w$. If S_w is nonsingular, it can be expressed as $S_w^{-1} S_B w = \lambda w$, and the solution of this problem is $w = S_w^{-1}(m_1 - m_2)$. In this way, the Fisher linear discriminant function is defined as

$$L(x) = \left(x - \frac{1}{2}(m_1 + m_2) \right)^T S_w^{-1}(m_1 - m_2) \tag{11}$$

and a new sample x is classified as belonging to the class W_1 if $L(x) > 0$ and it belongs to class W_2 if $L(x) \leq 0$.

The result of the classification is a binary image, the mask, which must be smoothed in order to obtain likely shapes for the lumen and plaque. By using morphological operations on the binary mask the borders are straightened and the disconnected parts eliminated. We perform a closing with a disk of radius 6 pixels, followed by an opening with a disk of radius 4. Finally, if there is any small isolated region left, it is suppressed. The final result determines the plaque and lumen zones in the vessel, which are measured in order to obtain a quantitative information. By performing these operations on successive frames of the IVUS sequence, one can estimate the volume of the plaque along the vessel, which is of interest for the diagnosis of CAV.

4. Ivus Analysis

The IVUS images were acquired using a Clearview Ultra (Boston Scientific Corp., Natick, MA, USA) with a catheter model Atlantis. The ultrasound frequency was 40MHz and the catheter pull-out speed was 0.5 mm/s. A sequence of 23 IVUS images was manually segmented by an expert who drew by hand the vessel and plaque contours and used a software to measure them. We have performed our analysis on the same sequence and have compared qualitative and quantitatively the results. Figure 4 shows the segmentations of two frames by the expert and by our software. In order to assess the performance of our software, we have compared the vessel and lumen areas of the analyzed sequence.

The differences of both measurements are normally distributed, specially the vessel area differences. The t-test reveals no significant differences between the expert and the software measurements. In order to visually assess the suitability of the software measurements, we present the Bland-Altman plots for the vessel and lumen areas obtained in both ways. In the diagram corresponding to the vessel areas, shown in Figure 5, it can be observed that the deviation is bigger for high values of the average, which corresponds to frames where the vase section is not well defined. The differences lie between the lines corresponding to the mean value plus or minus 1.96 times the standard deviation (the 95% confidence interval), except for two outliers.

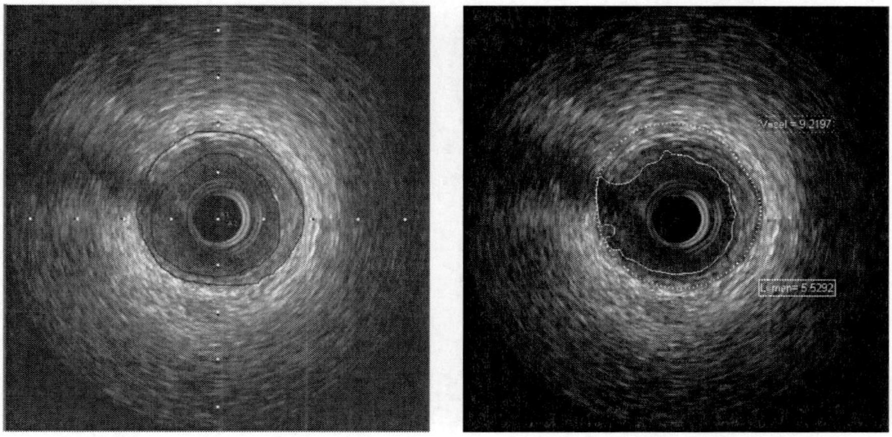

Figure 4. Left: expert segmentation. Right: software segmentation.

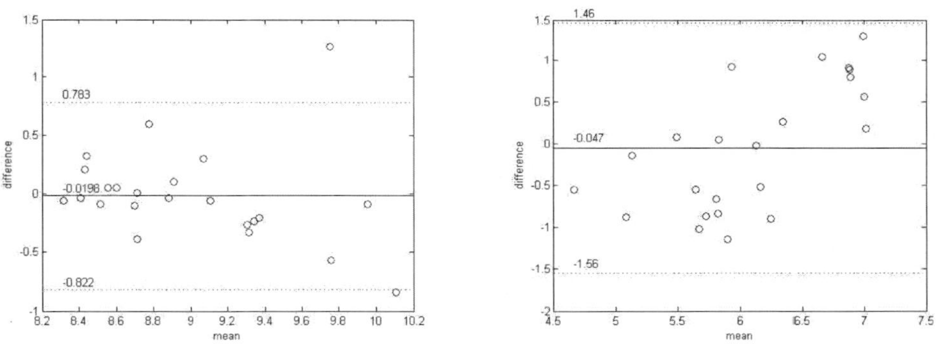

Figure 5. Bland-Altman diagram for the vessel (left) and lumen area (right) measurements.

Conclusion

We have presented a semi-automatic procedure for segmenting IVUS images of the coronary artery that is of interest in the diagnosis of cardiac allograft vasculopathy. The procedure has two phases. In the first phase, the vessel contour is detected. In the second phase, the interior of the vessel is segmented, identifying and measuring the lumen and the plaque. With the result of successive segmentations, a volumetric estimation of the vessel occupations is obtained, which is of interest for the diagnosis.

The procedure has to be supervised from time to time, in order to avoid an erroneous detection of the vessel, which would cause meaningless results in the segmentation phase. The variations in intensity or contrast in the images can also require and adjustment in the classification criteria. The minimization of these limitations will be the object of our future work in the area.

Acknowledgments

This research was supported by Ministerio de Ciencia y Tecnología MTM2011-28636-C02-02 and by Vicerrectorado de Investigación, Universitat Politècnica de València PAID-SP-2012-0498 and PAID-SP-2012-0474.

References

[1] Weis, M.; von Scheidt, W. *Circulation* 1997, 96, 2069-2077.

[2] Sanz-Requena, R.; Moratal, D.; García-Sánchez, D. R.; Bodí, V.; Rieta, J.J.; Sanchis, J.M.; *Comput. Med. Imaging Graph.* 2007, 31, 71-80.

[3] Mendizabal-Ruiz, G.; Rivera, M.; Kakadiaris, I. A. *IEEE Conference on Computer Vision and Pattern Recognition*, 2008, pp 1-8.

[4] Brusseau, E.; de Korte, C.L.; Mastik, F.; Schaar, J.; van der Steen, A. F. W. *IEEE Trans. Med. Imag.* 2004, 23, 5, 554-566.

[5] Gil, D.; Hernández, A.; Rodriguez, O.; Mauri, J.; Radeva, P. *IEEE Trans. Med. Imag.* 2006, 25, 6, 768-778.

[6] Canny, J. *IEEE Trans. Pattern Anal. Mach. Intell.* 1986, PAMI-8, 6, 679-698.

[7] Xu, C.; Prince, J.L.; *IEEE Trans. Image Process.* 1998, 7, 3, 359-369.

[8] Cohen, L. In *Computer Vision, Graphics, and Image Processing: Image Understanding*, 1991; 53, 2, 211-218.

[9] Duda, R.O.; Hart, P.E.; Stork, D.G. *Pattern Classification. 2nd Edition*; John Wiley & Sons, Inc. New York, 2001.

In: Mathematical Modeling in Social Sciences ...
Editors: J. C. Cortés López et al, pp. 261-272
ISBN: 978-1-63117-335-6

Chapter 25

ANALYSIS AND DETECTION OF V-FORMATIONS AND CIRCULAR FORMATIONS IN A SET OF MOVING ENTITIES

***Francisco Javier Moreno Arboleda*[*], *Jaime Alberto Guzmán Luna* and *Sebastián Alonso Gómez Arias*[†]**
Universidad Nacional de Colombia, Bogota, Colombia

Abstract

Diverse movement patterns are identifiable when a set of moving entities is studied. One of these patterns is known as a V-formation for it is shaped like the letter V. Informally, a set of entities presents a V-formation if the entities are located on one of their two characteristic lines. Another movement pattern is known as a circular formation for it is shaped like a circle. Informally, circular formations present a set of individuals grouped around a center in which the distance from these individuals to the center is less than a given threshold. In this chapter, we present a model to identify V-formations and circular formations with outliers. An outlier is an entity which is part of a formation but is distant from the formation. We present formal rules for our models and an algorithm for the detection of outliers. Our models were validated via NetLogo, a programming and modeling environment for the simulation of natural and social phenomena.

Keywords: Mobile object, movement patter, V-formation, circular formation, outlier

1. Introduction

Diverse movement patterns are identifiable when a set of moving entities is studied, e.g., a flock of birds [1] and a school of fish [2]. One of these patterns is known as a V-formation for it is shaped like the letter V. Another movement pattern is known as a circular formation for it is shaped like a circle.

[*] E-mail address: jaguzman@unal.edu.co.
[†] E-mail address: seagomezar@gmail.com.

Informally, a set of entities presents a V-formation if the entities are located on one of their two characteristic lines. The lines meet in a position where there is just one entity considered the entity leader [3]. Several authors have analyzed V-formations. In [4]and [5], there is an attempt to explain from a physical point of view the reasons why certain species of birds as Canadian geese (*Brantacanadensis*), red knots (*Calidriscanutus*) and plovers (*Calidrisalpina*), tend to fly this way.

Other authors try to simulate V-formations at a computational level. For instance, Nathan and Barbosa [6] propose a model based on rules that allows us to generate V-formations depending on specific parameters. The authors validated their model via NetLogo [7], a programming and modeling environment to simulate natural and social phenomena.

On the other hand, circular formations are a set of entities grouped around a common center in which the distance from these individuals to the center is less than given threshold, see Figure 1. Regarding circular formations the following works were identified. Reynolds[2] proposes that in fish these types of formations are the result of three basic rules that act upon individuals: i) Avoiding collisions with nearby individuals, ii) Trying to equal the speed of nearby individuals, and iii) Trying to remain nearby close individuals.

In [8], the authors present experiments with sets of moving data of different animal species. It was determined that despite being in different ecosystems, the sespecies follow similar behavior patterns. The authors also tried to model general grouping behaviors of fish, birds, insects, and even persons. One of these behaviors is circular formations in which they identified physical forces such as: attraction, repulsion, alignment and frontal interaction.

Source: authors' own presentation.

Figure 1. Circular formations in a school of fish.

On the other hand, researchers in the field of robotics and in control theory, inspired by social grouping phenomena and the movement patterns of birds and fish [9], [10] have developed applications for the coordination of multi-vehicle movement systems. Circular [6] and V-formations are among these patterns. Analyses of V-formations and of circular formations may be applied in fields as zoology to analyze the movement of birds [3] and fish. Romey [12] analyzes some formations in the military field and in the field of videogames where squadrons of airplanes, combat ships, and robots usually adopt the setypes of formations [3], [13]. In addition, V-formations usually appear in stock markets (prices of shares) [11].

This chapter is organized as follows. In Section 2, we present a method to detect V-formations. In Section 3, we present a method to detect circular formations. Then, in Section 4 we present a method to detect outliers in V-formations and in circular formations, i.e., formations which despite having some entities that do not tend to be grouped with the rest of the members are considered part of a V-formation or a circular formation. In Section 5, we present experiments, and finally, in Section 6 we conclude the chapter and propose future work.

2. V-Formations

In this section we present the definitions and the mathematic model to identify V-formations.

Let $F = <e_1, e_2, \ldots, e_n>$ be a list of moving entities in a point in time t. $e_k \in F$ is the entity leader, $1 < k < n$. F is a V-formation if:

i) Entities e_i, $1 \leq i \leq k$, *tend to form a straight line* l_1.
ii) Entities e_j, $k \leq j \leq n$, *tend to form a straight line* l_2.
iii) Straight lines l_1 *and* l_2 converge in position $(xpos(e_k, t), ypos(e_k, t))$. *xpos* and *ypos* are function that return the coordinates of an entity at t.
iv) $\angle ap_t > 0$ (the smallest angle defined by straight lines l_1 *and* l_2).

Regarding conditions i) and ii), to establish if a set of entities *tend* to form a straight line, we use Pearson correlation coefficient r [14]. Thus, given a set of points $\{(x_1, y_1), (x_2, y_2), \ldots, (x_n, y_n)\}$, r indicates how much they adjust to a straight line (degree of linearity). $r \in (-1, 1)$, if $|r| \approx 1$ then data tend to form a straight line, see Figure 2. r is calculated as shown in Equation (1).

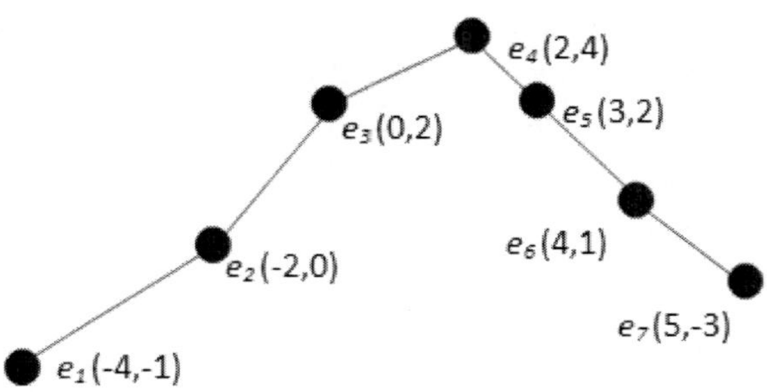

Source: authors' own presentation.

Figure 2. V-formation.

$$r = \frac{n \times (\sum_{i=1}^{n} x_i \times y_i) - \sum_{i=1}^{n} x_i \times \sum_{i=1}^{n} y_i}{\sqrt{[n \times \sum_{i=1}^{n} x_i^2 - (\sum_{i=1}^{n} x_i)^2] \times [n \times \sum_{i=1}^{n} y_i^2 - (\sum_{i=1}^{n} y_i)^2]}} \quad (1)$$

We can specify an h threshold to indicate the degree of linearity demanded for formation lines, i.e., $|r| \geq h$.

To obtain the equation for each straight line $y = mx + b$, characteristic of the formation (l_1 and l_2), we can apply the equations that correspond to the straight line which mostly adapts to a set of points $\{(x_1, y_1), (x_2, y_2), \ldots, (x_n, y_n)\}$, see equations (2) and (3).Table 1 presents the results for the formation in Figure 2.

$$m = \frac{n \times (\sum_{i=1}^{n} x_i \times y_i) - \sum_{i=1}^{n} x_i \times \sum_{i=1}^{n} y_i}{n \times \sum_{i=1}^{n} x_i^2 - (\sum_{i=1}^{n} x_i)^2} \tag{2}$$

$$b = \frac{\sum_{i=1}^{n} y_i - m \times \sum_{i=1}^{n} x_i}{n} \tag{3}$$

Table 1. Characteristic lines: results for the formation in Figure 2

Characteristic line	Entities that form the characteristic line	Equation of the characteristic line	Pearson coefficient	Entity	Coordinates of the entity	Coordinates of the entity calculated with the equation of the characteristic line
				e_1	(-4,-1)	(-4,-1.3)
l_1	e_1, e_2, e_3, e_4	$y = 2.1+0.85x$	0.99	e_2	(-2,0)	(-2,0.4)
				e_3	(0,2)	(0,2.1)
				e_4	(2,4)	(2,3.8)
				e_4	(2,4)	(2,4.3)
l_2	e_4, e_5, e_6, e_7	$y = 8.7-2.2x$	0.97	e_5	(3,2)	(3,2.1)
				e_6	(4,1)	(4,0.1)
				e_7	(5,-3)	(5,2.3)

Source: authors' own presentation.

Regarding condition iv), $\angle ap_t$ is calculated as follows: we obtain straight lines l_1 and l_2 of the formation, and we find the smallest angle among them as follows: a is the angle of the entity leader towards l_1, b the angle of the entity leader towards l_2, and $w = |a - b|$, then $\angle ap_t = w$ if $w \leq \pi$ and $\angle ap_t = 2\pi - w$, otherwise. For example a = 40.36, b = 294.44, and w =254.08; thus, $\angle ap_t = 105.91$.

3. Circular Formations

In this section we present the definitions and the mathematic model to identify circular formations.

3.1. Centroid of a Set of Points

The centroid (x_c, y_c) of a set of points $\{(x_1, y_1), (x_2, y_2), \ldots, (x_n, y_n)\}$ is obtained as follows. x_c is the average of the sum of ordinates and y_c is the average of the sum of abscissas [15], see equations (4) and (5).

$$x_c = \frac{\sum_{i=1}^{n} x_i}{n} \tag{4}$$

$$y_c = \frac{\sum_{i=1}^{n} y_i}{n} \tag{5}$$

We say that a set of moving entities exhibits a circular formation in a point in time *t* if

i) Distance *d* from each entity to the centroid is less than distance *R* (radius).
ii) The minimum number of the members of the formation is N_{min}.

A circular formation with eight members is shown in Figure 3.

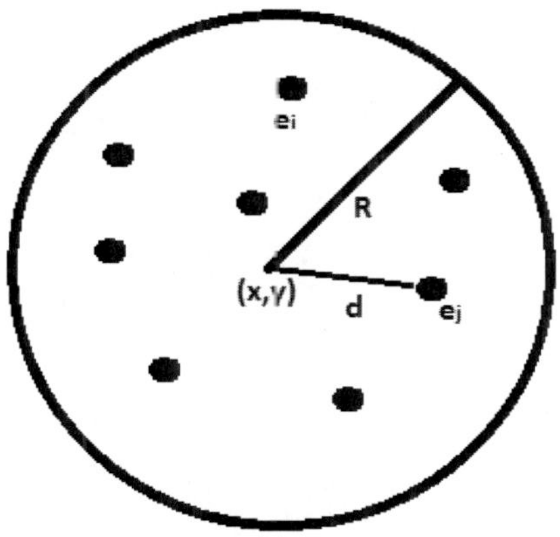

Source: authors' own presentation.

Figure 3. Circular formation with eight members.

4. Outliers

For a V-formation an outlier is an entity that is far from its characteristic lines; and for a circular formation, it is an entity found beyond the radius of the formation.

4.1. Outliers in V-formations

There are sets of entities that although tend to present a V-formation, they may have (in a point in time *t*) entities that are far from their characteristic straight lines. Therefore, this affects Pearson correlation coefficient. These entities are named outliers [16], [17].

There are many methods to detect outliers in different domains [18]. In List 1, we present an algorithm that receives a set of *m* entities (*lineMembers* array) which form the characteristic straight line of a formation in a point in time *t*. The algorithm determines if when we remove a maximum number of entities from the given set, Pearson coefficient surpasses a given μ_p threshold. For instance, if it is allowed to remove a maximum of three entities from a straight line of the formation, it is deemed that the entities in Figure 4 present a V-formation having three outliers. Then, the algorithm receives the minimum value of Pearson coefficient μ_p which must be met, and the maximum percentage of the entities (*percentage Outliers*) that may be removed from a set of entities. This percentage is calculated with regard to the total number of *m* entities.

Algorithm input parameters:

lineMembers= $[e_1, e_2, ..., e_8]$, $\mu_p = 0.99$, and *percentage Outliers* = 30%. Thus, *maxOutliers* = $\lceil 8*0.3 \rceil = \lceil 2.4 \rceil = 3$, i.e., a maximum of three outliers are allowed in the array of entities.

First, the algorithm calculates all the combinations of 1 element in 8. Table 2 shows some of the combinations generated, the content of the array *aux Members* and the value of Pearson coefficient (*newPearson*) that corresponds to the positions of the entities in this array.

Upon not finding the value of a *newPearson* that meets threshold $\mu_p = 0.99$, see Table 2, the algorithm calculates all the combinations of 2 elements in 8, results are shown in Table 3. When the algorithm evaluates the combination $\{e_2, e_4\}$, the *auxMembers* array = $[e_1, e_3, e_5, e_6, e_7, e_8]$ and the *newPearson* = 1. Since this value is greater than the demanded threshold ($\mu_p = 0.99$), the algorithm determines that entities e_2 and $e_4$are outliers. We then conclude that entities e_1, e_3, e_5, e_6, e_7, and e_8 tend to form a straight line when Pearson coefficient is greater than 0.99.

List 1. Algorithm to detect outliers

Example. Consider the set of entities in Figure 5.These entities form a straight line if entities e_2 and e_4 were not considered, i.e., e_2 and e_4 are outliers.

```
ALGORITHM: Detection of outliers on a characteristic line
INPUT: lineMembers = [e1, e2, ..., em] //Array of m entities
μp //Threshold for Pearson coefficient
percentageOutliers //Maximum percentage of outliers allowed in the lineMembers array
OUTPUT: outliers // Array of outliers
BEGIN
// Maximum number of outliers allowed in the lineMembers array
maxOutliers = ceil(m * percentageOutliers/100); // ceil function rounds to the next integer number
FOR k = 1 TO maxOutliersLOOP
    //Find all the combinations of k elements from the lineMembers array
    combinationsMatrix = combinations(k, lineMembers);
    FOR i = 1 TO size(combinationsMatrix) LOOP //For each combination of k elements
        outliers = combinationsMatrix(i); //Obtain the current combination
        /*Copy the lineMembers array in auxMembers but without the elements of the
        current combination */
        auxMembers = remove(lineMembers, outliers);
        //Calculate Pearson coefficient using positions (x,y) for each member in auxMembers
        newPearson = PearsonCoefficient(auxMembers);
        IF (newPearson ≥ μp) THEN
          /*If after eliminating from the lineMembers array the elements of the
        current combination, the μp threshold is met*/
            RETURN outliers;
        END IF
    END FOR
END FOR
PRINT "It was not possible to meet Pearson threshold: " μp;
END
```

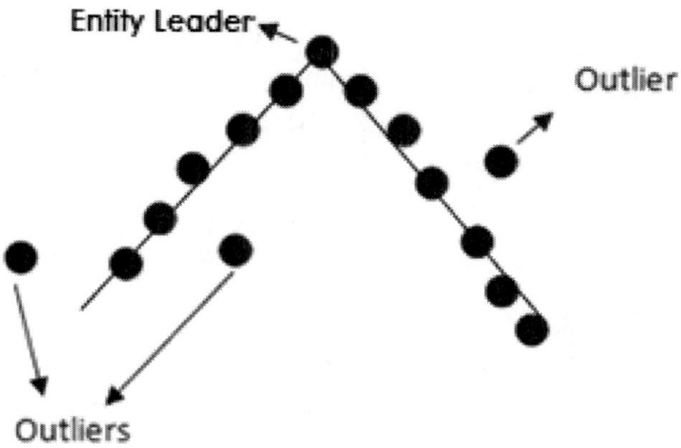

Source: authors' own presentation.

Figure 4. Example of a V-formation with three outliers.

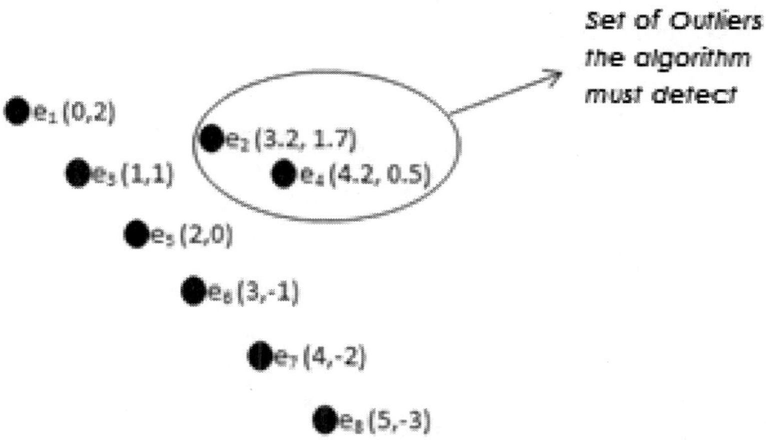

Source: authors' own presentation.

Figure 5. Characteristic straight line of a V-formation with two outliers.

Table 2. Combinations of an element and Pearson coefficient of entities in auxMembers

Combination	***auxMembers* array**	***newPearson***
$\{e_1\}$	$[e_2, e_3, e_4, e_5, e_6, e_7, e_8]$	0.90
$\{e_2\}$	$[e_1, e_3, e_4, e_5, e_6, e_7, e_8]$	0.90
…	…	…
$\{e_7\}$	$[e_1, e_2, e_3, e_4, e_5, e_6, e_8]$	0.90
$\{e_8\}$	$[e_1, e_2, e_3, e_4, e_5, e_6, e_7]$	0.90

Source: authors' own presentation.

Table 3. Combinations of two elements and Pearson coefficient of entities in *auxMembers* (in gray the combination that meets μ_p= 0.99)

Combination	*auxMembers* array	*newPearson*
$\{e_1, e_2\}$	$[e_3, e_4, e_5, e_6, e_7, e_8]$	0.93
$\{e_1, e_3\}$	$[e_2, e_4, e_5, e_6, e_7, e_8]$	0.91
...	....	...
$\{e_2, e_4\}$	$[e_1, e_3, e_5, e_6, e_7, e_8]$	1
	....	

Source: authors' own presentation.

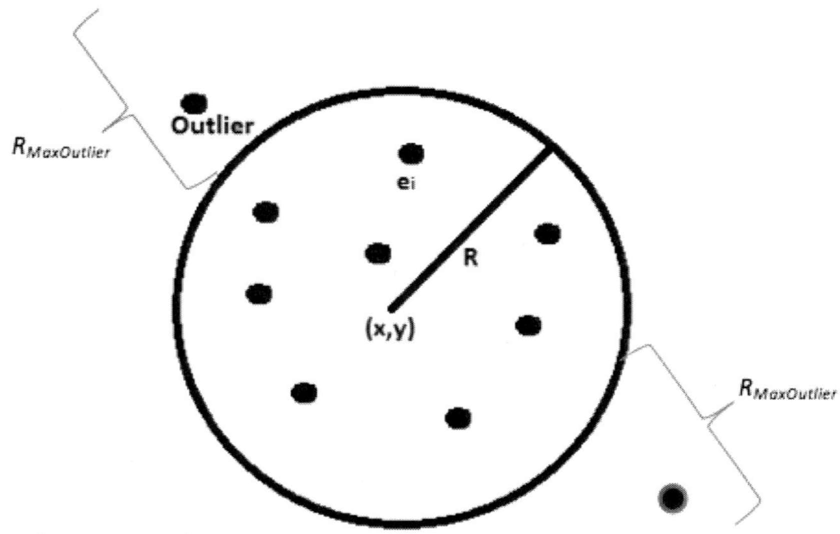

Source: authors' own presentation.

Figure 6. Circular formation with two outliers. MaxNumberOutliers = 2.

4.2. Outliers in Circular Formations

In a set of moving entities that tend to present a circular formation, there could be entities whose distance from the centroid is greater than *R*. These entities, named outliers, can be considered members of the formation which are temporarily away from it. To identify these types of entities, we introduce a parameter $R_{MaxOutlier}$, where $R_{MaxOutlier}>R$. An entity is considered an outlier in a time *t* if its distance *d* to the centroid is greater than *R* and less than $R_{MaxOutlier}$.

Because the separation of the entity outlier is temporary; an analyst may introduce a second parameter $t_{MaxTimeSeparation}$ to control the maximum continuous time of allowed separation. That is, if an entity separates from a circular formation in a time *t*, then, to be considered an outlier it will have to be reincorporated ($d \leq$ R) to the formation before $t + t_{MaxTimeSeparation}$. This same aspect may also be considered for outliers in V-formations.

An analyst may also specify a maximum number of outliers *MaxNumberOutliers*in the formation. This value may be calculated starting from a percentage (*PercentageOutliers*) with regard to the total number of individuals of the formation.

Experiments

For our experiments, we worked with NetLogo, it enables us to explore the relation between the behavior at the micro level of individuals and patterns at the macro level of groups. This is an approach that has been implemented in previous works [19], [20]. To generate V-formations in NetLogo, we used the model given in [6], which was conceived specifically for this goal. To generate circular formations, we used the model given in[7], which generates random formations of individuals in NetLogo.

Table 4.Parameters in Netlogo used to generate V-formations (on the left) and circular formations (on the right)

	V-formations		Circular formations
Number of individuals	15	Number of individuals	102
Parameters of vision		**Parameters of vision**	
Distance of the vision	9	Distance of the vision	3
Cone of vision	103°	Minimum separation	1
Cone of obstruction	43°	**Parameters of movement**	
Parameters of movement		Maximum turning angle	4.75°
Velocity	0,2	Maximum monitoring angle	2.50°
Velocity change factor	0,15	Maximum separation angle	3.5°
Vertical distance movement	9		
Minimum distance allowed from bird to bird	3,1		
Maximum turn allowed	8°		

Source: authors' own presentation.

Table 5. Parameters used to detect V-formations(on the left) and circular formations (on the right) in Netlogo

	V-formations		Circular formations
Minimum number of entities(N_{min})	3	Minimum number of entities(N_{min})	5
μ_p	> 0.92	Maximum distance (R) from centroid to entities	15 patches
Maximum percentage of outliers allowed (*PercentageOutliers*)	4%	Maximum percentage of outliers allowed (*PercentageOutliers*)	30%
		$R_{MaxOutlier}$	30 patches
		$t_{MaxTimeSeparation}$	400 ticks

Source: authors' own presentation.

Table 6. Results of the identification of V-formations(on the left) and circular formations (on the right) in Netlogo

	V-formations		Circular formations
Total number of formations identified in the 100 runs	318	Total number of formations identified in the 100 runs	332
Average number of formations identified in each run (200 ticks)	3	Average number of formationsidentified in each run (1200 ticks)	3
Average number of individuals in each formation	4	Average number of individuals in each formation	13
Average number of outliers	1	Average number of outliers	2

Source: authors' own presentation.

The parameters used to detect V-formations and circular formations are shown on Tables 4 and 5. Experimental results are shown on Table 6. 100 runs were done both to generate V-formations and circular formations. The time for each run was 200 ticks (a *tick* is a time measurement unit in Netlogo and at normal velocity equals 0.5 seconds. Nonetheless, in Netlogo, it is possible to change and change the velocity, so the value of a tick in seconds is relative.) A *patch* is the unit of measurement of distance Netlogo.

Conclusion

In this chapter we propose two formal models:

i) A model to identify V-formations with outliers.
ii) A model to identify circular formations with outliers.

Both models considered the location of the entities to determine if they form this type of formation. The rules of our model for V-formations are flexible, for they allow V-formations which are not necessarily aligned as it usually happens in the real world. Furthermore, we considered the possible presence of entity outliers both in V-formations and in circular formations, i.e., members of the formation which may be far from it during some periods. Results in Netlogo showed that our models identified this type of formations in an environment where they are generated.

Regarding future work, we plan to apply our V-formation model to the stock market where these types of formations tend to take place [11]. Moreover, we plan to extend our models to identify other types of patterns, e.g., identify *isolated entities*, i.e., entities which have not been considered members of a group, follow their own path and do not come together with other entities [21], *convergence*, i.e., to identify a group of entities which converge or move together towards a place; *divergence*, i.e., to identify a group of entities which get disperse or move away from a place [22], and *self-organization*, i.e., a group de entities that moves as a set yet it does not have a leader or an entity which may guide the rest of the members [23] or that the leader is not known by the members [24], [25].

Acknowledgments

This chapter presents preliminary results of the project "Apoyo al Grupo de Sistemas Inteligentes Web-SINTELWEB" with Quipúcode 205010011129, developed at the Universidad Nacional de Colombia, Sede Medellín.

References

[1] Dodge, S., Weibel, R., & Lautenschütz, A.K. Towards a taxonomy of movement patterns. *Inf. Vis.* 2008, vol. 7(3-4), 240-252.

[2] Reynolds, C.W. Flocks, herds and schools: A distributed behavioral model.*ACM SIGGRAPH Computer Graphics*. 1987, vol. 21, 25-34.

[3] Cattivelli, F., & Sayed, A.H. Self-organization in bird flight formations using diffusion adaptation.*3rd IEEE International Workshop on Computational Advances in Multi-Sensor Adaptive Processing (CAMSAP)*. Aruba,2009.

[4] Seiler, P., Pant, A., Hedrick, K. Analysis of bird formations.*41stIEEE Conference on Decision and Control,* Las Vegas, NV, 2002.

[5] Sewatkar, C.M., Sharma, A., & Agrawal, A. A first attempt to numerically compute forces on birds in v formation. *Artif. Life*.2010, vol. 16(3), 245-258.

[6] Nathan, A., Barbosa, V.C. V-like formations in flocks of artificial birds. *Artif. Life*.2008, vol. 14(2), 179-188.

[7] Wilensky, U., &Rand, W. Making models match: Replicating an agent-based model. *J. Artif. Soc. Soc. Simul.*2007, vol. 10(4), no pages.

[8] Lukeman, R. Li, Y.X.,& Edelstein-Keshet, L. Inferring individual rules from collective behavior. *Proc. Natl. Acad. Sci.*, 2010, vol. 107(28), 12576-12580.

[9] Wilde, S. Thoughts forming in a fish.2013. Available from: http://www.stuartwilde.com/2013/02/thoughts-forming-in-a-fish.

[10] Grandin, T. Understanding Flight Zone and Point of Balance for Low Stress Handling of Cattle, Sheep, and Pigs. 2011. Available from: http://www.grandin.com/behaviour/principles/flight.zone.html

[11] Rueda, A. (2002) *Para Entender la Bolsa: Financiamiento e inversión en el mercado de valores* (first edition). Miami, FL: Thomson Publishers.

[12] Romey, W.L. Individual differences make a difference in the trajectories of simulated schools of fish. *Ecol. Model*.1996, vol. 92(1), 65-77.

[13] Moshtagh, N., Michael, N., Jadbabaie, A.,& Daniilidis, K. Bearing-only control laws for balanced circular formations of ground robots. *Robotics: Science and Systems IV*. Zurich, 2008.

[14] Calderón-Meza, G.,& Sherry, L. Adaptive agents in NAS-wide simulations: A case-study of CTOP and SWIM. *Integrated Communications, Navigation and Surveillance Conference*, Herdon, VA, 2011.

[15] Hibbeler, R.C. (2004) *Mecánica vectorial para ingenieros: estática* (first edition). Mexico, DF: Pearson Educación.

[16] Hawkins, D.M, Bradu, D., &Kass, G.V. Location of several outliers in multiple-regression data using elemental sets.*Technometrics*.1984, vol. 26(3), 197-208.

[17] Ben-Gal, I. Outlier detection. In: Maimon, O. & Rokach, L. *Data Mining and Knowledge Discovery Handbook*. Washington, DC: Springer; 2005; 131-146.

[18] Papadimitriou, S., Kitagawa, H., Gibbons, P.B., &Faloutsos, C. LOCI: Fast outlier detection using the local correlation integral.*19th International Conference on Data Engineering*. Los Alamitos, CA, 2003.

[19] Andersson, M., Gudmundsson, J., Laube, P., &Wolle, T. Reporting leadership patterns among trajectories.*22nd Annual ACM Symposium on Applied Computing*, Seoul, 2007.

[20] Miller, B.W., Breckheimer, I., McCleary, A.L., Guzmán-Ramirez, L., Caplow, S.C., Jones-Smith, J.C., &Walsh, S.J. Using stylized agent-based models for population–environment research: a case study from the Galápagos Islands. *Popul. Environ.* 2010, vol. 31(6), 401-426.

[21] Laube, P. &Imfeld, S. Analyzing relative motion within groups of trackable moving point objects. *Second International Conference on Geographic Information Science*. London, 2002.

[22] Gudmundsson, J. van Kreveld, M. &Speckmann, B. «Efficient detection of motion patterns in spatio-temporal data sets», *12th Annual ACM International Workshop on Geographic Information Systems*. New York, NY, 2004.

[23] Canizo, J.A., Carrillo, J.A., &Rosado, J. Collective behavior of animals: Swarming and complex patterns. *Arbor*. 2010, vol. 186, 1035-1049.

[24] Wang, Z., &Gu, D. «Distributed cohesion control for leader-follower flocking», *Fuzzy Systems Conference (FUZZ-IEEE)*. London, 2007.

[25] Su, H., Zhang, N., Chen, M.Z., Wang, H., &Wang, X. Adaptive flocking with a virtual leader of multiple agents governed by locally Lipschitz nonlinearity. *Nonlinear Anal. Real World Appl*. 2013, vol. 14(1), 798-806.

In: Mathematical Modeling in Social Sciences ... ISBN: 978-1-63117-335-6
Editors: J. C. Cortés López et al.

Chapter 26

Analysis of Noise for the Sparse Givens Method in CT Medical Image Reconstruction

A. Iborra[1,*]***, M. J. Rodríguez-Álvarez***[2]***,***
A. Soriano[2]***, F. Sánchez***[2]***, M. D. Roselló***[1]***, P. Bellido***[2]***, P. Conde***[2]***,***
E. Crespo[2]***, A. J. González***[2]***, L. Hernández***[2]***, F. Martos***[2]***, L. Moliner***[2]***,***
J. P. Rigla[2]***, M. Seimetz***[2]***, L. F. Vidal***[2] ***and J. M. Benlloch***[2]
[1]Instituto de Matemática Multidisciplinar (IM2),
Universitat Politècnica de València, Valencia, Spain
[2]Instituto de Instrumentación para Imagen Molecular (I3M),
Centro Mixto CSIC - Universitat Politècnica de València - CIEMAT,
Valencia, Spain

Abstract

The analytical methods like Filtered Backprojection (FBP) have dominated the image reconstruction in Computed Tomography (CT) because they generate images of a reasonable quality with low cost in terms of computing time.

Considering that image reconstruction in CT can be modelled by a large sparse linear system of equations such as $Ax = b$, direct methods like QR decomposition might also be suitable but they are not commonly used because such kind of methods present various drawbacks with a great difficulty in resolution. The reconstruction of high resolution images requires very large systems of equations and a large amount of computer memory. But these methods let us speed up image reconstruction because heavy computational cost operations are precalculated once and each image reconstruction only involves a backward substitution process.

The previous mentioned model has to take into account the geometry of the scanner and the physical processes involved on the measurement. In order to reduce computational costs x-ray scattering is often disregarded and monoenergetic x-ray assumption is made. The numerical stability depends on the method used to solve the system. Unavoidable errors as finite precision arithmetic errors and electronic noise occur. The accumulation of these effects often renders $Ax = b$ a system of equations with no

*E-mail address: amibcar@upv.es

exact solutions. QR decomposition is a good choice to solve this kind of systems of equations because its solution is equivalent to the least squares solution.

In this chapter we analyze the noise in the reconstructed image x, as we increase the error assumed in the linear system. Noise analysis is made for simulated and real data from Albira μCT. The results obtained with both the simulated and real data show that the ratio between the number of pixels in the detector and the desired image resolution is the main factor related to the error in the reconstructed images, so the number of projections (radiation dose received by the patient) can be lowered without loss of image quality.

Keywords: medical imaging, image reconstruction, QR decomposition, image noise

1. Introduction

Each projection of the CT device gives the intensity of a radiation beam transmitted through the object and measured at a detector pixel (b_i). This intensity can be expressed as

$$Ax = b \tag{1}$$

where A is the matrix that models the CT geometry. x is a vector that contains each one of the elements of the object to reconstruct (voxels) x_j. Each voxel attenuates the x-ray radiation according to its density. When a beam of radiation passes through a percentage $a_{i,j}$ of voxel x_j produces an intensity measurement on the detector's pixel b_i.

When the system (1) is solved, the solution x is obtained in terms of the density of each reconstructed voxel. There are several ways to solve system 1 but we are interested in QR decomposition using Givens rotations because when the factorization is done, finding x only implies a backward substitution process [2].

$Ax = b$ and applying QR decomposition [3] is obtained

$$Rx = Q^T b \tag{2}$$

where $Q \in \mathbb{R}^{m\times m}$ is orthogonal, $R \in \mathbb{R}^{m\times n}$ is triangular, I is the identity matrix and $Q^T b \in \mathbb{R}^m$. The fact the model does not describe the exact reality and the system 1 is overdetermined ($m \geq n$) is sufficient to not seek an exact solution $x \in \mathbb{R}^n$. Instead we will seek

$$\min \|r\|_p = \min \|Ax - b\|_p \tag{3}$$

for a norm p. We will choose $p = 2$ because 2-norm is preserved under orthogonal transformation and $Q \in \mathbb{R}^{m\times m}$ produced by the QR decomposition is orthogonal so the problem 3 is equivalent to 4

$$\min \|s\|_2 = \min \|Rx - Q^T b\|_2 \tag{4}$$

that is, the same solution (or solutions) $x \in \mathbb{R}^n$ that minimizes the residual $\|r\|_2$ also minimizes the residual $\|s\|_2$ [3].

In practice one has to solve the perturbed linear system

$$A\hat{x} = \hat{b} \tag{5a}$$

$$\hat{x} = x + \delta x \tag{5b}$$

$$\hat{b} = b + \delta b \tag{5c}$$

where x is the real object and $\hat{x}$ is the system's solution. b would be the measurement caused by x if the system behaved exactly as described in A and $\hat{b}$ is the CT device measurement. Considering this, the system 4 becomes

$$\min ||s||_2 = \min ||R(x + \delta x) - Q^T(b + \delta b)||_2 \tag{6}$$

and one would like to say that if δb is small, then δx is also small. The condition number of A is related to the sizes of $\frac{||\delta b||}{||b||}$ and $\frac{||\delta x||}{||x||}$.

$$\frac{||\delta x||}{||x||} \leq \kappa(A)\frac{||\delta b||}{||b||} \tag{7}$$

If we build A with a low condition number, $\frac{||\delta x||}{||x||}$ will be small if $\frac{||\delta b||}{||b||}$ is small (see Figure 1).

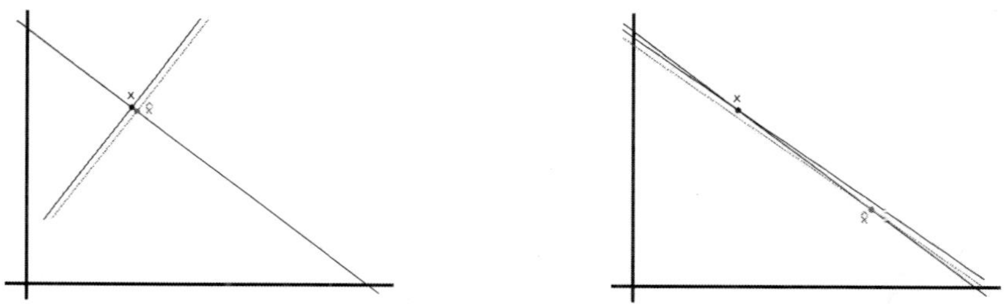

(a) Illustration of a well-conditioned system. (b) Illustration of an ill-conditioned system.

Figure 1. Well- and ill-conditioned systems. If the system is well-conditioned small perturbations on b imply small perturbations in x. If the system is ill-conditioned small perturbations on b do not imply small perturbations in x.

In order to obtain an accurate b with A, we propose to model the attenuation of x-rays as proportional to the density of the volume of the object they pass through and taking into account that the intensity of x-rays decays proportionally to the square of their source distance (cone beam factor [5]).

A CT measurement with an Albira μCT [4] of a phantom of well known densities was performed to prove that A characterizes b well enough (see Figure 2a). We reconstructed the measurement and generated a 3D image formed by the expected densities using the spatial information of the reconstruction (see Figure 2c). If the product of the matrix A and this phantom is done we obtain the theoretical value of the CT measurement, the values of the vector b (see Figure 2b).

The numerical analysis of the measurements is performed comparing the $\{b_i\}$ that are in the same projection p of the measurement ($\{b_i\}_p$). We compared the sum of each $\{b_i\}_p$ (see Figure 2d) and $\{b_i\}_p$ element-wise between real and generated measurements. In all cases there was little difference between real and generated measurements (less than 0.1% of relative error) that was caused mostly by the partial volume effect.

Once the measurement is generated has to be perturbed to model noise and physical effects that are not taken into account (as scattering). Let $r \in [0, 1]$ be the *level of perturbation* that we want to add to the measurement (from 0% to 100%). The perturbation is computed for each element b_i as follows:

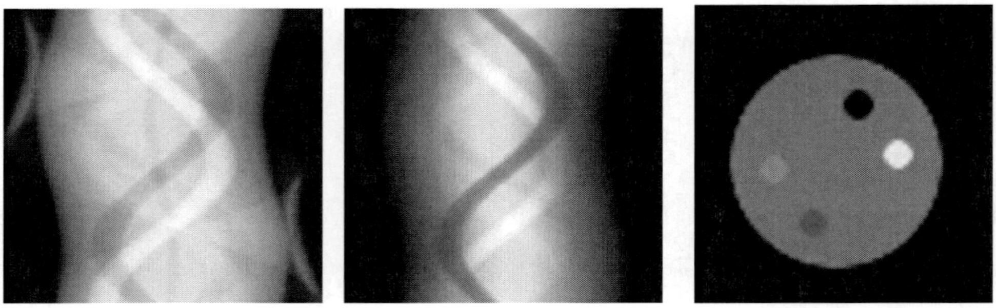

(a) Sinogram of a real CT measurement.

(b) Sinogram of a computer generated CT measurement.

(c) Central slice of the 3D computer generated phantom.

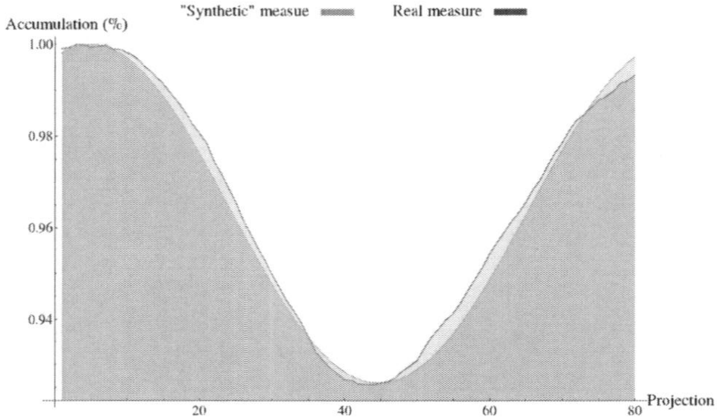

(d) Sum of all elements of each projection of the measurement for all projections. Comparing both real and computer generated CT measurements.

Figure 2. Sinograms of real (a) and computer generated (b) CT measurements. Figure (c) shows central slice of the 3D computer generated phantom of Figure (b). Comparison between real and computer generated CT measurements are shown in Figure (d).

- random generation of $p \in [-1, 1]$
- $n = b_i \, p$
- $\hat{b_i} = b_i + n \, r$

This way the obtained $\hat{b}$ models perturbations on data acquisition by detectors.

2. Results

The measurements were simulated varying the following parameters: 80, 100, 120, 140, 160, 180, 200, 220, 240, 260 projections taken, 60×60, 72×72, 84×84, 96×96, 108×108, 120×120, 132×132, 144×144, 156×156, 168×168, 180×180, 192×192 pixels per detector and 1%, 2%, 3%, 4%, 5%, 6%, 7%, 8%, 9%, 10% levels of perturbation. We generated a sample of 10 measurements for each combination of above parameters. If two measurement samples with same level of perturbation have different relative errors must be because of the condition number of A. Therefore we can establish a relationship between the CT model parameters and the condition number of the system matrix.

For the error measurement we will use the standard relative error $\frac{\|x-\hat{x}\|_2}{\|x\|_2}$ where x represents the ideal image (that is used to generate the measurement) and $\hat{x}$ represents the result of the reconstruction. Please note that $\|x - \hat{x}\|_2 = \|\delta x\|_2$ (see equation 5b) so the error must be under

$$\frac{\|x - \hat{x}\|_2}{\|x\|_2} \leq \kappa(A)\frac{\|\delta b\|_2}{\|b\|_2} \tag{8}$$

Figure 3a show the relative errors of a 84×84 pixels detector. The condition number doesn't depend on the number of projections. If we increase the number of pixels (to 132×132) as shown in Figure 3b the condition number of the system matrix drops.

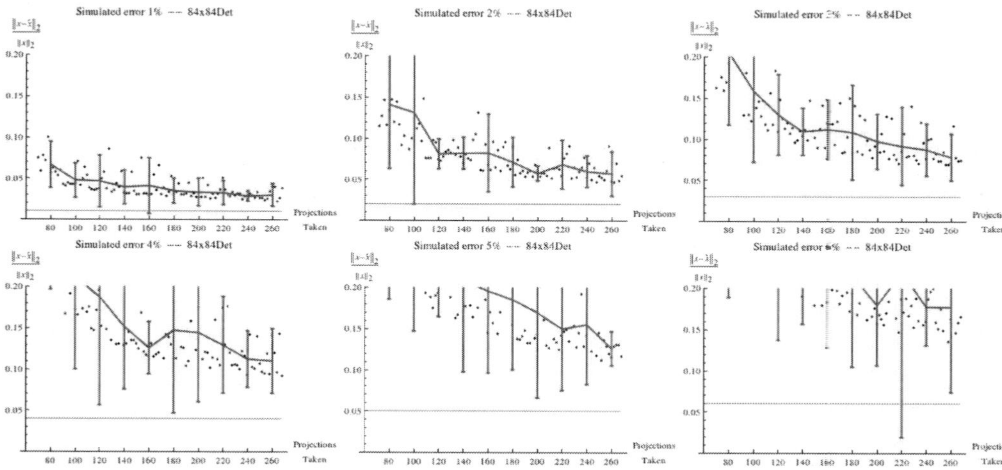

(a) Relative errors with a detector panel of 84×84 pixels.

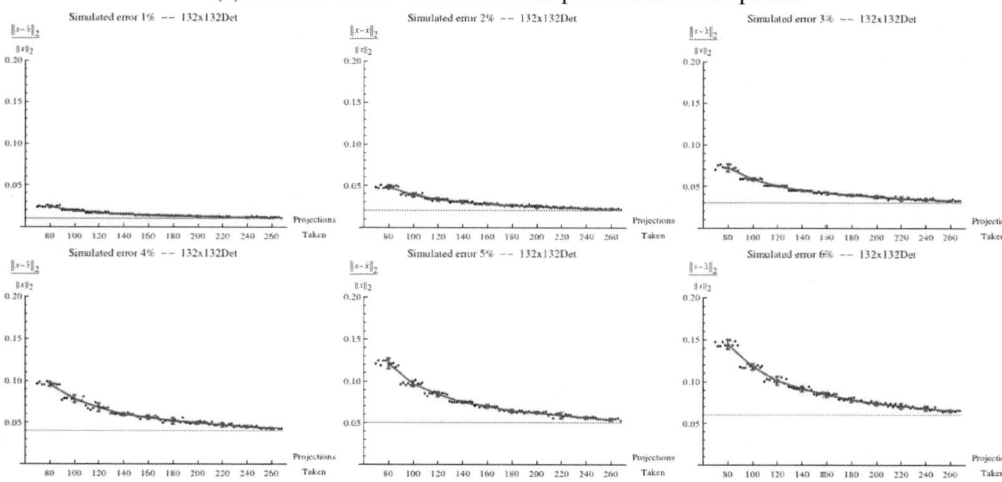

(b) Relative errors with a detector panel of 132×132 pixels.

Figure 3. Relative error in reconstructions as the number of projections increases. Black dots represent the relative error of each simulation for each system configuration. Each configuration has a vertical error bar showing 2σ of the sample (which contains 95% of the results). A line joins the mean relative error of each configuration. The level of perturbation is shown in a horizontal line.

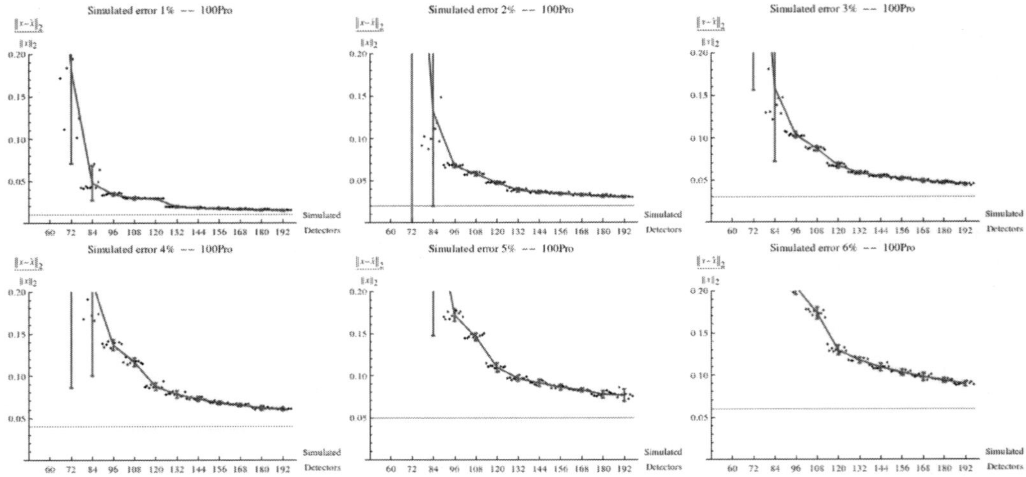

(a) Relative errors with 100 projections.

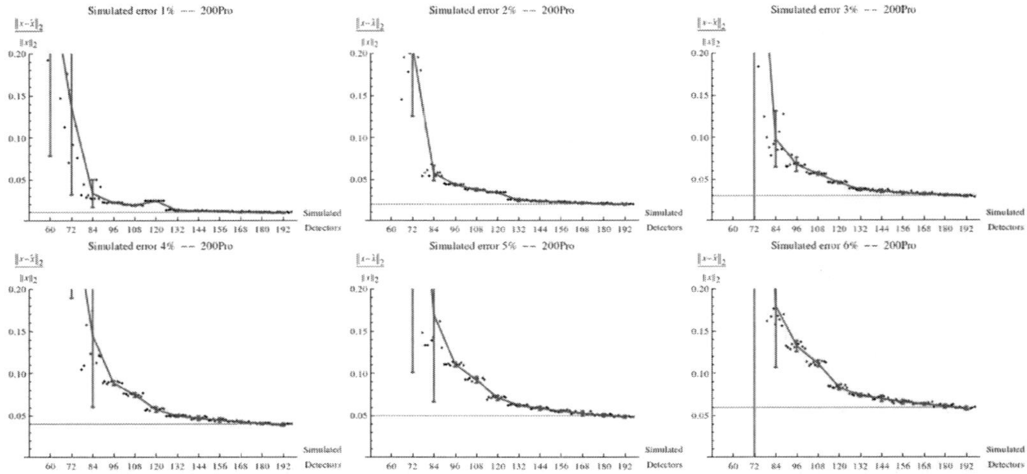

(b) Relative errors with 200 projections.

Figure 4. Relative error in reconstructions as the number of detector's pixels increases. Same legend used as in Figure 3.

As in the previous comparison, in Figures 4a and 4b we can see the evolution of relative error as the number of pixels increases for 100 and 200 projections. The intention of Figure 4 is to show how increasing the number of pixels decreases the condition number (at least an order of magnitude) but increasing the number of projection does not decrease it (or decrease it by units). The condition number of the system matrix depends mostly on the number of detector's pixels. It is true that increase the number of projection improves the condition of the system but only if it is reached a minimum number of pixels.

Previous results are obtained with a resolution of 1.3mm. We performed the same process for other resolutions (see Table 1). If we keep the pixel / voxel ratio between 2.2 and 2.5 we still face a well-conditioned system (taking into account that to solve a higher resolution problem implies a bigger problem with many more unknowns).

Table 1. Mean relative errors for configurations that match the pixel / voxel ratio and only 100 projections as the required resolution increases

Voxel Size	2.17mm	1.30mm	0.93mm
Detector / Voxel ratio	2.33	2.20	2.28
Relative error at 1% noise level	0.0134	0.0196	0.0215
Relative error at 2% noise level	0.0288	0.0383	0.0436
Relative error at 3% noise level	0.0441	0.0574	0.0670

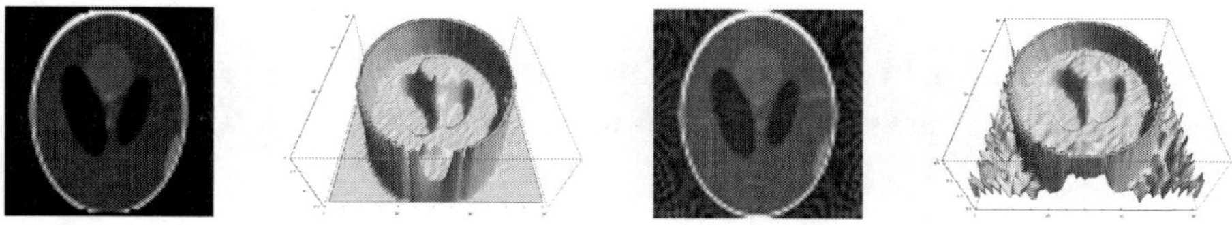

(a) Reconstruction of a Shepp-Logan measurement by QR decomposition.

(b) Density plot of Figure 5a.

(c) Reconstruction of a Shepp-Logan measurement by FBP.

(d) Density plot of Figure 5c.

Figure 5. Reconstructions of a measure perturbed by 1% by QR decomposition and FBP.

In Figure 5 we can see how QR decomposition reconstruct a measurement with low number of projections against FBP [1] (with a level of perturbation of 1% and a system resolution of 1.3mm). FBP's relative errors stay close to 2.5% while QR decomposition's relative error stays close to 1%.

Conclusion

Results show that above a detector's pixels / voxels ratio the condition number of the system matrix drops to a level that it can be say that the system is well-conditioned. This allows us to obtain reconstructed images with a relative error close to the level of perturbation introduced in the measurements.

As projections are not the main factor to a low condition number, QR decomposition needs fewer projections than dominant reconstruction methods in the field, like FBP. This can be exploited as modern CT devices have detectors with a high number of pixels and QR decomposition will allow decreasing the number of projections reducing patient dose.

We propose the reconstruction of CT images with QR decomposition as an alternative to FBP in cases that patient low radiation dose is needed.

References

[1] Feldkamp, L. A.; Davis, L. C.; Kress, J. W. Practical cone-beam algorithm. *J. Opt. Soc. Am. A*. 1984, 1, 612–619.

[2] Rodríguez-Álvarez, M. J.; Sánchez, F; Soriano, A.; Iborra, A. Sparse Givens resolution of large system of linear equations: Applications to image reconstruction. *Mathematical and Computer Modelling*. 2010, 52, 1258–1264.

[3] Golub, G. H.; Van Loan, C. F. *Matrix Computations*. JHU Press, 1996.

[4] Sánchez, F.; Orero, A.; Soriano, A.; Correcher, C.; Conde, P.; González, A.; Hernández, L.; Moliner, L.; Rodríguez-Álvarez, M. J.; Vidal, L. F.; Benlloch, J. M.; Chapman, S. E.; Leevy, W. M. ALBIRA: A small animal PET/SPECT/CT imaging system. *Med. Phys*. 2013, 40, 051906.

[5] Yao, W.; Leszczynsky, K. Analytically derived weighting factors for transmission tomography cone beam projections. *Phys. Med. Biol*. 2009, 54, 513–533.

In: Mathematical Modeling in Social Sciences ...
Editors: J. C. Cortés López et al.
ISBN: 978-1-63117-335-6

Chapter 27

AGENT-BASED MODEL TO DETERMINE THE EVOLUTION OF THE SEROPROTECTION AGAINST MENINGOCOCAL C OVER THE NEXT YEARS

L. Pérez-Breva[1,*], *R. J. Villanueva*[2,†], *J. Villanueva-Oller*[3,‡], *L. Acedo*[2,§], *F. J. Santonja*[4,¶], *J. A. Moraño*[2,‖], *R. Abad*[5,**], *J. A. Vázquez*[5,††] *and J. Díez-Domingo*[1,‡‡]
[1]Centro Superior de Investigación en Salud Pública - Fundación para el Fomento de la Investigación Sanitaria y Biomédica de la Comunidad Valenciana (CSISP-FISABIO), Valencia, Spain
[2]Instituto Universitario de Matemática Multidisciplinar, Universitat Politècnica de València, Spain
[3]Centro de Estudios Superiores Felipe II, Aranjuez, Spain
[4]Departamento de Estadística e Investigación Operativa, Universitat de València, Spain
[5]Instituto de Salud Carlos III, Majadahonda, Madrid, Spain

Keywords: Meningococcal C, Agent-based model, Seroprotection evolution, Prediction over the next years

---

*E-mail address: perez_lin@gva.es
†E-mail address: rjvillan@imm.upv.es
‡E-mail address: jvillanueva@pdi.ucm.es
§E-mail address: luiacrod@imm.upv.es
¶E-mail address: Francisco.Santonja@uv.es
‖E-mail address: jomofer@mat.upv.es
**E-mail address: rabad@isciii.es
††E-mail address: jvazquez@isciii.es
‡‡E-mail address: diez_jav@gva.es

1. Introduction and Motivation

Neisseria meningitidis is a major cause of morbidity and mortality during childhood in industrialized countries and has been responsible for epidemics in Africa and in Asia. This bacterium is the main cause of meningitis type C (MenC), an infection of the brain and spinal cord that can even infect the blood. Neisseria meningitidis is transmitted exclusively among humans, mainly during adolescence, by healthy carriers. Men C, even properly treated with specific antibiotics, there is up to 10% of mortality and 10% of survivors have sequelae. From 2000 the Conjugate Vaccine C (MCC) is used in campaigns with different strategies in the Community of Valencia (Spain). In 2006 is fixed the current vaccination schedule with three doses: 2, 6 and 18 months of age.

Recent studies on the MCC-vaccination have determined that levels of protection provided by this vaccine are lower than expected, in particular, in toddles. Doctors conjecture that, in $5 - 10$ years, there will be an increase of cases in children younger than a year because the herd immunity provided among the adolescents by the current vaccination schedule will disappear. Because of this, health experts in UK and Spain have decided to change the current vaccination schedule removing a dose in infants and adding it in adolescence.

In this work, we describe a seroprevalence study and present a dynamic agent-based model to analyse the evolution of the population protection given by the MCC vaccine in order to find out if the doctor's conjecture is correct.

The chapter is structured as follows. In Section 2, we describe the seroepidemiological study. In Section 3 we present the agent-based model. In Section 4, we present the prediction over the next few years and discuss the Doctor's conjecture.

2. Seroepidemiological Study in the Community of Valencia

The study was carried out from October 2010 to April 2012. 1800 samples were collected in twelve primary care centers and in three hospitals after getting the written informed consent. Exclusion criteria included: immunosuppression, severe medical illness, previous meningococcal disease and organ transplants. Blood samples were collected from subjects from 3 to 90 years of age, after getting the written informed consent from the subject (older than 17 years of age), and/or subject's parent(s) or legal guardian(s) in those younger than 18. Sera were stored frozen at -80 centigrade degrees, until they were sent to the Spanish National Reference Laboratory for Meningococci, Instituto Salud Carlos III (Madrid). Functional meningococcal serogroup C antibody levels were determined using the serum bactericidal activity (SBA) assay, with baby rabbit complement (rSBA). Titers of serum bactericidal antibody were expressed as the reciprocal serum dilution yielding 50% or greater killing after incubation for 60 min.

Only samples with rSBA 1:8 or higher were first qualified. Antibody levels were log transformed. Geometric mean titers (GMT) with 95% confidence intervals were calculated. Titers less than 1:8 were assigned a value of 2 for computational purposes, being a quarter of the value of the lowest limit of detection.

Meningococcal C conjugate vaccine (MCC vaccine) immunisation status was obtained from the Vaccine Information System (SIV) of the Region of Valencia, which is a population-based computerized vaccination registry put in place in 2000 [1]. SIV contains data on vaccine type, manufacturer, batch number, place and administration date, and number of doses administered of the approved immunisation series. All records can be linked by a unique personal identification number and SIV information is verified by consistency algorithms and quality.

Several meningococcal C conjugated vaccine (MCCV) programs in Valencian Community have been carried out. There are three types of vaccination

- Primary: when the child is vaccinated before the first year of life.
- Booster: second dose of the vaccine.
- Catch-up: especial campaign of vaccination.

In the Community of valencia, in 1997 one dose of plain polysaccharide vaccine (the previous to MCCV) was administered in subjects 18 months to 19 years (coverage 85%).

Also, in 2000, MCCV were scheduled at infants and a progressive catch up until 19 years with coverage over 90% in children less than 6 years, and decreased in older ages. In 2006 a booster dose was added to children born from 2005 with a coverage greater than 90%. In 2006 is fixed the current vaccination schedule with three doses: 2, 6 and 18 months of age.

As a consequence of the analysis of the database obtained from the 1800 samples under the described vaccination context, we were able to determine the present serological situation (Oct 2011), the seroprotection of unvaccinated individuals and the seroprotection evolution of vaccinated individuals depending on the way they were vaccinated (primary, booster or catch-up) and age. They can be seen depicted in Figures 1 and 2.

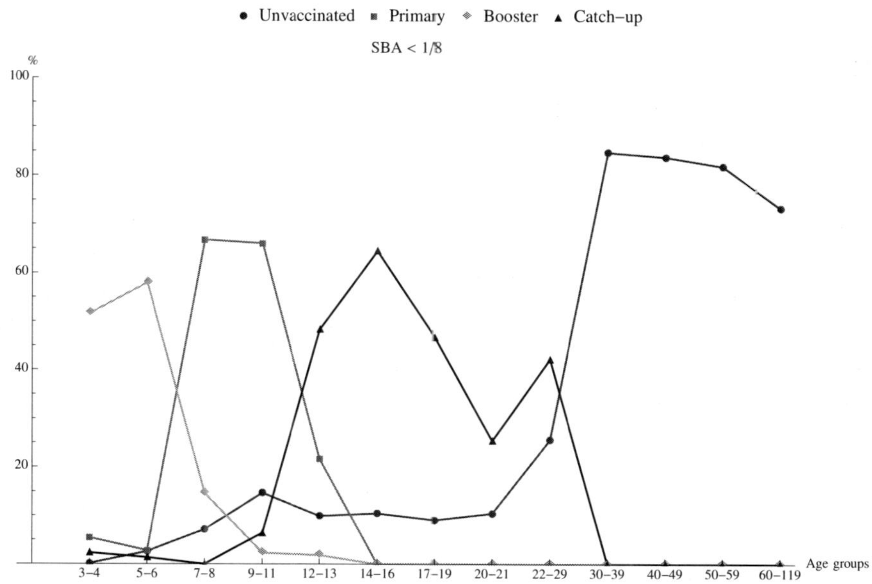

Figure 1. Percentage of unprotected (SBA $< 1{:}8$) by age group depending on if the individual is unvaccinated, primary, booster or catch-up.

Also, other interesting conclusion can be obtained:

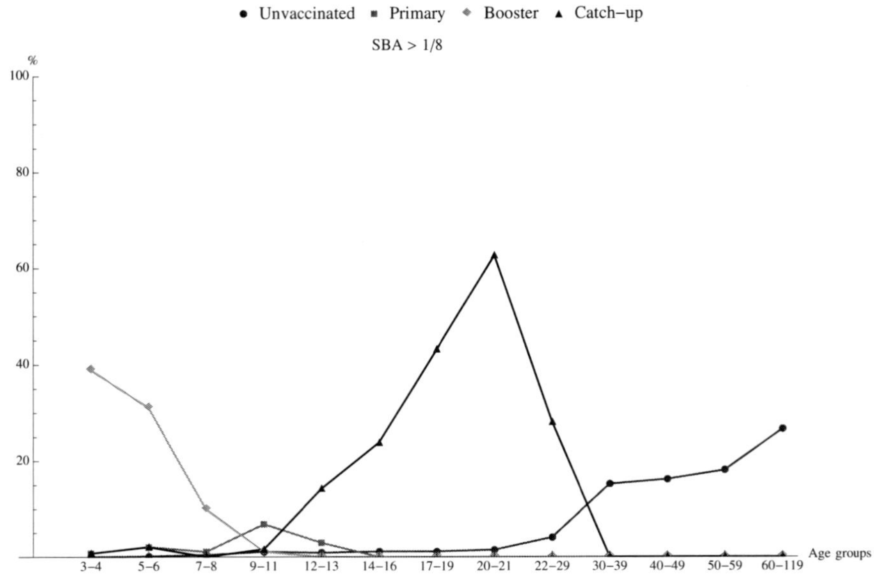

Figure 2. Percentage of protected individuals (SBA $\geq$ 1:8) by age group depending on if the individual is unvaccinated, primary, booster or catch-up.

- Subjects vaccinated in the catch up programme had higher levels of protection, even though a longer time elapsed since vaccination. Especially those vaccinated at an age older than 8 years.
- Subjects under 16 years of age had lower levels of seroprotection, as they were younger when the catch up or were routinely vaccinated in the first-second year of life.
- Seroprotection is highly related to the age at immunization and the time elapsed since vaccination.

As it was mentioned in the Introduction, recent studies on the MCC-vaccination have determined that levels of protection provided by this vaccine are lower than expected, in particular, in toddles (young children). Doctors conjecture that, in $5 - 10$ years, there will be an increase of cases in children younger than a year because the herd immunity provided among the adolescents by the current vaccination schedule will disappear.

The Joint Committee on Vaccination and Immunization of DH has recommended in January 2012 a change in the vaccination schedule for UK:

- An adolescent dose of MCC-vaccine should be introduced and a dose in infants should be removed.
- This change needs to ensure that coverage is high enough to maintain the herd immunity.

In Spain, the Grupo de Trabajo MENCC 2012 recently recommended a new the vaccination schedule 2 months, 12 months and 12 years old. In both cases, the new schedule will start in Jan 2014.

But, is this conjecture true? Using the data given by the database obtained from the 1800 samples in the Community of Valencia, we are going to develop a dynamic agent-based model in order to support this conjecture.

3. Agent-Based Model Building

To keep the model as close as possible to the real situation we opted for implementing an agent-based model. Agent-based models constitute a mainstream technique in modern epidemiological studies. In an agent-based model, every node or site represents an individual characterized by a dataset useful for the simulated evolution of the disease and seroprotection. Firstly the model should be age-structured because it is a well-known fact that incidence and transmission of meningococcal C disease depends strongly on age. We consider the following age groups:

1. 0-2 years = 0-35 months.
2. 3-4 years = 36-59 months.
3. 5-6 years = 60-83 months.
4. 7-8 years = 84-107 months.
5. 9-11 years = 108-143 months.
6. 12-13 years = 144-167 months.
7. 14-16 years = 168-203 months.
8. 17-19 years = 204-239 months.
9. 20-21 years = 240-263 months.
10. 22-29 years = 264-359 months.
11. 30-39 years = 360-479 months.
12. 40-49 years = 480-599 months.
13. 50-59 years = 600-719 months.
14. Older than 60 years = older than 720 months.

Adolescence is considered as the period between 12 years old (144 months) until 19 years old (239 months). In our model we do not distinguish among men and women because meningococcal C disease incidence or transmission does not depend on the sex of the individuals.

Our unit of time is a month because the average carriage time is initially considered as 3 months and it seems natural to set the time scale sufficiently small to follow the carriage evolution but also sufficiently large to avoid time-consuming computations associated with every update corresponding to every discrete time-step.

Our starting month is October, 2011 ($t = 0$) and the simulation ends at September, 2025 ($t = 167$). We consider a model with 1,000,000 sites.

Every site in the model is characterized by the following labels:

1. Label [site i] 1): the age in months.
2. Label [site i] (2): This label classifies the sites according to their state of seroprotection. We set it to 1 if rSBA is smaller than 1:8 and 2 if r SBA is greater than 1:8. The initial information will be obtained for the seroprotection study described in the previous section.
3. Label [site i] (3): This label classifies the individuals according to their vaccination status: we set it to 0 if the individual has never been vaccinated, P for primed, B for people which have received a booster dose and C for individuals participating in a catch-up campaign.
4. Label [site i] (4): age in months at the moment of the last vaccination.

Newborns are considered to be susceptible, unvaccinated and with rSBA $<$ 1:8.

A necessary requisite for starting any epidemiological simulation is the identification of the initial state of the individuals. The population distribution for the age of the individual was obtained from public data at the Valencian Institute of Statistics [2]. The assignment of ages to the sites was discussed elsewhere [3].

We started our simulation in October 2011. We distribute seroprotection and vaccination depending on data in Figures 1 and 2. Vaccination coverage is assumed to be 96% and the current vaccination strategy involves doses at 2, 6 and 18 months. The model evolution rules are described in the following

- FOR every month t (from Oct 2011 to Sep 2025)
  - FOR every individual $i = 1$ to $T = 1000000$
    1. ADD a month to his/her age
    2. IF this node i does not die
       - IF this node i has to be vaccinated (following the current schedule), UPDATE the type of vaccination, the age of the last vaccination and the SBA becomes greater than $1:8$
       - ELSE UPDATE his/her protection depending on his/her age and age and type of the last vaccination
    3. ELSE this node dies, it is *resurrected* as a unprotected unvaccinated newborn.

The objective is to simulate the future dynamics on Meningococcal C transmission since October, 2011. With this model we will be able to see if the Doctor's conjecture is correct.

4. Model Simulations and Predictions

Now we discuss the implementation of a simulation program for the evolution of seroprotection status and the effect of vaccination strategies in the Community of Valencia, Spain. The code relies heavily on the results of the seroprevalence study previously described.

Agent-based models constitute a mainstream technique in modern epidemiological studies and, although they may require a huge computational effort, they provide more detailed characterization of individuals and the possibility of micromanage the evolution of the clinical history of each individual represented by a site.

Computational power was obtained by means of the open-source BOINC software for distributed computing. Servers were located at the FALUA Laboratory for Distributed Computing [5]. BOINC is better known for their applications to large computational projects such as SETI@HOME [6] and it has been used also in epidemiological studies such as malaria control [7].

Previous experience in the simulation of other infectious-contagious diseases (respiratory syncytial virus) was also an advantage [8].

Thus, we run the agent-based model starting in Oct 2011 [1]. The initial situation can be seen in Figure 3.

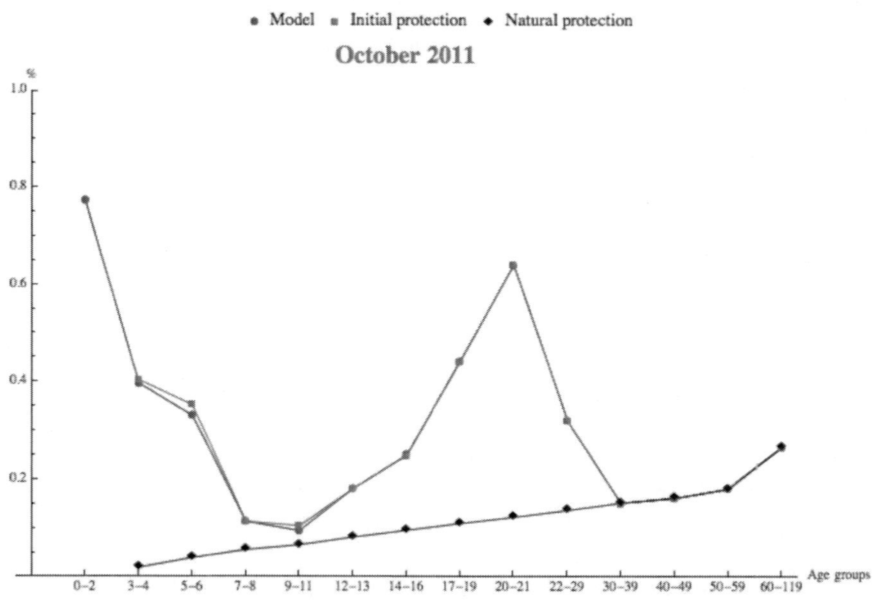

Figure 3. Initial (Oct 2011) percentage of protected individuals (SBA $\geq$ 1:8) per age group.

As the running progresses, the adolescents lose their protection over the time. In October 2015 (Figure 4), the protection of teenagers $12 - 19$ years old is below 20%.

Thus, this model supports the experts' conjecture about the dramatic reduction of adolescent protection expected over the next few years if the vaccination schedule does not

[1] In fact, we should carry out several simulations using distributed computing environment and calculate the mean of all of the output. The mean is what we show in the figures.

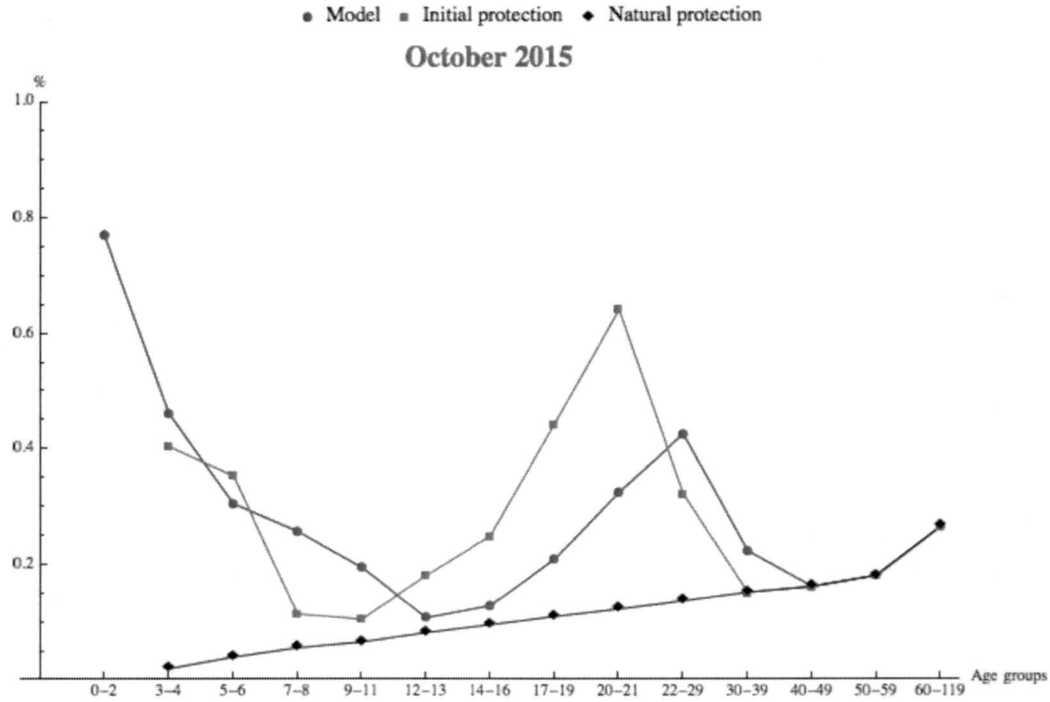

Figure 4. Simulated prediction of the percentage of protected individuals (SBA $\geq$ 1:8) per age group in October 2015. The adolescents are loosing their protection over the time.

change. In Figure 5 we can see that adolescent $14 - 19$ years old, the most important transmitters, have minimal protection.

Conclusion

The seroprotection study provided a snapshot of the antibody persistence in the Valencian population in 2011. We fitted an evolution curve for seroprotection levels. Individuals with rSBA$\geq$1:8 are considered protected against Neisseria meningiditis C disease. Since early studies, it is known that SBA levels decrease very fast in children but persist longer in adolescents. A recent study [4] supports these results and shows that seroprotection wanes slower as the age of the vaccinated individual increases. A remarkable difference among primed (children under 1 year of age) and those who received the 2005 catch-up dose is also found.

The vaccinated seroprotection wanes fastest for children under one year of age; for children from one year until sixteen years of age the period of seroprotection is considerably larger and increases steadily with age.

Thus, we build an agent-based model to study the evolution of the serprotection over the time per age groups. A Public Health goal is to maintain adolescents and young adults well protected without reducing the protection in the younger than 4 years of age. However, with the current vaccination schedule, the simulations of the agent-based model show that the adolescents, the most important carriers and transmitters of MenC, lose their protection

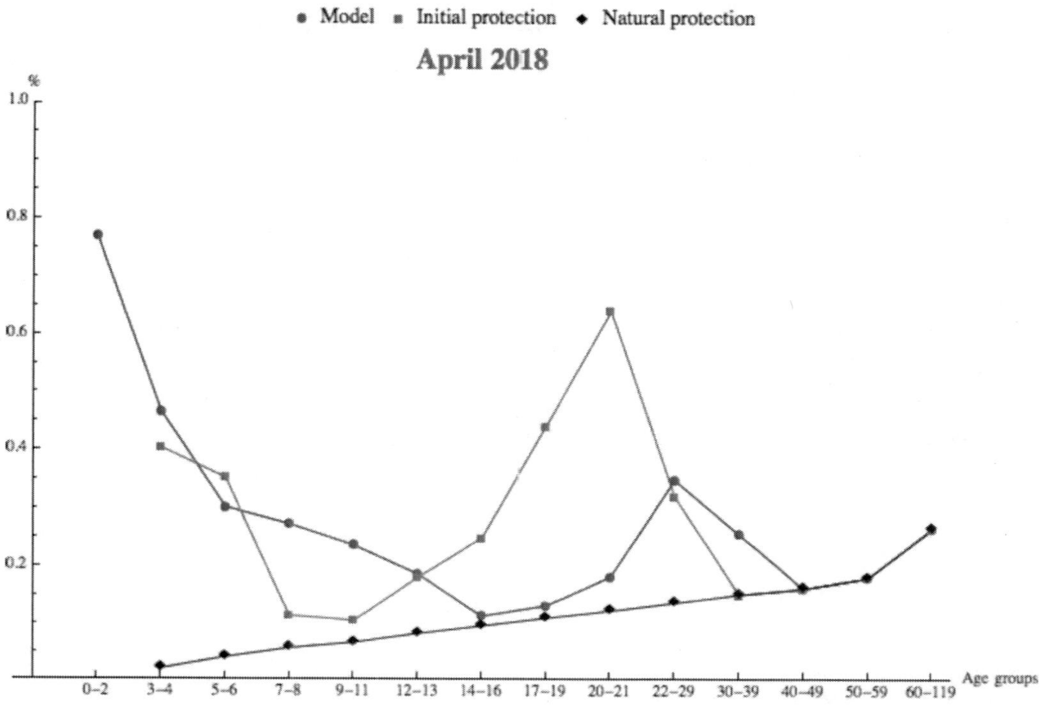

Figure 5. Simulated prediction of the percentage of protected individuals (SBA $\geq$ 1:8) per age group in October 2018. The adolescents $14 - 19$ years old, the most important transmitters, have minimal protection.

over the time in such a way that in October 2015, the protection of teenagers $12 - 19$ years old is below 20%.

Then, the presented model supports the Doctor's conjecture and suggests, as the Joint Committee on Vaccination and Immunization of DH and the Spanish Grupo de Trabajo MENCC 2012 proposed, a change in the vaccination schedule.

References

[1] Puig Barberà J, Pérez Vilar S, Pérez Breva L, Pastor Villalba E, Martín Ivorra R, Díez Domingo J. Validity of the Vaccine Information System to ascertain influenza vaccination status in hospitalized adults in Valencia, Spain. *Poster presentation in The 4th International Conference & Exhibition on Influenza Vaccines for the World- IVW 2012*. Valencia, Spain. 9-12 October 2012.

[2] http://www.ive.es

[3] Acedo L, Moraño JA, Villanueva RJ, Villanueva Oller J, Díez Domingo J. Using random networks to study the dynamics of respiratory syncytial virus in the Spanish region of Valencia, *Mathematical and Computer Modelling*, 54 (2011): 1650-54.

[4] Ishola DA, Borrow R, Findlow H, Findlow J, Trotter CL, Ramsay ME. *Prevalence of serum bactericidal antibody to serogroup C Neisseria meningitidis in England a decade after vaccine introduction. Clin Vacc Immun* 2012; 19 (8):1126-30.

[5] http://falua.cesfelipesegundo.com

[6] http://setiathome.berkeley.edu/

[7] http://www.malariacontrol.net/

[8] Acedo L, Moraño JA, Villanueva RJ, Villanueva Oller J, Díez Domingo J. Using random networks to study the dynamics of respiratory syncytial virus in the Spanish region of Valencia, *Mathematical and Computer Modelling*, 54 (2011): 1650-54.

In: Mathematical Modeling in Social Sciences ...
Editors: J. C. Cortés López et al.
ISBN: 978-1-63117-335-6

Chapter 28

APPLYING CLUSTERING BASED ON RULES FOR FINDING PATTERNS OF FUNCTIONAL DEPENDENCY IN SCHIZOPHRENIA

***Karina Gibert*[1,*] *and Luis Salvador Carulla*[2,3]**
[1]Knowledge Engineering and Machine Learning group
Dpt. Statistics and Operations Research,
Universitat Politècnica de Catalunya, Barcelona, Spain
[2]PSICOST Research Association, Spain
[3]Center for Disability Research and Policy, Faculty of Health Sciences
University of Sydney, Australia

Abstract

In 1996 Fayyad described the *Knowledge Discovery process* as an integral process including prior expert knowledge, preprocessing, data mining and knowledge production to produce understandable patterns from data. *Clustering based on rules* (ClBR) is a particular data mining method suitable for profiles discovery. ClBR is an hybrid AI and Statistics technique, which combines some Inductive Learning (from AI) with hierarchical clustering (from Statistics) to extract knowledge from complex domains in form of typical profiles. It has the particularity to embed the prior expert knowledge existent about the target domain in the clustering process itself, guaranteeing more comprehensible profiles. In this paper, the results of applying this technique to a sample of patients with mental disorders are presented and their advantages with regards to other more classical analysis approaches are discussed. The final step of knowledge production is supported by post–processing tools, like Class panel graphs (CPG) and Traffic Lights panels (TLP), which were appreciated by domain experts as powerful, friendly and useful tools to transform raw clustering results into understandable patterns suitable for later decision-making.

It was confirmed that functional impairment (FI) in schizophrenia and other severe mental disorders show a different pattern than FI in physical disability or in ageing population. Understanding the patterns of dependency in schizophrenia and getting criteria to recognize them is a key step to develop both elegibility criteria and services

*E-mail address: karina.gibert@upc.edu

for functional dependency in this particular population. This research was related with the implantation of the Spanish Dependency Low, in Catalonia, acting from 2007.

Keywords: Data mining and Knowledge Discovery, clustering based on rules, decision support and Knowledge management, class panel graph, prior expert knowledge, schizophrenia, clinical test, dependency

1. Introduction

Nowadays it is well known that *Knowledge Discovery from Databases (KDD)* [Fayyad et al. 96] provides a good framework to analyze complex phenomena, to obtain novel and valid knowledge that can improve the background *doctrine corpus*. In this work a specific KDD method named Clustering Based on Rules is used to find patterns of functional dependency in a schizophrenic population. The importance of characterizing dependency in severe mental disorders' patients is related with a new legal frame where dependent persons receive attention by health systems.

In 1998, the European Council recommended the member states to develop services for people with dependency. The European Council defined *Dependency* as a condition where, due to the lack or loss of physical, psychological or intellectual functions, the person needs assistance and/or significant aids to perform daily living activities related to self-help and autonomy. Thus, it can be said that dependency is understood as a limitation on functionalities and thus, one can talk about *functional dependency* [Salvador-Carulla et al. 10].

Persons with functional dependency form a population characterized by high special needs, including the aged, and disabled (either physical or psychological). However, when the global concept of dependency was used as a reference to deploy dependency services in different European countries and elegibility criteria to access those services were developed, it became clear that severe mental disorders did not fit well into the functional dependency model developed for physical and ageing populations. Among mental disorders, schizophrenia is a major cause of functional impairment [Prince et al. 07]. In [Ustun et al. 99], relationship between disability and physical and mental conditions was studied, and positive symptoms of schizophrenia (active psychosis) were ranked the third most disabling condition, higher than paraplegia and blindness, by the general population. In the Global Burden of the Disease study [WHO 01], schizophrenia accounted for 1.1% of total disability-adjusted life years (DALYs) and 2.8% of years lived with disability (YLDs). The use of services and the economic cost of schizophrenia to society are also high [Haro et al. 06]. However the functional impairment related to schizophrenia an other mental disorders widely differs from functional impairment in physical disabilities or ageing. Daily living activities such as grooming or moving around are impaired in the later groups. However, impairment in schizophrenia concerns social isolation, difficulty in medication compliance and behavioral problems which need monitoring from carers among other distinct impairments. Most of them produce dependency, even if patients are physically able to perform daily activities by themselves.

Spain was the first Mediterranean country to adopt a policy on dependency by also taking into account severe mental disorders. The *Law for the Promotion of personal autonomy and care for persons with dependency (LPAD, 39/2006, 14th december)* was approved by the Spanish government in 2006, to be enacted from 2007 on, by regional dependency agencies. In Catalonia, the catalan government created a dependency agency named PRODEP to lead the deployment of the law and the dependency services to be implemented. Being aware of the special characteristics of functional dependency in mental health population, PRODEP funded a specific project to adapt the dependency concept to persons with severe mental disorders (schizophrenia) [Salvador-Carulla et al. 06]. Also, the elegibility criteria for accessing dependency services and benefits derived by the Law should guarantee that dependency was correctly detected in mental patient's population. Obtaining a clear definition of dependency patterns in schizophrenia and proper eligibility criteria is specially relevant for decision-making related to the implantation of the Low.

This work contributed to update the know-what and know-how about dependencies in schizophrenia, by confirming that dependency in schizophrenic persons do not follow the same patterns that in ageing or physical impaired populations. Characteristics of functional dependency in schizophrenia were elicited and this was a relevant support to improve the official instrument designed to assess dependency, by adapting it to properly detect dependency in persons with mental disabilities.

Under the KDD approach many different problems can be addressed, provided that the proper Data Mining technique is used. For the particular application faced in this work, clustering methods are suitable, as they permit to identify distinguishable groups of similar individuals, which eventually admit a generic solution for each group. It has been seen in [Gibert, Sonicki 99] that classical clustering techniques cannot well recognize certain domain structures, producing some non-sense classes, and providing results difficult to be interpreted by the experts. In fact, this arises when dealing with *ill-structured domains* (ISD) [Gibert, Corts 98], as is the case of severe mental health disorders. ISD are characterized by [Gibert, Corts 94]: i) numerical and qualitative information coexists (see [Gibert, Corts 97], [Gibert et al. 05], [Gibert et al. 13]); ii) there exists some relevant semantic additional (but partial) knowledge to be regarded.

Clustering based on rules (ClBR) [Gibert, Corts 94] is a technique described bellow especially introduced by Gibert to improve clustering results on ISD by taking into account the prior expert knowledge existing on the target domain. In fact, a main advantage is to guarantee the meaningfulness of the resulting classes. In previous works [Gibert, Sonicki 99] [Gibert et al. 03] [Gibert et al. 13] the improvements in results related with ClBR instead of using other classical clustering techniques have been discussed.

In this study, patterns of dependency in schizophrenia were identified by using the Clustering Based on Rules (ClBR) methodology over a real database.

The database contains information about a sample of patients with severe mental disorders, in particular schizophrenia, regarding different aspects: clinical, socio-demographic characteristics, results on psychometric batteries of tests about functional impairment, information about the use of private or public Health services and about the amount of support required from their carers (usually relatives). Main idea is to induce from data homogeneous groups in schizophrenic population as well as their distinctive characteristics, contributing to provide an operational definition of functioning in schizophrenia. On the one hand, this is useful for better understanding dependency patterns in our immediate environment; on the other hand, discovered profiles support proper decisions about planning allocation of resources derived from the application of LPAD to the psychic disabled persons.

However, *KDD* is, as proposed by Fayyad [Fayyad et al. 96], the high level process combining *DM* methods with different tools for extracting *knowledge* from data. In fact, Fayyad's proposal, pointed to a new paradigm in KDD research: *"Most previous work on KDD has focussed on [...] DM step. However, the other steps are of considerable importance for the successful application of KDD in practice."* From this point of view, KDD includes prior and posterior analysis tasks as well as the application of DM algorithms. This work fits in this integral approach by using a data mining method (the ClBR) that permits prior expert knowledge on the target phenomenon be introduced into the system to guide the classes' construction process. Also some tools to assist interpretation of results and reporting (Class Panel Graph[Gibert et al. 05]). have been used.

Advantages of the proposed approach with regards to other more classical analysis are discussed. Advantages of proper pre and post processing of data are also stressed.

2. Methodology

2.1. Preprocessing

First, descriptive statistics was done. Very simple statistical techniques [Tukey 77] were used to describe data and to get preliminary information about. Next, data cleaning, including missing data treatment or outlier detection was performed. It is a very important phase, since the quality of final results directly depends on it. Decisions were taken on the basis of descriptive statistics and background knowledge of the experts. A selection of relevant variables among the whole battery of scales was also done together with the experts. Redundant items through different scales were eliminated.

2.2. Data Mining

Data was analyzed using two methods: i) A hierarchical clustering was performed, using chained reciprocal neighbors method [Murtagh 83], with Ward criterion [Ward, J.H. 63] and the Gibert's mixed metrics [Gibert, Corts 97], since both numerical and categorial variables were considered. ii) A Clustering based on rules (ClBR), described bellow, was used on the same data set. In this chapter, just an intuitive idea is given (see details in [Gibert, Corts 98] and [Gibert, Corts 94]). It is a hybrid AI and Statistics technique which combines inductive learning (AI) and clustering (Statistics) especially designed to extract knowledge in form of typical profiles from certain complex domains like ISD. A Knowledge Base (KB) expressing the existent prior domain knowledge is considered to properly bias the clustering on the database. It is implemented in the software KLASS [Gibert, Nonell 08] [Gibert, Nonell, 05b] and it has been successfully used in several real applications [Gibert et al. 13], or [Gibert, Sonicki 99] [Comas et al. 01], [Gibert et al. 03], [Gibert et al. 08], [Gibert et al. 12]. Our experience is that ClBR perform better than using any statistical clustering method by itself, since an important property of the method is that semantic constraints implied by the KB are hold in final clusters; what guarantees interpretability of the resulting classes. Also, it uses to be better than pure inductive learning methods, since it reduces the effects of missing some implicit knowledge in the KB:

1. Build a (KB) with additional prior knowledge provided by the expert, which can even be a partial description of the domain.
2. Evaluate the KB on data. Induce an initial partition over data from it; build a residual class (RC) with the data not included in this partition.
3. Independent hierarchical clustering for every rules-induced class (RIC).
4. Generate prototypes of each rules-induced class.
5. Build the extended residual class as the union of RC with the set of prototypes of RIC, conveniently weighted by the number of objects they represent.
6. Perform a weighted hierarchical clustering of the extended residual class.
7. In the resulting dendrogram, substitute every rules-induced prototype by its hierarchical structure, obtained in 3, integrating a single hierarchy.

For both methods, clustering results can be graphically represented in a dendrogram. Final number of classes was determined on best horizontal cut (maximizing the ratio of between-classes inertia versus within-classes inertia). This identifies a partition of the data.

2.3. Post-Processing

Interpretation of the classes use to be difficult and time consuming and requires much human guidance. It is critical to bridge the gap between data mining and effective decision support [Gibert et al. 13]. Very recently, A Sandy Pentland, the Head of MediaLab Enterpreneurship of MIT, pronounced a keynote in the Campus Party Europa Sept 4th 2013 where, appart from stressing the importance of data era and a lack of data scientists to face the future needs, he specifically referred to the need of general literacy about data interpretation.

Here, some interpretation-oriented tools have been used to complete the KDD process by producing understandable knowledge [Fayyad et al. 96] on the basis of the data mining results. For the particular case of clustering, this means to produce understandable profiles on the basis of discovered clusters itself, by post-analysing clusters composition. In this works, the conceptualization of the classes was performed by means of close interaction between experts and data miners, by using the Class panel graph (CPG) [Gibert et al. 08b] as a support, combined with a significance analysis of variables with respect to clusters. The CPG displays a compact overview of conditional distributions of the variables through the classes, in such a way that characteristic behaviours of variables in especial classes is quickly identified. On the other hand, the relevance of differences between classes is assessed using ANOVA, Kruskall-Wallis or χ^2 independence test, depending on the required assumptions hold by each of the variables. Experts use the CPG to get a meaningful description of classes by identifying which variables indicate particularities of every class regarding the others and making a conceptualization process which leads to a class-labeling proposal regarding the semantic entity represented by each class.

Later, the Traffic Lights Panel (TLP) was designed [Gibert et al. 08b] and it was also produced for this application and discussed with the experts to confirm the previous interpretation. The TLP is a symbolic abstraction of the CPG where the central trend of every variable in each class is represented by means of a color-coding inspired in traffic lights and related with the semantics of the variable, and some latent concept relying behind the discovered cluster. In this case, severity of disorder. Thus, for this particular application, red color are associated to more severity whereas green color with less severity.

3. Application to Functional Dependency in Schizophrenia

3.1. Data

The analysis is performed on the database PSICOST-II, a naturalistic study of assisted prevalence pre-post with a follow up of two years and three waves of data collection (beginning of study, first year, second year) of a representative sample of six-month's prevalence [Ochoa et al. 12]. There were 306 patients between 18 and 65 years with diagnoses of schizophrenia DSM-IV [APA 00]. Patients were in contact with mental health services in 4 small healthcare areas in Spain which represented different socio-economical contexts regarding familiar rent, construction levels and mental health services provided. For 205 patients it was also possible to interview the main care giver. Four persons were specially trained to interview the patients for evaluating a battery of assessment scales over the patient. Independent interviews with psychiatrist, and main care giver as well as a revision of the clinical history of the patient were performed, provided the informed consent of the patient. The assessment scales used for the evaluation concerned disease (PANSS [Kay et al. 86], [Prudo, Blum 87]), quality of life (EuroQol [Brooks 96]), functioning (through the scales GAF [Endicott et al. 76][Goldman et al. 92] DAS [Janca et al. 96]), familiar help requirements (ECFOS [Vilaplana et al. 07] about daily activities performances, behaviour, economic management, etc), health services use (CECE [Vzquez-Polo et al. 05]).

The dominant profile in the sample is a man (68%), single (77%), with primary school (49%), getting a pension (62%) and mainly leaving with parents (67%). Disease started about 24 years and stay for more than 14 years, (sample of long duration schizophrenia).

3.2. Clustering

Clustering of the 306 patients was made under the two approaches mentioned before:

i) With classical hierarchical clustering, 5 classes emerged. However, most of the variables shown no significant differences *vs* classes and their interpretation was confusing and psychiatrists could not learn too much from the results. Patients with different levels of dependency were mixed in the different clusters and it was no possible to understand the underlying clustering criteria.

ii) Several iterations of the ClBR process conformed the prior knowledge acquisition phase, and experts could elicit their implicit knowledge. The knowledge provided by the experts concerned some clear situations of dependency or autonomy (as an example, they stated that patients with bad levels of functioning (GAF), high familiar support requirements

Table 1. List of significant variables

Variable id	Meaning
ANYEVOL	The years of evolution of disorder
ESCOLAR2	Educational level at second year of study
SUBTEV2	Evolutive subtype at second year of study (specify the schizophrenia diagnosis with a subtype describing the curse of the disorder)
TIPOCON2	Indicates the cohabitative frame of the patient
GAFCLA2	Clinical Functioning component from GAF instrument at second year of study. Measures the severity of the disorder
DASFFA2	Familiar functioning component from DAS instrument at second year of study. Measures disability in familiar functioning
URGCON2	Use of emmergency services in private hospitals receiving public funds at second year of study
PROFECON2	Indicates the professional condition of the patient
ASISVIS2	Estimated level of asistance to programmed visits with the health profesionals at second year of study
CASOIN2	Records the patient's reasons to drop out the study at some point before the second year of study.
A3C	Load in domestic tasks from ECFOS instrument at second year of study. It measures the hours x week that care giver employs to help the patient in development of domestic tasks
A7C	Load due to affection in money management from ECFOS instrument at second year of study. It measures the hours x week that care giver employs to manage the money for the patient
A8C	Load in time scheduling from ECFOS instrument at second year of study. It measures the hours x week that care giver employs to help the patient in organizing his/her time
MEDIANAA	Median load of all items from block A of ECFOS instrument at second year of study. Average load per week of care giver in helping to daily leaving activities
B1A	Behavioural problems item from ECFOS instrument at second year of study. It measures how many times in the last 30 days the care giver had to act to prevent, avoid or solve the consequences of innappropiate behaviours of the patient
B5A	Suicide item from ECFOS instrument at second year of study. It measures how many times in the last 30 days the care giver had to do something to make the patient forget about suicide or to avoid suicide attempts.
MAXIMOB	Maximum load of care giver in items of Block B from ECFOS at second year of study. It measures the carer cost of most reincident behavioural problem of the patient.

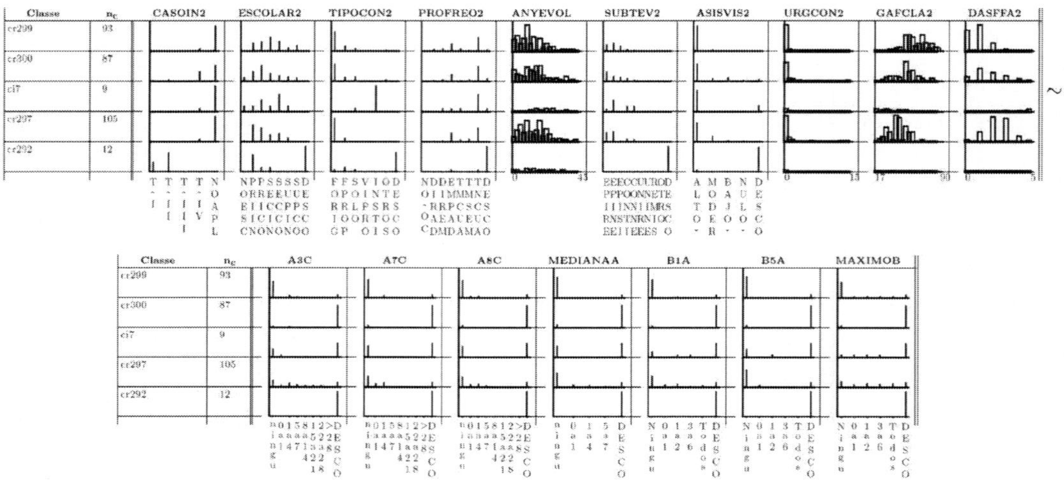

Figure 1. Panel graph of most relevant variables.

in daily activities (ECFOS section A) and behavioral problems (ECFOS section B) are patients in ill conditions; in another rule they stated that patients able to work and with high levels of functioning (GAF) are in good condition). Finally 5, classes with different patterns of dependency were found.

3.3. Postprocessing and Profiles Interpretation

The profiles obtained by ClBR were postprocessed by detecting significant variables 1 according to the statistical tests mentioned before, and providing the CPG for significant variables for both the classical clustering and the one obtained by ClBR. A clearer conceptual interpretation of the classes is possible from the experts' point of view, looking at results provided by ClBR [Salvador-Carulla et al. 06]. Fig. 1 shows the CPG with significant variables in this case. On the basis of this variables conceptualization of the classes was performed by the experts. Fig. 2 shows the corresponding traffic lights panel which provides a higher symbolic abstraction where technical skills are no more required by the experts to understand the profiles proposed by the clustering. Experts confirmed that the TLP provided information aligned with the CPG and is was much friendly for health profesionals in the wide spectrum. The interpretation provided is the following:

Autonomous (c299) : 93 persons with the better conditions: They are autonomous and can do tasks by themselves; they can work; they require little support from their carers (less than 4 h. a week) and they do not make intensive use of health-care services. This group has the greater educational level: 28 started secondary school and 18 could finish it; other 18 started higher education and 11 could finish it. Younger disease than other groups (13 years in average)

Singles (C300) : 87 persons whose main characteristic is to live alone. They tend to complete primary school. They have been ill for 15 years in average. They have intermediate scores in the assessment scales. But their condition is not good; probably

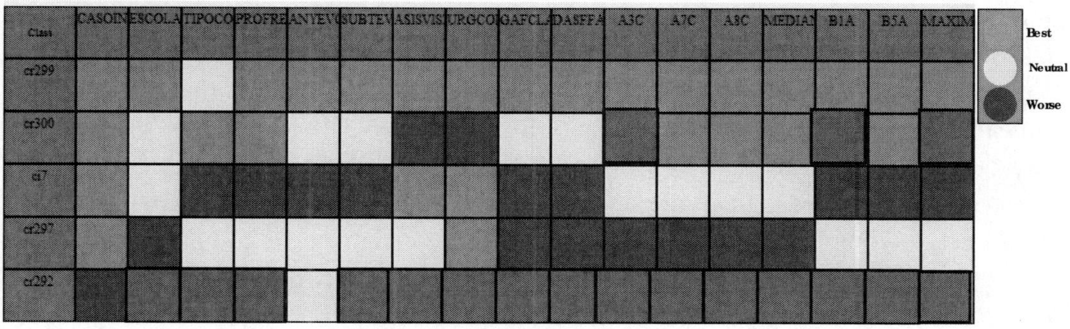

Note: Violet color indicates complete missing data in the cell.

Figure 2. Traffic Lights Panel of most relevant variables.

they would require higher supervision, but they have not. Treatment adherence and contacts with doctors are very low. They show a healthcare pattern very different from other groups, in fact they use the services in an inappropriate way: they may fail to the scheduled visits with professionals, whereas they may overuse emergency services (up to 15 times), probably because they feel bad and they are alone. When ECFOS could be evaluated, they show low familiar support requirements (less than 7 h. a week), mainly focused on domestic tasks. They do not generate family burden due to behavioral problems.

Institutionalized (ci7) : This group includes all the persons (a total of 9) which along the first year of the study ended-up to long-term residential care. They completed secondary school, they tend to be slightly older than the other groups, but not significantly, and they have a longer course of disease (23 years in average), they have non-contributive pension, they have an evolutive subtype of episodes with residual inter-episodic symptoms and severe negative symptoms. They show worse functioning levels that other groups and higher levels of severity as well as more self-harm attempts. They provoke family burden due to behavioral problems and they use to require support in daily activities.

Dependents (c297) : It is a class of 105 patients with high levels of dependency. They couldn't finish primary school. This are the patients in worse conditions; thus they make a high use of health care. This is the group requiring higher support from carers, up to 28 hours a week.

Uncomplete (c292) : They are 12 patients which dropped out the study by different reasons. Only socio-demographic and clinical data is available.

4. Discussion and Conclusion

Clustering techniques allow detecting different groups of patients of different dependency profiles. The analysis of the data under method i) only provided a confusing partition of patients difficult to understand. Facing such a complicated phenomenon as dependency, concerned with a lack of clear patterns and difficulties for establishing relationships between patient characteristics and patient needs of support, requires to take into account as much prior expert knowledge as possible, even if it is a partial description of the phenomenon. Mental disorders fit in the definition of ISD stated in [Gibert, Corts 98] and ClBR shown better results in general. ClBR incorporates the additional prior knowledge provided by experts by means of logical rules; often a partial description of the domain is provided, as ISD are so complex that it is usually impossible to announce a complete domain-KB (this is a bottleneck for using pure AI methods in ISD). Here, the KB expressed 5 rules with antecedents involving no more than 4 variables of the whole set of 75 measurements available. None of the classical statistical methods support expert knowledge influencing the analysis. ClBR is a hybrid technique which sensibly improved results by integrating clinical knowledge inside the analysis, which produces classes with proper interpretation [Gibert et al. 13]. Finally, a set of 5 classes was recommended by the system. Several tools were used to assist the interpretation of final classes, ensuring the understandability of the proposed model. Among them, CPG appeared as a successful support for the conceptualization process, but TLP was perceived as an excellent much friendly complement for end users with non deep statistical skills. From the medical point of view, ClBR provided a set of classes which fit well with different patterns of increasing degrees of dependency. All the patients that dropped out the study appear in a single group. Patients with dependency are subdivided in three different profiles: those who ended in long-term residential care, those who leave alone and those in such an ill condition that they cannot leave alone, but stay at home. Particularly interesting to the experts, elicitation of the special situation of the **Singles** group, which do not have extremely high dependency regarding daily leaving activities or functioning, but they show behavioral problems, probably because, leaving alone and being not properly supervised, whereas they should, they end up loosing treatment adherence and making an irrational use of services, *i. e.* missing scheduled visits and using emergency service as their main care resource.

The use of ClBR+CPG-TLP produces meaningful classes and sensibly improves, from a semantics point of view, the results of classical clustering, according to our opinion that hybrid techniques combining AI and Statistics are more powerful for KDD than pure ones. This work contributed to increase the knowledge about the dependency situations in the population with severe mental disorders. A clearer knowledge about how dependency behaves in schizophrenic population was achieved and this provided to the policy makers the inputs for a better resource allocation and planning of dependency services derived from the LPAD. Indeed, from this results, assigning specific *packages of care/support* according to the dependency profile of the patient becomes possible, thus bridging the gap between data mining and decision-making. An independent group of experts was asked to manually elaborate a second profiles proposal, to validate the methodology. Later, predictors of the different profiles will be identified to properly assign resources and benefits to LPAD applicants with severe mental health problems.

Acknowledgments

This research was supported and leaded by PRODEP, the specific program of the Generalitat de Catalunya for encouraging and structuring the promotion to the personal autonomy and the attention to persons with dependencies. Thanks to Hospital Sant Joan de Deu to provide the database and contributing to the interpretation of the results. Thanks also to APPS, for partially financing the research.

References

[APA 00] *Diagnostic and Statistical Manual of Mental Disorders*. DSM-IV-TR. APA (American Psychiatric Association): Washington, US, 2000.

[Brooks 96] Brooks *Health Policy*. 1996, 37, 53-72.

[Comas et al. 01] Comas, J. and Dzeroski, S. and Gibert, K. and Roda, I. and Sànchez-Marrè, M. *AI Communications*. 2001, 14(1), 45–62.

[Endicott et al. 76] Endicott, J., Spitzer, R.L., Fleiss, J.L., Cohen, J *Archives of General Psychiatry* 1976, 33, 766-771.

[Fayyad et al. 96] Fayyad, U. and Piatetsky-Shapiro, G. and Smyth, P. In *Data Mining to KDD: An overview*. Fayyad et al.; Ed. AAAI/MIT Press: US. 1996.

[Gibert et al. 08] Gibert, K. and Garca-Rudolph, A. and Garca-Molina, A. and Roig-Rovira, T. and Bernabeu, M. and Tormos, J.M. *Medical Archives* 2008, 62(3), 132-135.

[Gibert et al. 08b] Gibert K, A. Garca-Rudolph, G. Rodrguez-Silva *Acta Informatica Medica* 2008, 16(4), 178-182.

[Gibert et al. 12] Gibert K, D. Conti, D. Vrecko *Environmental Engineering and Management Journal* 2012, 11(5), 931-944.

[Gibert, Corts 94]: Gibert K, Corts U *Selecting Models from Data.*, Cheeseman P. and Oldford RW.; Ed.; LNS 89; Springer: NY, US, 1994, 351–360. Springer.

[Gibert, Corts 97]: Gibert, K., Corts, U. *Mathware and Soft Computing, 1997, 4(3)*, 251–266.

[Gibert, Corts 98]: Gibert K, Corts U. *Computación y Sistemas*, 1998, 1(4),213–227.

[Gibert et al. 05]: Gibert, K., Nonell, R., Colillas, MM., Velarde, JM. *Neural Network World*, 2005, 4/05, 319–326.

[Gibert, Nonell, 05b] Gibert K., Nonell, R. In *Procs 3rd World Conf on Computational Statistics and Data Analysis*; Limassol, Cy, 2005, pp 90 (Cyprus).

[Gibert, Nonell 08] Gibert, K, Nonel, R. In *Proc. of the iEMSs IVth Int'l Congress of Environmental Modeling and Software (DM-TES'08 Workshop)*; Barcelona, Sp, vol III, pp 1965-1966.

[Gibert et al. 03]: Gibert, K., Rodas, J., E. Rojo and U. Cortés *Medicinska Informatica*, 2003, 6, 15–21.

[Gibert et al. 13] Gibert K, Rodrguez-Silva G, Annicchiarico R *Mathematical and Computer Modelling*, 2013, 57(7-8), 1633-1639.

[Gibert, Sonicki 99]: Gibert, K., Sonicki, Z *AMSDA*, 1999, 15(4), 319–324.

[Gibert et al. 13] Gibert, K., Valls, A., Batet, M *Knowledge and Information Systems* 2013 (in press) DOI: 10.1007/s10115-013-0663-5.

[Goldman et al. 92] Goldman, Howard H., Andrew E. Skodol, and Tamara R. Lave. *American Journal Psychiatry*, 1992, 149 (9), 1148-56.

[Haro et al. 06] Haro JM, Salvador-Carulla L, et al. *Acta Psychiatrica Scandinavica,* 2006, (111(Suppl. 432)), 29–38.

[Janca et al. 96]: Janca, A., Kastrup, ML, Katschnig, H., Lopez-Ibor, J.J. Jr, Mezzich, J.E., Sartorius, N *Soc Psychiatry Psychiatr Epidemiol*, 1996, 31, 349-54.

[Kay et al. 86]: Kay, S.R., Opler, L.A., Fiszbein, A. *Social Behav Sci Doc*, 1986, 17, 28–29.

[Murtagh 83] Murtagh, F. *The Computer Journal* 1983, 26(4), 354-359.

[Ochoa et al. 12] Ochoa S. Salvador-Carulla, L., Villalta-Gil, V., Gibert, K., Haro, JM. *European Journal of Psychiatry* 2012, 26(1), 1-12.

[Prince et al. 07]: Prince M, Patel V, Saxena S et al. *Lancet*, 2007, 8;370(9590), 859-77.

[Prudo, Blum 87] Prudo R, Blum HM. *British Journal of Psychiatry*, 1987, 150, 345-54.

[Salvador-Carulla et al. 06]: Salvador-Carulla L, Gibert K *et al. Estudio DEFDEP: Definición operativa de dependencia en personas con discapacidad psíquica* PRODEP: Barcelona, Spain, 2006, vols. 1 and 2.

[Salvador-Carulla et al. 10] L. Salvador-Carulla, K. Gibert, S. Ochoa *Atencin Primaria*, 2010, 42, 344-345.

[Tukey 77] : J.W. Tukey. *Exploratory Data Analysis*. Addison-Wesley.

[Ustun et al. 99] :stn TB, Rehm J, Chatterji S et al. *LANCET*, 1999, 354 (9173), 111–115.

[Vzquez-Polo et al. 05]: F Vazquez-Polo, M Negrn, JM Cabass, E Snchez, JM Haro, L Salvador-Carulla *J Ment Health Policy Econ*, 2005, 8(3), 153-65.

[Vilaplana et al. 07]: Vilaplana, M., Ochoa, S., Martinez, A., et al. *Actas Esp Psiquiatr*, 2007, 35(6), 372-81.

[Ward, J.H. 63] *J. Am. Statis. Ass.*, 1963, 58, 236–244.

[WHO 01]: *The World Health Report 2001 – Mental Health: New Understanding, New Hope*. WHO: Geneva, Sw, 2001.

In: Mathematical Modeling in Social Sciences ...
Editors: J. C. Cortés López et al.
ISBN: 978-1-63117-335-6

Chapter 29

MODELING MATHEMATICAL FLOWGRAPH MODELS IN RECURRENT EVENTS. AN APPLICATION TO BLADDER CARCINOMA

B. García-Mora*, C. Santamaría†, G. Rubio‡ and J. Camacho§
Instituto Universitario de Matemática Multidisciplinar,
Universitat Politècnica de València, Valencia, Spain

Abstract

Mathematical Flowgraph models have proven to be very useful in providing an efficient approach for the analysis of time–to–event data. They essentially provide time-to-event distributions and also data for intermediate events to the final event of interest. This technique is able to model complex systems with large number of states. Solving a Flowgraph model we use a mixture of parametric distributions. An integral transform allows us to reduce a complicated problem to a much simpler one in algebraic equations. We apply this methodology to the evolution of bladder carcinoma in a three–state illness model of a recurrence–progression process. The probability of being free of progression at a given time is determined and applied to different risk groups of patients according to common characteristics.

Keywords: Statistical Flowgraph Model, Survival analysis, Bladder carcinoma, Erlang distribution, Phase–Type distribution

1. Introduction

Mathematical Flowgraph models consist of a model structure for representing Multistate Stochastic Processes (MSP) in which *nodes* represent system states and directed *links* represent the transitions between them. They provide time-to-event distributions and also for intermediate events to the final state, such as for example, recurrences, progressions or

*E-mail address: magarmo5@imm.upv.es
†E-mail address: crisanna@imm.upv.es
‡E-mail address: grubio@imm.upv.es
§E-mail address: fcamacho@imm.upv.es

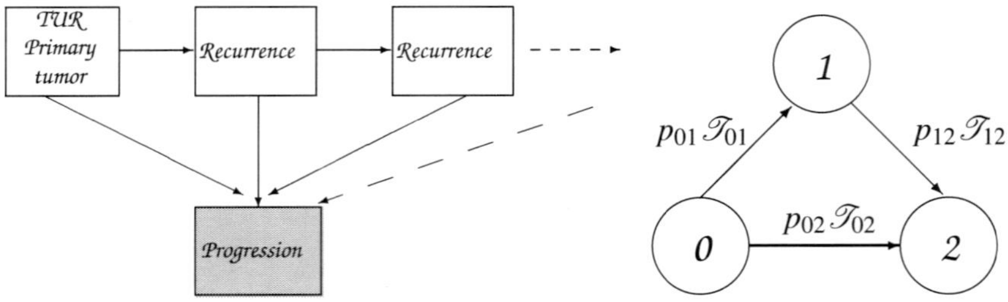

Figure 1. Multi–state Stochastic Process examples: a) Evolution of the bladder cancer; b) Transmittances for the Flowgraph model in bladder carcinoma.

death in the evolution of a disease. The *objective* is to predict the time until a given event and to calculate quantities of interest such as *cumulative distribution functions* for the time to occurrence of that event.

Flowgraph models were developed to model semi–Markov processes and there are many applications in the engineering framework [1]. Recently, they have expanded their applications in the field of medicine [2], providing richer models in this framework that allow specification of *recurrences* or *progressions* (see Figure 1a).

The objective of this chapter is to discuss the application of this methodology to the evolution of bladder carcinoma. Previously, we have developed several models trying to capture different aspects of the evolution of this cancer [3, 4] but now our aim is, in a step further, to explore the evolution of this disease by means of this type of model.

Transitional Cell Carcinoma (*TCC*) is the 11th most common cancer worldwide, accounting for 3–4% of all malignancies. Approximately 75-85% of patients present a *superficial* (*TCC*), which can be managed with a surgical endoscope technique, that is, the transurethral resection (*TUR*). This generally has a favorable prognosis, although *recurrence* rates are 30-80% and *progression* to muscle invasive tumor is 1-45%. We are interested in predicting the risk of *recurrences* and tumor *progression*.

In the Section 2 of this chapter we review the Phase–type and Erlang distributions needed to build the model. Next, we introduce the Flowgraph model. In Section 3 we present the important features of our approach. In Section 4 risk groups of patients are defined according to common characteristics. Finally, conclusions are discussed in Section 5

2. Preliminary Statistics

Survival analysis deasl with the analysis of data, taking times from well-defined *time-origin* until the occurrence of some particular event or *end-point*. Let T be the random variable associated with the survival time in that period. The *Survival Function* is

$$S(t) = \mathrm{P}(\mathrm{T} \geq t) = 1 - F(t) = 1 - \mathrm{P}(\mathrm{T} < t)$$

where $F(t)$ is the distribution function of T. On the other hand, in survival analysis data are frequently censored, which means that the event of interest may not be observed in the follow–up period, although these times must be taken account because this allows us to know if that the individual has been free of undergoing the event during the period of study.

2.1. The Phase–Type Distribution

The distribution, $F(.)$ on $[0, \infty[$, of the time until the absorption state in a Markov process with one absorbing state, is a *phase–type* (PH)-distribution [5] given by

$$F(t) = 1 - \alpha \exp(Tt)\mathbf{e}, \; t \geq 0 \tag{1}$$

where (α, α_{m+1}) is an initial probability vector and T is a matrix of order *m* representing transition rates of the m transient states , with negative diagonal entries and non–negative off–diagonal entries.This satisfies $-T\mathbf{e} = T^0 \geq 0$, with T^0 representing the absorbing rates from the transient states and $\mathbf{e} = (1, 1, \ldots, 1)' \in \mathbb{R}^{m \times 1}$.

The Laplace Transform of a (PH)-distribution is given by

$$L(s) = \alpha_{m+1} + \alpha(sI - T)^{-1}T^0, \text{ for } Re(s) > 0 \tag{2}$$

2.2. The Erlang Distribution

This is a particular case of the *phase–type* distribution. The representation (α, T) of order *r* of the Erlang distribution denoted by $E[r, \mu]$ is

$$\alpha = (1, 0, \ldots, 0)_{1 \times r} \qquad T = \begin{pmatrix} -\mu & \mu & & & \\ & -\mu & \mu & & \\ & & \ddots & \ddots & \\ & & & -\mu & \mu \\ & & & & -\mu \end{pmatrix}_{r \times r} \tag{3}$$

The initial vector indicates that the lifetime begins in the first phase, and matrix T indicates that only transitions from one phase to the following are allowed. We will specifically deal with a linear combination of three Erlang proposed in [6] and given by

$$G(t) = p_1 F_1(t) + p_2 F_2(t) + p_3 F_3(t), \tag{4}$$

with $p_1 + p_2 + p_3 = 1$, $p_i > 0$, $i = 1, 2, 3$. The three Erlang distributions are denoted by $E[r_1, \mu_1]$, $E[r_2, \mu_2]$, $E[r_3, \mu_3]$, with $\mu_i > 0$ and r_i a positive integer, $i = 1, 2, 3$. If $r_1 = 1$, $r_2 = 3$, $r_3 = 5$ the representation as *phase-type* distribution of G is (α, T) where

$$\alpha = (p_1 \quad p_2 \quad 0 \quad 0 \quad p_3 \quad 0 \quad 0 \quad 0 \quad 0) \tag{5}$$

$$T = \begin{pmatrix} -\mu_1 & 0 & 0 & 0 & 0 & 0 & 0 & 0 & 0 \\ 0 & -\mu_2 & \mu_2 & 0 & 0 & 0 & 0 & 0 & 0 \\ 0 & 0 & -\mu_2 & \mu_2 & 0 & 0 & 0 & 0 & 0 \\ 0 & 0 & 0 & -\mu_2 & 0 & 0 & 0 & 0 & 0 \\ 0 & 0 & 0 & 0 & -\mu_3 & \mu_3 & 0 & 0 & 0 \\ 0 & 0 & 0 & 0 & 0 & -\mu_3 & \mu_3 & 0 & 0 \\ 0 & 0 & 0 & 0 & 0 & 0 & -\mu_3 & \mu_3 & 0 \\ 0 & 0 & 0 & 0 & 0 & 0 & 0 & -\mu_3 & \mu_3 \\ 0 & 0 & 0 & 0 & 0 & 0 & 0 & 0 & -\mu_3 \end{pmatrix} \tag{6}$$

2.3. The Flowgraph Model

The first step is to define the necessary *nodes* and *transition* (links) according to the evolution of the bladder carcinoma. We focus only on the earlier states of this disease with the *progression* (final state) either *primary tumor* (state 0) or from the *first recurrence* (state 1).

Nodes are connected by directed line segments called *branches* and each one of them is characterized in terms of two elements (see Figure 1b):

- p_{ij} the probability on entry to state i with the transition in state j (the probability of direct progression from the *TUR* is p_{02} and the probability of recurrence is p_{01}).
- $F_{ij}(x)$ the cumulative distribution function (CDF) of the time x spent in state i, given that a transition to j occurs.

The *branches* are labeled with a *transmittance*, which consists of the product $p_{ij}\mathcal{T}_{ij}$ (see Figure 1b) where $\mathcal{T}_{ij}$ is The Laplace transform of the cumulative distribution function $F_{ij}(x)$. Both the p_{ij} and the F_{ij} are based on data analysis. As we have to decide a probability model family for each waiting time distribution $F_{ij}(x|\theta_{ij})$, firstly, we compute the empirical distributions for each transition $i \to j$ by means of the Kaplan Meier estimator [7] and secondly we will approximate them selecting a mixture of three Erlang distributions for $F_{ij}(x|\theta_{ij})$ given by (4). Note that the CDF is easily computed from the expression (1) with the probability vector α and matrix T given by (5)–(6).

Next, the parameters θ_{ij} of the probability distribution, in our case the parameters p_i and μ_i of each transition, will be estimated to minimize the following expression

$$\|K_{ij}(t) - G_{ij}(t)\|, \tag{7}$$

where K_{ij} is the empirical distribution of the Kaplan-Meier estimator for the transition $i \to j$ and G_{ij} is the mixture distribution (4). In order to estimate the parameters p_i and μ_i, and to obtain a suitable mixture we use a non-negative least squares fit (Lawson–Hanson algorithm [8]).

2.4. Data

The database was obtained from La Fe University Hospital of Valencia (Spain). It records clinical-pathological information from 957 patients, monitored between January 1995 and January 2010. The primary tumor is categorized as stage Ta or T1, according to the World Health Organization (WHO). After removal of the tumor by *TUR*, this may recur at a similar stage, which we call *recurrence*; or it may progress to muscle invasive stages T2, T3 or T4, which we call *progression*. The other *pathological* characteristic, *Grade*, is categorized from G1 to G3, from low aggressive to highly aggressive.

434 patients underwent a recurrence, 24 a progression, and 499 had censored times (some patients have no recurrence at all), so 63 patients were lost. From the remaining 371 patients, 17 underwent a progression, 226 a recurrence and times of the remaining 128 patients were censored. All computations were made in *R* [9] with the packages *expm*, *Matrix* and *survival*.

3. Results

We determine the empirical, K_{ij}, and parametric distributions, G_{ij}, for each transition $i \to j$ according to the procedure described above. It can be seen that the selected linear combination of three Erlang distributions presents a good approximation to the empirical distribution, see Figures 2 a), b) y c).

As we want to calculate the first passage distribution of transition from state 0 to state 2, irrespective of the path that was taken (directly with the transition $0 \to 2$ or undergoing a recurrence, transition $0 \to 1 \to 2$), we start computing the *transmittance* of the three transitions in the graph (see Figure 1b), which is the product $p_{ij}\mathcal{T}_{ij}$. Note the Laplace transform (2) for each F_{ij} is computed with the vector α and matrix T in the mixture of three Erlang distributions (5)-(6). On the other hand, the estimation of the probabilities p_{ij} is based on the sample data: $p_{01} = 0.3967742$, $p_{02} = 0.02507837$ and $p_{12} = 0.03252033$.

Having calculated a *transmittance* for each transition $i \to j$, the objective is now to reduce the flowgraph model to a *single transmittance* for the first passage between the states 0 and 2. In our case, the possible path from state 0 to state 2 is $0 \to 2$ and, at the same time, $0 \to 1 \to 2$. For this we use two Manson's rules [10]:

- The transmittance of transitions *in series* is the product of the series transmittances. In our case is the path $0 \to 1 \to 2$. This means that

$$\mathcal{T}_{02}(s) = p_{01}\mathcal{T}_{01}(s)p_{12}\mathcal{T}_{12}(s) = 0.4\mathcal{T}_{01}(s)0.03\mathcal{T}_{12}(s) = 0.012\mathcal{T}_{01}(s)\mathcal{T}_{12}(s)$$

- The transmittance of transitions *in parallel* is the sum of the parallel transmittances.

$$\mathcal{T}_{02} * (s) = p_{01}\mathcal{T}_{01}(s)p_{12}\mathcal{T}_{12}(s) + p_{02}\mathcal{T}_{02}(s) = 0.012\mathcal{T}_{01}(s)\mathcal{T}_{12}(s) + 0.03\mathcal{T}_{02}(s)$$

$\mathcal{T}_{02}*$ refers to the first passage from the state 0 to state 2, irrespective of the path.

The transition from state 0 to state 2 may not occur, i.e. patient may only suffer recurrences, or even no recurrence. In this case, the probability of taking the considered path

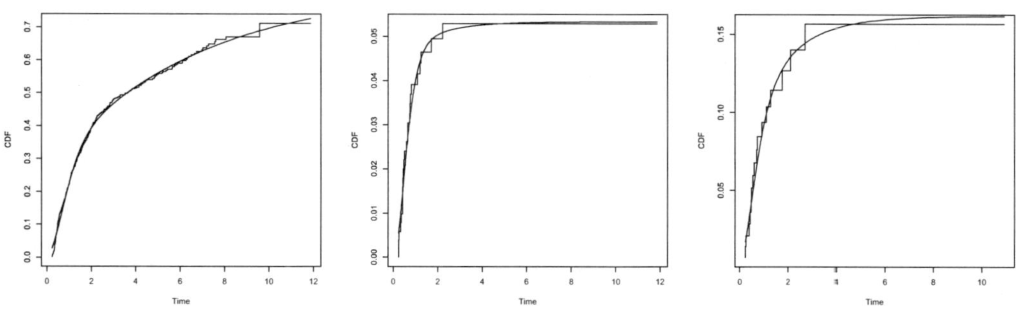

Figure 2. Erlang mixture (*smooth line*) and empirical distributions (*step function*) for a) transition $0 \to 1$; b) transition $0 \to 2$; c) transition $1 \to 2$.

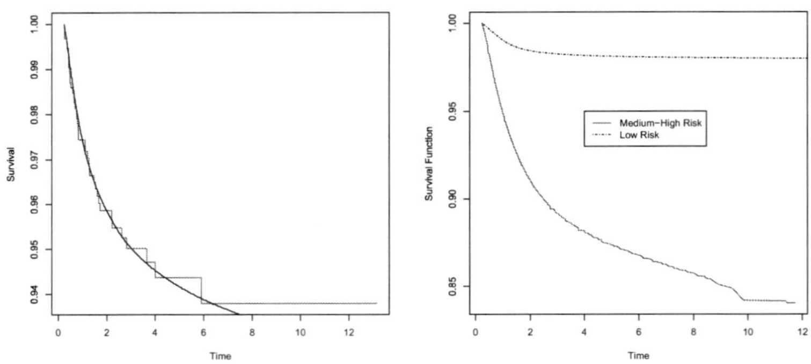

Figure 3. a) Survival function model for progression (*smooth line*) and empirical survival function for progression (*step function*). Time in years. b) Probability of being free of progression for low risk group and medium-high risk group.

is $p_{01}p_{12} + p_{02}$, and so we must divide the preceding by this probability to obtain the true Laplace transform. Then, the final expression is

$$L(s) = \frac{p_{01}\mathcal{T}_{01}(s)p_{12}\mathcal{T}_{12}(s) + p_{02}\mathcal{T}_{02}(s)}{p_{01}p_{12} + p_{02}} \tag{8}$$

3.1. Transform Inversion

As our real interest is to recover the probability distribution function we require the *inversion* of the Laplace transform (8). For this we use a variant of the inversion algorithm EULER [11]. In this way we can obtain the survival function with regard to *progression*, that is shown in the Figure 3a) , jointly with the empirical survival function.

Note we have obtained a parametric model to predict the probability of being free of *progression* at a given time. This procedure may be easily used to define risk groups, simply by calculating the survival functions of patients grouped according to common characteristics. Then the monitoring and the treatment of patients could be adjusted according to their risk.

4. Risk Groups

The joint evolution of the two possible processes in bladder cancer (*recurrence process* and *progression process*) was modeled by means of a non-parametric penalized likelihood method for estimating hazard functions in a general joint frailty model for recurrent events and terminal event in [12]. Three variables (*age*, *stage* and *grade*) were obtained as significant covariates in the progression process. A score scale ranging from 0 (best prognosis) to 4 (worst prognosis) was elaborated (Table 1), based on the coefficients of the fitted variables in the resulting model. According to Kaplan–Meier and Log–Rank test we establish two risk groups: *low risk group* for patients with a score of zero or one point and *medium-high risk group* for patients with more than one point. The statistical difference between both risk groups was contrasted ($p - value < 0.01$).

Table 1. Score for the progression process

	Progression process
Age	
≤ 66 years	**0**
≤ 66 years	**1**
Stage	
Ta	**0**
T1	**1**
Grade	
G1-G2	**0**
G3	**2**

Table 2. Probabilities to free survival for progression according to total score (Table 1).

	Probability "Free of progression" 1 YEAR	3 YEARS	5 YEARS
Low Risk (0-1 points)	99.0	98.3	98.1
Medium-High Risk (> 1 points)	94.9	89.2	86.8

Once these risk groups have been established, the flowgraph methodology was apply in each risk group. The survival function or the probability of being free of progression at one, three and five years is obtained for each group, see Table 2 and Figure 3b).

Conclusion

Flowgraph Models offer a comprehensive framework of theory and computational methods capable of modeling highly complex systems with large number of states (multiple recurrences). The use of integral transformations reduces the solution of solving a Flowgraph model from a complicated problem to a much simple one in algebraic equations. The versatility of the Erlang distribution allows us to present the expressions in an algorithmic and computationally tractable form.

Flowgraph models are suitable for modeling the evolution of the bladder cancer. These models can incorporate covariates such as the authors proposal in [1]. This versatility, along with the inclusion of molecular biomarkers and the clinical–pathological factors of the patients, allow us to increase the model predictability in a not too distant future.

References

[1] A. V. Huzurbazar and B. Williams. Incorporating covariates in flowgraph models: Applications to recurrent event data. *Thecnometrics*, 52:198–208, 2010.

[2] C. L. Yau and A. V. Huzurbazar. Analysis of censored and incomplete survival data using flowgraph models. *Stat Med*, 21:3727–43, 2002.

[3] C. Santamaría, B. García-Mora, G. Rubio, and S. Luján. An analysis of the recurrence-progression process in bladder carcinoma by means of joint frailty models. *Math Comput Model*, 54:1671–75, 2011.

[4] B. García-Mora, C. Santamaría, G. Rubio, and J. L. Pontones. Computing survival functions of the sum of two independent markov processes: an application to bladder carcinoma treatment. *Int J Comput Math*, 2013. DOI:10.1080/00207160.2013.765560.

[5] M. F. Neuts. *Matrix Geometric Solutions in Stocastic Models. An Algoritmic Approach*. The Johns Hopkins University Press, Baltimore, 1981.

[6] R. Pérez-Ocón and M. C. Segovia. Modeling lifetimes using phase-type distributions. In 3rd editionh Taylor & Francis, editor, *Risk, Reliability and Societal Safety, Proceedings of the European Safety and Reliability Conference 2007 (ESREL 2007)*, 2007.

[7] E. L. Kaplan and P. Meier. Nonparametric estimation from incomplete observations. *J Amer Statist Assoc*, 53:457–481, 1958.

[8] K. M. Mullen and Ivo H. M. van Stokkum. *nnls: The Lawson-Hanson algorithm for non-negative least squares (NNLS)*, 2012. R package version 1.4.

[9] R Development Core Team. *R: A Language and Environment for Statistical Computing*. R Foundation for Statistical Computing, Vienna, Austria, 2010. http://www.R-project.org.

[10] S. J. Manson. Feedback theory-some properties of signal flowgraphs. In *Proc IRE*, 1953.

[11] D. H. Collins and A. V. Huzurbazar. Prognostic models based on statistical flowgraphs. *Appl Stochastic Models Bus Ind*, 28:141–51, 2012.

[12] S. Luján. *Modelización matemática de la multirrecidiva y heterogeneidad individual para el cálculo del riesgo biológico de recidiva y progresión del tumor vesical no músculo invasivo*. PhD thesis, Universitat València, 2012.

In: Mathematical Modeling in Social Sciences ... ISBN: 978-1-63117-335-6
Editors: J. C. Cortés López et al.

Chapter 30

NUMERICAL SOLUTION OF AMERICAN OPTION PRICING MODELS USING FRONT-FIXING METHOD

V. Egorova*, R. Company† and L. Jódar‡
Instituto Universitario de Matemática Multidisciplinar,
Universitat Politècnica de València, Valencia, Spain

Abstract

This chapter deals with the numerical solution of the American option valuation problem formulated as a parabolic partial differential equation. The opportunity of early exercise for American options leads to a free boundary problem that has an additional difficulty derived by the moving computational domain. By using the front-fixing method the free boundary is involved to a transformed equation that results non-linear. An explicit finite difference scheme is proposed for the numerical solution. Numerical examples showing the conditional stability as well as comparison with other authors are included.

Keywords: American option, Exercise boundary, Black-Scholes equation Front-fixing method

1. Introduction

Black and Scholes show the opportunity for describing option pricing by partial differential equations [1]. American options can be exercised at each moment before expiration date. Therefore value of American option is associated with a partial differential equation with the moving boundary. This free boundary depends on time, it is a priori unknown and has to be found in the solution process. American option pricing is a subject of intensive research. Several approaches since analytical approximations to numerical methods have been developed for valuation of American options.

The main problem in American option pricing is unknown behaviour of the exercise boundary. Close to maturity it can be analytically approximated. Green's theorem is used

*E-mail address: veeg@doctor.upv.es (Corresponding author)
†E-mail address: rcompany@imm.upv.es
‡E-mail address: ljodar@imm.upv.es

to convert the boundary value problem for the price of the option into an integral equation for the optimal exercise boundary in [6]. A comparison of the different numerical and analytical approximation is provided in [7].

There are methods based on integral representation. The Mellin transform is used for European and American options pricing in [10]. Later this approach is extended in [2]. The extension is non-trivial in the sense that the original Mellin transform will not work for American call options due to convergence problems.

Options with early exercise possibility can be priced with finite difference methods. It can be considered as a linear complementary problem. Another ways to obtain value of the option are projected successive over relaxation (PSOR) method [11] and penalty method [9], [11], [12].

Another numerical approach is the front-fixing method. It comes from numerical solution of Stefan's problem. The idea of the method is to fix the computational domain by involving the free boundary to the equation. This method is proposed in [9], [13] and [14].

At time $\tau = T - t$ American option price for asset price $S > B(\tau)$ satisfies the Black-Scholes equation [4]:

$$\frac{\partial P}{\partial \tau} = \frac{1}{2}\sigma^2 S^2 \frac{\partial^2 P}{\partial S^2} + rS\frac{\partial P}{\partial S} - rP, \qquad S > B(\tau), \qquad 0 < \tau \leq T, \tag{1}$$

where:

- τ denotes the time to maturity.
- S is the asset's price.
- $P(S, \tau)$ is the option price at the time instant τ for asset's price S.
- $B(\tau)$ is the exercise price. If asset price $S > B(\tau)$ the optimal strategy is to exercise the option, if $S < B(\tau)$ the optimal strategy is to hold the option. This value is a-priori unknown.
- σ is a volatility of the asset.
- r is the risk free rate.

Equation (1) is subject to the following boundary and initial conditions

$$P(S, 0) = \max(E - S, 0), \qquad S \geq 0, \tag{2}$$

$$\frac{\partial P}{\partial S}(B(\tau), \tau) = -1, \tag{3}$$

$$P(B(\tau), \tau) = E - B(\tau), \tag{4}$$

$$\lim_{S \to \infty} P(S, \tau) = 0, \tag{5}$$

$$B(0) = E, \tag{6}$$

where E is the strike price.

The free boundary problem is the equation (1) together with the boundary and initial conditions (2)- (6). It can be solved numerically, if the value of $B(\tau)$ is determined.

In this chapter an explicit finite difference scheme for the valuation of American options is proposed based on the front-fixing method. Numerical results are included showing desirable positivity, stability and monotonicity properties as well as the comparison with other techniques in the literature.

2. Front-Fixing Method

We use the methodology of solving free boundary problems by known front-fixing transformation (Landau, 1950). In that case the unknown free boundary is involved into the differential equation on the fixed domain.

Let us consider the dimensionless transformation:

$$p(x,\tau) = \frac{P(S,\tau)}{E}, \quad S_f(\tau) = \frac{B(\tau)}{E}, \qquad x = \ln\frac{S}{S_f(\tau)}. \tag{7}$$

Under this transformation (7) the problem (1) - (6) can be rewritten in normalized form:

$$\frac{\partial p}{\partial \tau} = \frac{1}{2}\sigma^2\frac{\partial^2 p}{\partial x^2} + \left(r - \frac{\sigma^2}{2}\right)\frac{\partial p}{\partial x} - rp + \frac{S'_f}{S_f}\frac{\partial p}{\partial x}, \qquad x > 0, \qquad 0 < \tau \leq T, \tag{8}$$

where S'_f denotes the derivative of S_f with respect to τ. The new boundary and initial conditions are:

$$p(x,0) = 0, \qquad x \geq 0, \tag{9}$$

$$\frac{\partial p}{\partial x}(0,\tau) = -S_f(\tau), \tag{10}$$

$$p(0,\tau) = 1 - S_f(\tau), \tag{11}$$

$$\lim_{x\to\infty} p(x,\tau) = 0, \tag{12}$$

$$S_f(0) = 1. \tag{13}$$

3. Finite-Difference Approximation

The equation (8) is a non-linear differential equation on the domain $(0;\infty) \times [0,T]$. In order to solve numerically (8), one has to consider a bounded domain. Let us introduce $x_{\max}$ large enough to translate the boundary condition (12), i.e $p(x_{\max};\tau) = 0$. Then the problem (8)-(13) can be study on a fixed domain $[0;x_{\max}] \times [0,T]$.

We introduce the computational grid of M space points and N time levels with respective step sizes h and k:

$$h = \frac{x_{max}}{M+1}, \tag{14}$$

$$k = \frac{T}{N}, \tag{15}$$

$$x_j = hj, j = 0,..,M+1, \tag{16}$$

$$\tau^n = kn, n = 0,..,N. \tag{17}$$

The approximate value of $p(x, \tau)$ at the grid points is denoted by:

$$p_j^n \approx p(x_j, \tau^n). \tag{18}$$

Then we use a two-level in time explicit scheme with spatial centred differences for the numerical solution:

$$\frac{p_j^{n+1} - p_j^n}{k} = \frac{1}{2}\sigma^2 \frac{p_{j-1}^n - 2p_j^n + p_{j+1}^n}{h^2} +$$

$$\left(r - \frac{\sigma^2}{2}\right) \frac{p_{j+1}^n - p_{j-1}^n}{2h} - rp_j^n + \frac{S_f^{n+1} - S_f^n}{kS_f^n} \frac{p_{j+1}^n - p_{j-1}^n}{2h}. \tag{19}$$

By denoting

$$\mu = \frac{k}{h^2}, \qquad \lambda = \frac{k}{h},$$

the scheme (19) can be rewritten in form:

$$p_j^{n+1} = ap_{j-1}^n + bp_j^n + cp_{j+1}^n + \frac{S_f^{n+1} - S_f^n}{2hS_f^n} \left(p_{j+1}^n - p_{j-1}\right), \tag{20}$$

where

$$a = \frac{\mu}{2} \left(\sigma^2 - \left(r - \frac{\sigma^2}{2}\right) h\right), \tag{21}$$

$$b = 1 - \sigma^2\mu - rk, \tag{22}$$

$$c = \frac{\mu}{2} \left(\sigma^2 + \left(r - \frac{\sigma^2}{2}\right) h\right). \tag{23}$$

From the boundary conditions we can obtain:

$$p_0^n = 1 - S_f^n, \tag{24}$$

$$\frac{p_1^n - p_{-1}^n}{2h} = -S_f^n. \tag{25}$$

To exclude the auxiliary point p_{-1}^n we need an additional boundary condition. Consider the equation (8) at the point $x_0 = 0$:

$$-\frac{dS_f}{d\tau} = \frac{1}{2}\sigma^2 \frac{\partial^2 p}{\partial x^2} - \left(r - \frac{\sigma^2}{2}\right) S_f(\tau) - r\left(1 - S_f(\tau)\right) - \frac{dS_f'}{d\tau}, \tag{26}$$

$$\frac{1}{2}\sigma^2 \frac{\partial^2 p}{\partial x^2} + \frac{\sigma^2}{2} S_f(\tau) - r = 0. \tag{27}$$

Second order discretization of the equation (27) is following:

$$\frac{\sigma^2}{2} \frac{p_1^n - 2p_0^n + p_{-1}^n}{h^2} + \frac{\sigma^2}{2} S_f^n - r = 0. \tag{28}$$

The value of p_{-1}^n can be excluded by using (25) and (28). We can obtain the connection of the free boundary S_f and option value p at the n-th time level, $n \geq 1$:

$$p_1^n = \alpha - \beta S_f^n, \tag{29}$$

where

$$\alpha = 1 + \frac{rh^2}{\sigma^2}, \tag{30}$$

$$\beta = 1 + h + \frac{1}{2}h^2. \tag{31}$$

At the point x_1 we have two equations for option price - finite difference scheme and boundary condition:

$$p_1^{n+1} = ap_0^n + bp_1^n + cp_2^n + \frac{S_f^{n+1} - S_f^n}{2hS_f^n}\left(p_2^n - p_0^n\right), \tag{32}$$

$$p_1^{n+1} = \alpha - \beta S_f^{n+1}. \tag{33}$$

Let us express the free boundary S_f using the equations above:

$$S_f^{n+1} = \frac{\alpha - \left(ap_0^n + bp_1^n + cp_2^n - \frac{p_2^n - p_0^n}{2h}\right)}{\frac{p_2^n - p_0^n}{2hS_f^n} + \beta}. \tag{34}$$

By denoting

$$d^n = \frac{\alpha - \left(ap_0^n + bp_1^n + cp_2^n - \frac{p_2^n - p_0^n}{2h}\right)}{\frac{p_2^n - p_0^n}{2h} + \beta S_f^n}, \tag{35}$$

the free boundary motion in time can be written in form:

$$S_f^{n+1} = d^n S_f^n, \quad 0 \leq n \leq N - 1. \tag{36}$$

Under the expression (36) we can consider S_f^{n+1} as a known value. Let us denote:

$$\tilde{a}^n = a - \frac{S_f^{n+1} - S_f^n}{2hS_f^n}, \tag{37}$$

$$\tilde{c}^n = c + \frac{S_f^{n+1} - S_f^n}{2hS_f^n}, \tag{38}$$

then the numerical scheme for the problem (8) - (13) can be rewritten for any $n = 0, .., N - 1$:

$$S_f^{n+1} = d^n S_f^n, \tag{39}$$

$$p_0^{n+1} = 1 - S_f^{n+1}, \tag{40}$$

$$p_1^{n+1} = \alpha - \beta S_f^{n+1}, \tag{41}$$

$$p_j^{n+1} = \tilde{a}^n p_{j-1}^n + b p_j^n + \tilde{c}^n p_{j+1}^n, \quad j = 2, .., M, \tag{42}$$

$$p_{M+1}^{n+1} = 0, \tag{43}$$

with the initial conditions

$$S_f^0 = 1, \quad p_j^0 = 0, \quad 0 \le j \le M+1. \tag{44}$$

4. Application of the Proposed Methodology

In this section we represent results of the numerical experiments. We use the explicit scheme which is conditionally stable. This property of the numerical solution is proved empirically.

4.1. Example 1

We consider the same problem as in [9]. The front-fixing method is used, but with another transformation:

$$x = \frac{S}{S_f(t)}, \qquad p(x,t) = P(S,t) = P(xS_f(t), t). \tag{45}$$

$$r = 0.1, \tag{46}$$

$$\sigma = 0.2, \tag{47}$$

$$T = 1, \tag{48}$$

$$x_\infty = 2. \tag{49}$$

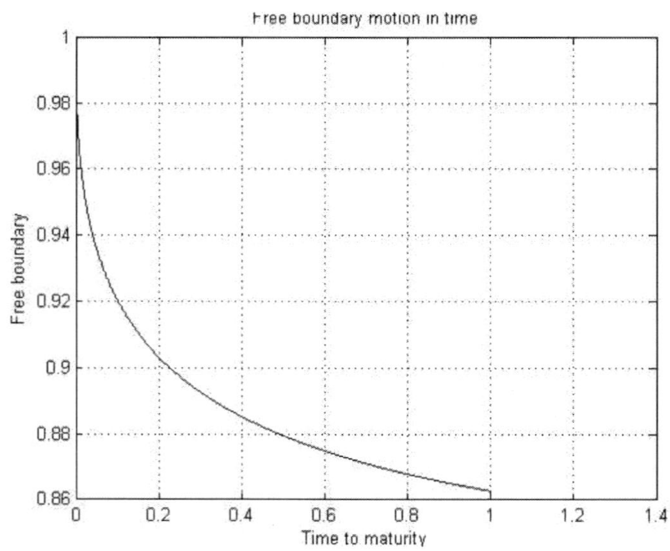

Figure 1. Free boundary motion in time for example 1.

We consider fixed space step

$$h = 0.001,$$

and change time step to check the condition for a stability. The connection between space step h and time step k is defined by γ. By the numerical tests it is shown that the considered scheme is stable for $\gamma \leq 24$ while the condition for a stability of the scheme in [9] is $\gamma \leq 6$. The results are represented in Table 1.

Table 1. Comparison with front-fixing method under transformation (45)

Method	$S_f(T)$
Implicit (in Ref. [9])	0.8615
Explicit (in Ref. [9])	0.8622
Explicit (proposed)	0.8628

The front-fixing method with the logarithmic transformation has a good advantage: more weak condition for the stability. It means that we can choose time step larger than for explicit method in [9]. It reduces the computational costs. Fig. 1 shows the evolution of the free boundary S_f depending on time to maturity τ.

4.2. Example 2

Let's compare our method with another approach [10], based on the Mellin's transform.

The parameters of the problem:

$$r = 0.0488 \tag{50}$$

$$\sigma = 0.3 \tag{51}$$

$$T = 0.5833 \tag{52}$$

$$\tag{53}$$

To compare results of the explicit front-fixing method (FF explicit) with the Mellin's transform [10], we have to multiply our dimensionless value on $E = 45$ (Table 2).

4.3. Example 3

Close to maturity exercise boundary for the American put without dividend yield can be analytically approximated. This approximation can be used only near expire date. Green's theorem is used to convert the boundary value problem for the price of the option into an integral equation for the optimal exercise boundary in [6]:

$$S_f(\tau) \sim E\left[1 - \sqrt{2\sigma\tau \log\left(\frac{\sigma^2}{6r\sqrt{\frac{\pi\tau\sigma^2}{2}}}\right)}\right] \tag{54}$$

Table 2. Comparison with the Mellin's transform method

Method	$S_f(T)$
Mellin's transform (in Ref. [10])	32.77
FF (explicit)	32.7655

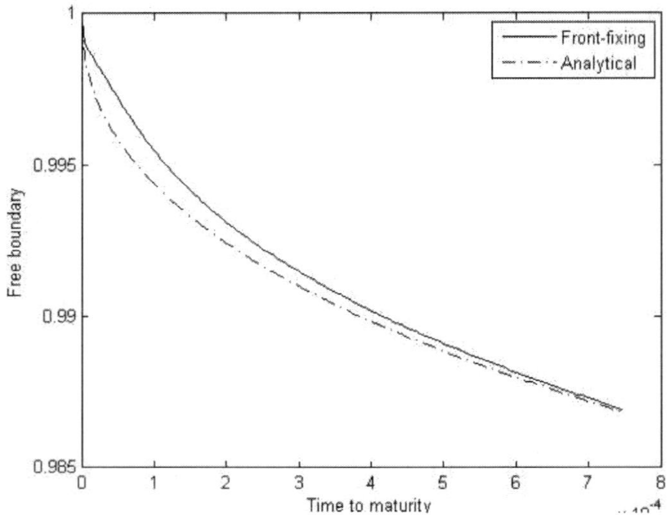

Figure 2. Comparison with the analytical approximation Kuske (example 3) near the maturity.

On the Fig. 2 the value of the free boundary in 10 first points is shown. We can see, that near the maturity, front-fixing method and analytical approximation (54) give closed results.

4.4. Example 4

We can also compare our explicit front-fixing method (FF) for American put with other numerical methods shown in [11] (Table 3). The problem parameters:

$$r = 0.08, \tag{55}$$

$$\sigma = 0.2, \tag{56}$$

$$T = 3, \tag{57}$$

$$E = 100. \tag{58}$$

There are considered several methods:

Table 3. Comparison with other methods

S	**True**	MBM	HW	OCA	OS	PM	FF
90	11.6974	11.6889	11.6974	11.6975	11.6922	11.7207	11.6898
100	6.9320	6.9203	6.9320	6.9321	0.9319	6.9573	6.9243
110	4.1550	4.1427	4.1548	4.1550	4.1548	4.1760	4.1468
120	2.5102	2.4996	2.5101	2.5102	2.5101	2.5259	2.5089

- The finite difference moving boundary method of Muthuraman (MBM) [8];
- Han-Wu algorithm (HW) transforms the Black-Scholes equation into a heat equation on an infinite domain [3];
- The OCA method uses an optimal compact scheme for the heat equation [11];
- Ikonen and Toivanen [5] proposed an operator splitting technique (OS) for solving the linear complementarity problem.
- Penalty method (PM) is considered in [9] and [12].

We can see that front-fixing method is more accurate than penalty method and moving boundary method.

Conclusion

We considered and tested the front-fixing method for an American put option pricing model. The main idea of the method is to involve free boundary to the equation to eliminate the motion of the computational domain. This method allows define the free boundary and option price together without additional costs.

We used explicit scheme for the numerical solution. This scheme is conditionally stable that was shown by the numerical tests.

The method was compared with other ones. It has an advantage: there is not iterative algorithm and we can choose any small space step to obtain the accurate enough result without large computational costs.

Acknowledgments

This chapter has been partially supported by the European Union in the FP7-PEOPLE-2012-ITN program under Grant Agreement Number 304617 (FP7 Marie Curie Action, Project Multi-ITN STRIKE-Novel Methods in Computational Finance).

References

[1] Black, F., Scholes, M. (1973) The Pricing of Options and Corporate Liabilties. *Journal of Political Economy*, 81. 637–654.

[2] Frontczak, R. and Schöbel, R. (2009) On modified Mellin transforms, Gauss-Laguerre quadrature, and the valuation of American call options. *Tbinger Diskussionsbeitrag*, 320.

[3] Han, H., Wu, X. (2003) A fast numerical method for the Black-Scholes equation of American options. *SIAM J. Numer. Anal.* 41. 2081–2095.

[4] Hull, J.(2000) *Options, Futures and Other Derivatives*. NJ, Prentice-Hall.

[5] Ikonen, S., Toivanen, J. (2004) Operator splitting method for American option pricing. *Applied Mathematics Letters*, 17. 809–814.

[6] Kuske, R. A., Keller, J. B. (1998) Optimal Exercise Boundary for An American Put. *Appl. Math. Finance*, 5. 107–116.

[7] Lauko, M., Sevcovic, D. (2010) *Comparison of Numerical and Analytical Approximations of the Early Exercise Boundary of the American Put Option*. Available at SSRN: http://ssrn.com/abstract=1547783.

[8] Muthuraman, K. (2008) A moving boundary approach to American option pricing. *Journal of Economic Dynamics and Control*, 32. 3520–3537.

[9] Nielsen, B.F., Skavhaug, O., Tvelto, A. (2002) Penalty and front-fixing methods for the numerical solution of American option problems. *Journal Comp. Finan.,* 5 (2002), (4, Summer).

[10] Panini, R., Srivastav, R.P. (2004) Option pricing with Mellin Transforms. *Mathematical and computer modelling*, 40. 43–56.

[11] Saib, A.A.E.F., Tangman, Y.D., Thakoor, N., Bhuruth, M. (2011) On Some Finite Difference Algorithms for Pricing American Options and Their Implementation in Mathematica. *Proceedings of the 11th International Conference on Computational and Mathematical Methods in Science and Engineering*, CMMSE 2011 26-30 June 2011.

[12] Toivanen, J. (2010) Finite Difference Methods for Early Exercise Options. *Encyclopedia of Quantitative Finance*.

[13] Wu, L., Kwok, Y.-K.(1997) A Front-Fixing method for the Valuation of American Option. *The Journal of Financial Engineering*, Vol. 6, N2. 83–97.

[14] Zhang, J., Zhu, S.(2009) A Hybrid Finite Difference Method for Valuing American Puts. *Proceedings of the World Congress of Engineering* 2009, Vol II.

In: Mathematical Modeling in Social Sciences ...
Editors: J. C. Cortés López et al.
ISBN: 978-1-63117-335-6

Chapter 31

ESTIMATION OF THE COST OF ACADEMIC UNDERACHIEVEMENT IN HIGH SCHOOL IN SPAIN OVER THE NEXT FEW YEARS

J. Camacho[1,*]***, R. Cervelló-Royo***[2,†]***, J. M. Colmenar***[3,‡]
and A. Sánchez-Sánchez[1,§]
[1] Instituto Universitario de Matemática Multidisciplinar,
Universitat Politècnica de València, Valencia, Spain
[2] Departmento de Económicas y Ciencias Sociales,
Universitat Politècnica de València, Valencia,Spain
[3]Centro de Estudios Superiores Felipe II,
Universidad Complutense de Madrid, Campus de Aranjuez,
Madrid, Spain

Abstract

High rates of academic underachievement have strong negative effects on the economic situation of families and the Spanish Government, mainly, in the current economic crisis which is affecting especially to Spain. We quantify the large costs that would entail for the Spanish society the high rates of academic underachievement in the Spanish *Bachillerato* in the coming years based on the predictions given with 95% confidence intervals in [1]. These predictions allow to provide an estimation of the investment could be made in this educational level by both, the Spanish Government and families, paying special attention on the groups of students who abandon and do not promote. According to our estimations over the next few years, these amounts of money, on average, would have been ranging between 47 348 373,89 and 83 499 397,50 euros for both, the Spanish Government and families investment.

Keywords: Academic underachievement, Abandon, High school, Economic costs, Predictions with confidence intervals

*E-mail address: fcamacho@mat.upv.es
†E-mail address: rocerro@esp.upv.es
‡E-mail address: jmcolmenar@ajz.ucm.es
§E-mail address: alsncsnc@posgrado.upv.es

1. Introduction

In recent years, there has been increased awareness of the importance of education in society by both governments and society in general. Education largely determines the professional life of an individual. It has an impact on the ease of getting and keeping a job and also influences on the conditions and characteristics of the job. This important issue has led to the educational experts and policy makers to focus their attention on the evolution of the academic results of students. There are many contributions which show us the increasing concern about young students' academic performance in worldwide [2, 3, 4], mainly, focusing on the bad academic results which could have serious negative influences on the country's economic development. Obviously, the better education of the population, the greater benefits could be brought by the population to the country.

The interest about academic underachievement in Spain is increasing and completely justified, not only by the high rates but also it is becoming a major social and political concern [5, 6, 7], especially in the unemployment and its serious consequences. This issue is of primer importance in the current context of economic crisis affecting particularly Spain. In fact, when the economic crisis started around the year 2008 affecting negatively on the international labor market, in Spain, the unemployment rates were twice higher than the rest of the European Countries [8]. Moreover, in 2012, the 80% of the Spanish people that had finished their higher studies, accessed to the work market while the Spanish population who had only got ESO or lower educational levels was around 27% [9].

Furthermore, high rates of academic underachievement have strong negative effects on the economic situation of families and the Spanish Government, mainly, in the current economic crisis in Spain. On the one hand, families must spend a lot of money on each of their children's education and it has been increasing as time goes reaching values that, on average, are around 1 300 euros per year for each Spanish student [10, 11]. There are many needs that parents have to deal with such as the school fees, books, uniform and, in some cases, accommodation each academic year, money that would be invested again if the student is not able to promote during the academic year. In the same way, each year the Spanish State Management spends a high percentage of its budget on education [4], a waste of large amounts of money if the rates of academic underachievement are increasing [7].

Taking into account the above-mentioned reasons, in this chapter, we will pay special attention on the Spanish *Bachillerato* educational level, mainly, because *Bachillerato* is a milestone in the career training of students because it represents a period to make important decisions about academic and professional future [12]. This educational level is made up of two stages (First and Second Stage of *Bachillerato*) and, when students finish *Bachillerato*, they can decide whether to continue their higher studies (university or professional training) or to access the work market. This is of paramount importance for society because, although the percentage of high school academic underachievement has slightly reduced over the last years, nowadays it seems to be at a worrying steady-level [13, 1], contributions in which it is analyzed the Spanish *Bachillerato* academic underachievement based on the transmission of academic habits among students in the same academic level.

In this chapter we propose to quantify the large costs that would entail for the Spanish society the high rates of academic underachievement in the Spanish *Bachillerato* in the coming years based on the predictions given with 95% confidence intervals in [1]. These predictions will allow to provide an estimation of the investment could be made in this educational level by both, the Spanish Government and families. We will pay special attention on the groups of students who abandon and do not promote during their corresponding academic year whose academic attitude could lead a high economic costs for the Spanish Government and families.

This chapter is structured as follows. In Section 2 and Section 3 we will quantify the economical cost that will have to support both, the Spanish Government and the Spanish families by 95% confidence intervals for the next few years, respectively. Finally, conclusions are given in Section 4

2. Estimation with 95% Confidence Intervals of the Cost of the Academic Underachievement in *Bachillerato* for the Next Few Years for the Spanish Government

In this section, we will pay special attention on the predictions of the percentage of Spanish *Bachillerato* students who may abandon or not promote in the next few years (see Tables 1 and 2 [1]). These predictions will allow us, with the required suitable economical data, to predict the cost to the academic underachievement in this educational level for the Spanish Government.

To perform estimations as close as possible, we will follow the next steps:

Step 1 We obtain the average Spanish Government cost of each *Bachillerato* student during the academic years 1999 – 2000 to 2008 – 2009.

Step 2 We predict the Spanish Government investment in each Spanish *Bachillerato* student during the academic years 2009 – 2010 to 2014 – 2015 using the cost of each *Bachillerato* student given in Step 1.

Step 3 We predict the number of *Bachillerato* students registered during the academic years 2009 – 2010 to 2014 – 2015. This is required to obtain the number of *Bachillerato* students that will not promote and abandon at that period using the corresponding percentages estimated in Tables 1 and 2 [1].

Step 4 We compute the total of the Spanish Government investment in *Bachillerato* students that will not promote and abandon during the academic years 2009 – 2010 to 2014 – 2015 using the predictions given in Step 2 and 3.

First, we obtain the Spanish Government cost of each *Bachillerato* student during the academic years 1999 – 2000 to 2008 – 2009 **(Step 1)**. For that, we collect the total investment in education (in euros) and calculate the percentage of the Spanish Government investment amount of money expended in *Bachillerato* educational level, in both, state and private high schools all over Spain from academic year 1999 – 2000 to 2008 – 2009 [4]. These available data let us know the total the Spanish Government investment in *Bachillerato* at that period of time. Furthermore, we also know the number of students registered during the mentioned period given in [14].

Table 1. The 95% confidence interval prediction corresponding to the First and Second Stage of *Bachillerato*, in both, state and private high schools all over Spain during academic years $2009-2010$ to $2014-2015$. Each row shows the rate of girls/boys who promote ($G_i|B_i$) and do not promote ($\overline{G}_i|\overline{B}_i$) for each level $i=1,2$.

Group	Time (t)	Mean	Median	Confidence interval
G_1	$2009-2010$	0.20205	0.20414	[0.17993 , 0.21227]
$\overline{G}_1$	$2009-2010$	0.06851	0.06866	[0.06512 , 0.07041]
G_2	$2009-2010$	0.18859	0.19092	[0.16987 , 0.19554]
$\overline{G}_2$	$2009-2010$	0.07847	0.07875	[0.07564 , 0.08020]
B_1	$2009-2010$	0.16101	0.15807	[0.15286 , 0.18575]
$\overline{B}_1$	$2009-2010$	0.06853	0.06834	[0.06574 , 0.07340]
B_2	$2009-2010$	0.16176	0.16099	[0.15632 , 0.17254]
$\overline{B}_2$	$2009-2010$	0.07100	0.07097	[0.06852 , 0.07391]
G_1	$2010-2011$	0.20126	0.20346	[0.17870 , 0.21222]
$\overline{G}_1$	$2010-2011$	0.06719	0.06734	[0.06333 , 0.06920]
G_2	$2010-2011$	0.18999	0.19241	[0.16938 , 0.19734]
$\overline{G}_2$	$2010-2011$	0.07787	0.07816	[0.07471 , 0.07969]
B_1	$2010-2011$	0.16165	0.15852	[0.15283 , 0.18840]
$\overline{B}_1$	$2010-2011$	0.06646	0.06624	[0.06346 , 0.07173]
B_2	$2010-2011$	0.16557	0.16470	[0.15965 , 0.17752]
$\overline{B}_2$	$2010-2011$	0.06994	0.06991	[0.06728 , 0.07287]
G_1	$2011-2012$	0.19969	0.20220	[0.17673 , 0.21202]
$\overline{G}_1$	$2011-2012$	0.06607	0.06630	[0.06179 , 0.06830]
G_2	$2011-2012$	0.19053	0.19355	[0.16803 , 0.19898]
$\overline{G}_2$	$2011-2012$	0.07744	0.07770	[0.07392 , 0.07949]
B_1	$2011-2012$	0.16306	0.15924	[0.15291 , 0.19165]
$\overline{B}_1$	$2011-2012$	0.06453	0.06430	[0.06137 , 0.07001]
B_2	$2011-2012$	0.16957	0.16870	[0.16287 , 0.18219]
$\overline{B}_2$	$2011-2012$	0.06903	0.06899	[0.06613 , 0.07221]
G_1	$2012-2013$	0.19730	0.20043	[0.17223 , 0.21172]
$\overline{G}_1$	$2012-2013$	0.06510	0.06536	[0.06044 , 0.06749]
G_2	$2012-2013$	0.19021	0.19409	[0.16952 , 0.20041]
$\overline{G}_2$	$2012-2013$	0.07715	0.07760	[0.07322 , 0.07948]
B_1	$2012-2013$	0.16530	0.16126	[0.15361 , 0.19406]
$\overline{B}_1$	$2012-2013$	0.06277	0.06247	[0.05941 , 0.06848]
B_2	$2012-2013$	0.17378	0.17278	[0.16616 , 0.18630]
$\overline{B}_2$	$2012-2013$	0.06826	0.06826	[0.06513 , 0.07170]
G_1	$2013-2014$	0.19497	0.19850	[0.17036 , 0.21135]
$\overline{G}_1$	$2013-2014$	0.06416	0.06444	[0.05934 , 0.06691]
G_2	$2013-2014$	0.18999	0.19112	[0.16989 , 0.20172]
$\overline{G}_2$	$2013-2014$	0.07686	0.07736	[0.07242 , 0.07916]
B_1	$2013-2014$	0.16736	0.16410	[0.15372 , 0.19705]
$\overline{B}_1$	$2013-2014$	0.06109	0.06072	[0.05772 , 0.06713]
B_2	$2013-2014$	0.17783	0.17656	[0.16943 , 0.19088]
$\overline{B}_2$	$2013-2014$	0.06756	0.06760	[0.06426 , 0.07116]
G_1	$2014-2015$	0.19361	0.19730	[0.16837 , 0.21089]
$\overline{G}_1$	$2014-2015$	0.06315	0.06351	[0.05798 , 0.06610]
G_2	$2014-2015$	0.19077	0.19262	[0.16786 , 0.20304]
$\overline{G}_2$	$2014-2015$	0.07645	0.07694	[0.07180 , 0.07888]
B_1	$2014-2015$	0.16821	0.16442	[0.15377 , 0.19860]
$\overline{B}_1$	$2014-2015$	0.05940	0.05898	[0.05586 , 0.06554]
B_2	$2014-2015$	0.18140	0.17994	[0.17236 , 0.19516]
$\overline{B}_2$	$2014-2015$	0.06680	0.06685	[0.06333 , 0.07060]

Table 2. Descriptive analysis of the percentage of abandon in Spanish *Bachillerato* during the academic years from $2009 - 2010$ to $2014 - 2015$.

Academic Year	2009 – 2010	2010 – 2011	2011 – 2012	2012 – 2013	2013 – 2014	2014 – 2015
Mean	1.25	1.23	1.21	1.19	1.18	1.16
Median	1.25	1.23	1.21	1.19	1.18	1.16
Percentile 2.5	1.20	1.17	1.15	1.13	1.12	1.10
Percentile 97.5	1.30	1.27	1.25	1.23	1.23	1.21

Table 3. Investment per Spanish student in the First and Second Stage of *Bachillerato*, in both, state and private high schools all over Spain from academic year $1999 - 2000$ to $2008 - 2009$ by the Government [4].

t	Academic Year	Euros
1	1999 – 2000	2 610,70
2	2000 – 2001	2 796,50
3	2001 – 2002	2 991,48
4	2002 – 2003	3 384,28
5	2003 – 2004	3 691,93
6	2004 – 2005	3 972,37
7	2005 – 2006	4 224,20
8	2006 – 2007	4 569,65
9	2007 – 2008	5 130,38
10	2008 – 2009	5 146,88

The aforementioned data allow us to work out (dividing the Spanish Government investment in *Bachillerato* by the number of *Bachillerato* students registered in each academic year) the amount of money in euros that the Spanish Government has invested in each *Bachillerato* student in recent years. The obtained results can be seen in Table 3. Notice that these figures have progressively increased as time goes on ranging between 2 610,70 euros in $1999 - 2000$ until 5 146,88 euros in $2008 - 2009$.

Then, we need to predict the Spanish Government investment in each *Bachillerato* student during the academic years $2009 - 2010, \ldots, 2014 - 2015$ **(Step 2)**. To do that, we are going to use statistical techniques, in particular, time series analysis [15, 16, 17]. This statistical technique provides tools for selecting a model in order to forecast future events. In our case, the application of these techniques will return predictions of the investment in each *Bachillerato* student over the next few years taking into account the known Spanish Government investment the previous years (Table 3). We will address our approach using *Statgraphics Plus for Windows 5.1* software [18]. This powerful statistical tool provides the user five different forecasting models: Random Walk with Trend, Linear Trend, Simple Moving Average, Simple Exponential Smoothing and Brown's Linear Exponential Smoothing. Then, the models are validated by their corresponding Root Mean Square Error (RMSE) and Percentage of the Mean Absolute Error (MAPE). Finally, it is selected the model that best fit the available data and provide us the predictions with 95% confidence

Table 4. The indicators (RMSE and MAPE) considered for the validation of the different models in order to determine the model that best fit the data in Table 3. The best is the *Linear Trend Model.*

Model	RMSE	MAPE
Random walk with trend	151.187	2.56029
Linear trend	104.440	1.78572
Simple moving average of 3 terms	633.230	14.7076
Simple exponential smoothing with alpha 0.999	315.820	6.49393
Brown's Linear Exponential Smoothing with alpha 0.853	222.497	4.35632

intervals, both analytically and graphically. The model that best fit our data is the *Linear Trend Model* because it returns the minimum Root Mean Square Error (RMSE=104.44) whose corresponding Mean Absolute Percentage of Error is 1.79, as can be seen in Table 4 (see [18, 19, 20, 21]). Therefore, the obtained equation which allows us to predict the Spanish Government investment in euros in each *Bachillerato* student over the next few years is stated as follows:

$$G_t = -601795.0 + 302.144t, \tag{1}$$

where G_t is the estimation of the investment at time $t = 1, 2, 3, \dots$ where $t = 1$ corresponds to the academic year $1999 - 2000$, $t = 2$ to the academic year $2000 - 2001$ and so on.

According to the time series model stated, in Table 5, we show the obtained estimations with 95% confidence intervals given by *Statgraphics Plus for Windows 5.1* (see [18, 19, 20, 21]) of the cost in euros that the Spanish Government would invest in each *Bachillerato* student during the academic years from $2009 - 2010$ to $2014 - 2015$. Graphically, this results can be seen in Figure 1.

Table 5. The prediction of euros invested by the Spanish Government in each Spanish student in the First and Second Stage of *Bachillerato*, in both, state and private high schools during the academic years from $2009 - 2010$ to $2014 - 2015$.

t	Academic Year	Prediction (Euros)	95% Confidence interval (Euros)
11	$2009 - 2010$	5 513,62	[5 221,95 , 5 805,29]
12	$2010 - 2011$	5 815,77	[5 509,97 , 6 121,56]
13	$2011 - 2012$	6 117,91	[5 796,43 , 6 439,39]
14	$2012 - 2013$	6 420,05	[6 081,52 , 6 758,58]
15	$2013 - 2014$	6 722,20	[6 365,47 , 7 078,92]
16	$2014 - 2015$	7 024,34	[6 648,42 , 7 400,26]

The next step is to predict the number of *Bachillerato* students registered during the academic years $2009-2010$ to $2014-2015$ **(Step 3)**. Due to the predictions of *Bachillerato* students are given in percentages (see Tables 1 and 2 [1]), we need to estimate the number of students registered in both First and Second Stage of *Bachillerato* to be able to estimate the number of them who do not promote and abandon over the next few years using our

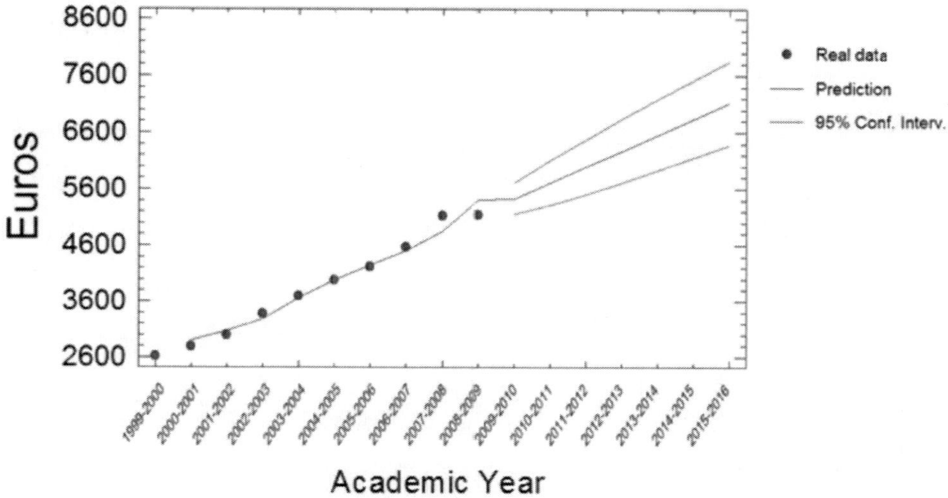

Figure 1. Graph of the prediction of euros invested by the Spanish Government in each Spanish student in the First and Second Stage of *Bachillerato*, in both, state and private high schools during the academic years from $2009-2010$ to $2014-2015$.

predictions. To do that, we will again use the time series models mentioned above following the same procedure as it was shown previously, applied, in this case, to the number of *Bachillerato* students in the specific period of time given in Table 6 [14].

Table 6. Number of Spanish student in the First and Second Stage of *Bachillerato* in both, state and private high schools, all over Spain from academic year $1999-2000$ to $2008-2009$ [14].

Academic Year	Number of *Bachillerato* Students
$1999-2000$	766 964
$2000-2001$	738 407
$2001-2002$	676 107
$2002-2003$	654 655
$2003-2004$	626 926
$2004-2005$	613 581
$2005-2006$	604 806
$2006-2007$	595 571
$2007-2008$	584 693
$2008-2009$	629 247

In this case, using *Statgraphics Plus for Windows 5.1*, the time series model selected that best fit our data in Table 6 is the *Random Walk with Trend Model*. This has the least Root Mean Square Error (RMSE) and Percentage of the Mean Absolute Error (MAPE), as can be seen in Table 7 (see [18, 17]).

As regards to the definition of the *Random Walk with Trend Model* (see [18, 17]), we consider Y_t as the observed number of *Bachillerato* students in a specific academic year at time t and $F_t(k)$ the obtained forecast. Despite the *Statgraphics Plus for Windows 5.1*

Table 7. The indicators (RMSE and MAPE) considered for the validation of the different models in order to determine the model that best fit the data in Table 6. The best is the *Random Walk with Trend Model*

Model	RMSE	MAPE
Random walk with trend	27978.4	2.70597
Linear trend	32784.3	3.83946
Simple moving average of 3 terms	43745.2	6.28963
Simple exponential smoothing with alpha 0.999	30496.7	3.49021
Brown's Linear Exponential Smoothing with alpha 0.853	29404.0	3.209

software only gives us the predictions if all the required assumptions are fulfilling, we will also confirm them analyzing statistically if the obtained white noise in this process follows a normal distribution, as is required. In order to check this, we apply the *Shapiro-Wilks normality test* which, with a significative p-value at significance level of 0.05 (p-value=0.407). The p-value confirms that the white noise follows a univariate normal distribution. This fact is also supported by having a closed mean and median (-13 345 and -15 302, respectively) and the kurtosis is 3.198, approximately 3, value considered as a reference to data following a univariate normal distribution [22].

Then, as the model is stated, in Table 8 we show the estimation with 95% confidence intervals of the number of Spanish *Bachillerato* students during the academic years from $2009-2010$ to $2014-2015$ (see [18]).

Table 8. Estimations with 95% confidence intervals of the number of Spanish students in the First and Second Stage of *Bachillerato* in both, state and private high schools, all over Spain from academic year $2009-2010$ to $2014-2015$.

Academic Year	Number of estimated *Bachillerato* Students	
	Predicted	95% Confidence Intervals
$2009-2010$	613 945	[562 245 , 665 646]
$2010-2011$	598 643	[525 528 , 671 759]
$2011-2012$	583 341	[493 793 , 672 889]
$2012-2013$	568 039	[464 638 , 671 441]
$2013-2014$	552 738	[437 132 , 668 343]
$2014-2015$	537 436	[410 796 , 664 076]

Finally, we compute the Spanish Government total investment in *Bachillerato* students that will not promote and abandon during the academic years $2009-2010$ to $2014-2015$ **(Step 4)**. To obtain them, we take into account the Spanish Government investment in each *Bachillerato* student given in Table 5 and the estimated number of *Bachillerato* students in Table 8. After some algebraic operations (simply multiplications of the extremes of the intervals obtained in each mentioned tables), Table 9 collects the estimated number of students who will not promote and abandon and their corresponding costs that would have for the Spanish Government in the next few years.

As we can see, if expectations are fulfilled and educational measures are not taken, the

Spanish Government would lose a huge amount of money in groups of *Bachillerato* students who, most of them, would not promote and abandon the year or access to the labor market without sufficient qualification to perform works requiring improved training. Notice that, for example, this investment could be ranging between 38 225 011,05 and 71 646 592,10 euros in the academic year $2013-2014$.

Table 9. Estimation with 95% confidence intervals of the number of *Bachillerato* students who do not promote and abandon in the First and Second Stage of *Bachillerato*, in both, state and private high schools all over Spain from academic year $2009-2010$ to $2014-2015$ and their corresponding cost for the Spanish Government also given with 95% confidence intervals.

Academic year	Estimated number of *Bachillerato* students who will not promote and abandon	Estimated Spanish Government investment (in euros)
2009 – 2010	[8 293 , 10 636]	[43 306 812,55 , 61 747 912,30]
2010 – 2011	[7 561 , 10 502]	[41 661 939,75 , 64 294 039,41]
2011 – 2012	[6 978 , 10 362]	[40 449 413,28 , 66 728 551,64]
2012 – 2013	[6 450 , 10 186]	[39 226 440,83 , 68 848 080,60]
2013 – 2014	[6 005 , 10 121]	[38 225 011,05 , 71 646 592,10]
2014 – 2015	[5 541 , 9 902]	[36 842 317,83 , 73 278 632,94]

3. Estimation with 95% Confidence Intervals of the Investment in Education by Spanish Families of *Bachillerato* Students in the Next Few Years

In the previous section, we have estimated the cost that would have for the Spanish Government the predicted negative academic results of *Bachillerato* students. However, Government not only has to make those educational investments but also students' families. Undoubtedly, families have a very important role on their children's education, in fact, most of students depend heavily on their parents for their studies, parents that, with their efforts, try to support and provide them the best conditions to develop their children's knowledge. That effort is commonly shown through their understanding, their care and, of course, with financial support. This financial support that, especially in periods of economical crisis as we are suffering at this moment, is really difficult for most families.

In this section, we will show that high rates of academic underachievement (including the abandon rates) not only have negative economical consequences on the Spanish Government but also to Spanish families, in particular, of *Bachillerato* students. To address this approach, we will estimate the Spanish families investment following the same procedure as it is shown in Section 2 For sake of clarity, in this case, we will follow the next steps:

Step 1 We obtain the Spanish families cost of each *Bachillerato* student during the academic years $1999-2000$ to $2008-2009$.

Table 10. Spanish families investment, on average, per Spanish student in the First and Second Stage of *Bachillerato* in both, state and private high schools, all over Spain from academic year from $1999-2000$ **to** $2008-2009$ **[4].**

Spanish families Investment per *Bachillerato* student		
t	Academic Year	Euros
1	$1999-2000$	889,21
2	$2000-2001$	900,85
3	$2001-2002$	951,66
4	$2002-2003$	1 008,23
5	$2003-2004$	1 028,53
6	$2004-2005$	1 067,29
7	$2005-2006$	1 131,23
8	$2006-2007$	1 156,78
9	$2007-2008$	1 173,82
10	$2008-2009$	1 141,92

Step 2 We predict the Spanish families investment in each *Bachillerato* student during the academic years $2009-2010$ to $2014-2015$ using the cost of each *Bachillerato* student given in Step 1.

Step 3 We compute the Spanish families total investment in *Bachillerato* students that will not promote and abandon during the academic years $2009-2010$ to $2014-2015$ using the predictions given in the previous step (Step 2) and in the Step 3 shown in Section 2

First of all, we need to obtain the Spanish families cost of *Bachillerato* student during the academic years $1999-2000$ to $2008-2009$ **(Step 1)**. For this, we collect the Spanish families investment over the total registered students in the non-university Spanish education during the corresponding academic years $1999-2000$ to $2008-2009$ given in [4]. Furthermore, we know the total number of non-university Spanish students registered [14]. These available data allow us to work out (dividing the Spanish families total investment over the total of registered non-university students by the corresponding number of non-university students) the Spanish families investment on each non-university Spanish student. Unfortunately, it has not been possible to get this information corresponding only to the *Bachillerato* educational level. As a consequence, we will consider these figures as a reference to determine, on average, the cost of a Spanish *Bachillerato* student for their families. Thus, in Table 10, we show, on average, the assumed Spanish families investment in each *Bachillerato* student during the academic years $1999-2000$ to $2008-2009$.

Then, we predict the Spanish families investment in each *Bachillerato* student during the academic years $2009-2010$ to $2014-2015$ **(Step 2)** using the cost of each *Bachillerato* student given in Step 1. These predictions, as it was developed in the previous section, have also been obtained by best time series model that fit the available data in Table 10 using, again, *Statgraphics Plus for Windows 5.1* software. After applying them, we consider that the model that best fit our data is the *Linear Trend Model* because it returns the minimum

Table 11. The indicators (RMSE and MAPE) considered for the validation of the different models in order to determine the model that best fit the data in Table 10. The best is the *Linear Trend Model*

Model	RMSE	MAPE
Random walk with trend	29.1247	2.04206
Linear trend	27.7048	1.70595
Simple moving average of 3 terms	74.3422	6.37513
Simple exponential smoothing with alpha 0.999	39.2763	2.9957
Brown's Linear Exponential Smoothing with alpha 0.999	29.95	2.31063

Root Mean Square Error (RMSE=27.705) whose corresponding Mean Absolute Percentage of Error is 1.71 (see Table 11). As a consequence, *Statgraphics Plus for Windows 5.1* software provides, by the model selected, 95% confidence intervals predictions of the Spanish families investment in each *Bachillerato* student over the next few years (see [18, 19, 20]). The obtained results can be seen in Table 12 and, graphically, in Figure 2.

Table 12. The prediction of euros Spanish families will invest in each Spanish student in the First and Second Stage of *Bachillerato*, in both, state and private high schools during the academic years from $2009 - 2010$ **to** $2014 - 2015$**.**

t	Academic Year	Prediction (Euros)	95% Confidence interval (Euros)
11	2009 – 2010	1 232,24	[1 154,86 , 1 309,61]
12	2010 – 2011	1 266,29	[1 185,17 , 1 347,40]
13	2011 – 2012	1 300,34	[1 215,06 , 1 385,62]
14	2012 – 2013	1 334,39	[1 244,59 , 1 424,19]
15	2013 – 2014	1 368,44	[1 273,81 , 1 463,07]
16	2014 – 2015	1 402,49	[1 302,77 , 1 502,21]

Finally, we compute the Spanish families total investment in *Bachillerato* students that will not promote and abandon during the academic years $2009 - 2010$ to $2014 - 2015$ **(Step 3)**. To obtain them, we use the estimated number of *Bachillerato* students that will not promote and abandon (see Table 9) and the cost for the Spanish families of each *Bachillerato* student during the academic years $2009 - 2010$ to $2014 - 2015$ (see Table 10). After some algebraic operations (simply multiplications of the extremes of the intervals obtained in each mentioned tables), in Table 13, we show the estimation of the Spanish families total investment in education during the academic years from $2009 - 2010$ to $2014 - 2015$.

Notice that these values could be ranging between 7 649 301,83 and 14 807 905,66 euros in the academic year $2013 - 2014$. No negligible amount of money if we consider the economic difficult situation of most Spanish families as a result of the severe economic crisis in Spain is immersed.

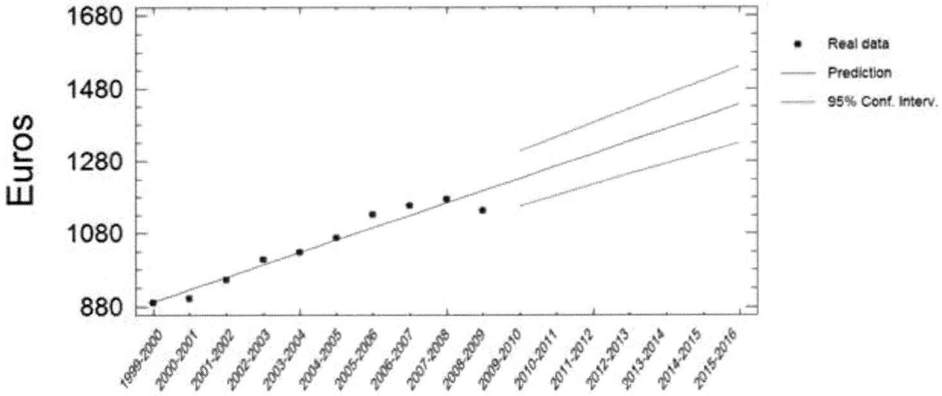

Figure 2. Graph of the prediction (in euros) the Spanish families will invest in each *Bachillerato* student during the academic years from 2009 − 2010 to 2014 − 2015.

Table 13. 95% **confidence intervals of the Spanish families cost in the group of *Bachillerato* students with academic underachievement over the next few years.**

Academic Year	Estimated Spanish families investment (in euros)
2009 − 2010	[9 577 515,21 , 13 929 654,41]
2010 − 2011	[8 961 297,64 , 14 151 586,96]
2011 − 2012	[8 479 092,15 , 14 358 567,46]
2012 − 2013	[8 027 735,83 , 14 507 891,88]
2013 − 2014	[7 649 301,83 , 14 807 905,66]
2014 − 2015	[7 219 319,24 , 14 875 138,87]

Conclusion

In this chapter, we quantify the important social problem of the academic underachievement, we take advantage of our predictions of the Spanish academic performance to propose an estimation of the Spanish Government and families investment in the *Bachillerato* students over the next few years, paying special attention on the groups of students who abandon and do not promote during their corresponding academic year. According to our results, notice that, for example, in the academic year 2013 − 2014, the Spanish Government would have invested in students with academic underachievement a large amount of money, ranging between 38 225 011,05 and 71 646 592,10 euros and, in case of the Spanish families, the costs will range between 7 649 301,83 and 14 807 905,66 euros. According to our predictions over the next few years (the total number of Bachillerato student and the cost per *Bachillerato* student for the Spanish Government and families given in Tables 6, 5,

12, respectively), these amounts of money, on average, would have been ranging between 47 348 373,89 and 83 499 397,50 euros for both the Spanish Government and families investment.

From our expectations and if new and innovative educational measures are not taken, the Spanish Government and families would lose a huge amount of money in groups of *Bachillerato* students who, most of them, would have to repeat a year or access to the labor market without sufficient qualification to perform works requiring improved training.

References

[1] Cortés, J.C. & Sánchez-Sánchez, A. & Santonja, F.J. & Villanueva, R.-J. (2013). Non-parametric probabilistic forecasting of academic performance in Spanish high school using an epidemiological modelling approach. *Applied Mathematics and Computation*, 221, 648-661.

[2] United Nations Educational, Scientific and Cultural Organization. (2012). *Youth and Skills: Putting Education to Work*. EFA Global Monitoring Report: UNESCO Publishing.

[3] EuroStat. Comision European. Education statistics at regional level [online]. 2013 [2013-01-11]. Available from: http://epp.eurostat.ec.europa.eu/statistics_explained/index.php/Education_statistics_at_regional_level.

[4] Instituto Nacional de Evaluación Educativa. Gobierno de España. Sistema estatal de indicadores de la educación. [Education indicator of the Spanish Government]. 2012 [2013-01-11]. Available from: http://www.mecd.gob.es/inee/publicaciones/indicadores-educativos/Sistema-Estatal.html#SEIE_2011_2.

[5] Eckert, H. (2006). Entre el fracaso escolar y las dificultades de inserción profesional: la vulnerabilidad de los jóvenes sin formación en el inicio de la sociedad del conocimiento. [Between academic underachievement and employability difficulties: the vulnerability of young people without training in the beginning of the knowledge society]. *Revista de Educación*, 341, 35-55.

[6] Psacharopoulos, G. (2007). The costs of school failure: A feasibility study. *European Expert Network on Economics of Education (EENEE).*

[7] Calero Martínez, J. & Gil Izquierdo, M. & Fernández Gutiérrez, M. (2011). *Los costes del abandono escolar prematuro (Recurso electrónico): una aproximación a las pérdidas monetarias y no monetarias causadas por el abandono prematuro en España. [The costs of early school abandon (Electronic resource): an approach to the monetary and nonmonetary losses caused by early abandon in Spain].* Investigación. IFIIE (Instituto de Formación del Profesorado, Investigación e Innovación Educativa). Gobierno de España); 191: Ministerio de Educación, Subdirección General de Documentación y Publicaciones.

[8] Instituto Nacional de Estadística. Mujeres y hombres en España. [Women and men in Spain]. 2010 [2013-10-11]. Available from: http://www.ine.es.

[9] Instituto Nacional de Evaluación Educativa. Ministerio de Educación. Gobierno de España. Panorama de la Educación. Indicadores de la OCDE 2012. [Education at a Glance. OECD Indicators 2012]. 2012 [2013-10-11]. Available from: http://www.mecd.gob.es/dctm/inee/internacional/panorama2012.pdf?documentId=0901e72b81415d28.

[10] Instituto Nacional de Estadística. Encuesta sobre Gasto de los Hogares en Educación. (Módulo Piloto de la Encuesta de Presupuestos Familiares 2007). [Survey of Household Spending on Education. (Module Pilot Household Budget Survey 2007)]. 2009 [2013-10-11]. Available from: http://www.ine.es/prensa/np541.pdf.

[11] Instituto Nacional de Estadística. Encuesta sobre Gasto de los Hogares en Educación. (Módulo Piloto de la Encuesta de Presupuestos Familiares (Curso 2011/2012)). [Survey of Household Spending on Education. (Module Pilot Household Budget Survey (Course 2011/2012)]. 2012 [2013-10-11]. Available from: http://www.ine.es/prensa/np763.pdf.

[12] Marchesi, A. & Lucena, R. (2003). La Representación Social del Fracaso Escolar. [The Social Representation of Academic Underachievement]. In Marchesi, A. & Gil, C.H. (Eds.), *El Fracaso Escolar: Una Perspectiva Internacional. [Academic Underachievement: An International Perspective]*: Alianza Editorial.

[13] Camacho, J. & Cortés, J.C. & Micle, R.M. & Sánchez-Sánchez A. (2013). Predicting the academic underachievement in a high school in Spain over the next few years: A dynamic modeling approach, Mathematical and Computer Modelling, 57, 7-8, 1703-1708.

[14] Ministerio de Educación. Gobierno de España. Enseñanzas no universitarias. Alumnado matriculado. [Non-university education. Registered students]. 2013 [2013-01-11]. Available from: http://www.mecd.gob.es/horizontales/estadisticas/no-universitaria/alumnado/matriculado.html

[15] Brockwell, P.J. & Davis, R.A. (2002). *Introduction to Time Series and Forecasting*. Springer Texts in Statistics: Springer.

[16] Brockwell, P. J. (2008). Time Series Analysis. In *Encyclopedia of Statistics in Quality and Reliability*: John Wiley and Sons, Ltd.

[17] Box, G.E.P. & Jenkins, G.M. & Reinsel, G.C. (2008). *Time Series Analysis: Forecasting and Control*. Wiley Series in Probability and Statistics: Wiley.

[18] Statgraphics.Net. Statgraphics tutorials. 2013 [11/01/2013]. Available from: http://www.statgraphics.net/wp-content/uploads/2011/12/tutoriales/Pronosticos.pdf.

[19] Fernández, S.M. (2001). *Guía completa de Statgraphics: Desde MS-DOS a Statgraphic Plus. [Statgraphics Complete Guide: From MS-DOS to Statgraphic Plus]*: Díaz de Santos.

[20] Nyblom, J. (1986). Testing for Deterministic Linear Trend in Time Series. *Journal of the American Statistical Association*, 81, 394, 545-549.

[21] Muth, J.F. (1960). Optimal Properties of Exponentially Weighted Forecasts. *Journal of the American Statistical Association*, 55, 290, 299-306.

[22] Hair, J.F. & Anderson, R.E. (2010). *Multivariate data analysis*: Prentice Hall.

In: Mathematical Modeling in Social Sciences ...
Editors: J. C. Cortés López et al.
ISBN: 978-1-63117-335-6

Chapter 32

A Finite Difference Scheme for Options Pricing Modeled by Lévy Processes

R. Company, M. Fakharany* and L. Jódar
Instituto Universitario de Matemática Multidisciplinar,
Universitat Politècnica de València, Valencia, Spain

Abstract

In this chapter, we use a new discretization strategy to generate schemes for option pricing modeled by Lévy process of finite and infinite activity. The aim of this discretization is to improve the accuracy of the numerical solutions and guarantees that these solutions are nonnegative. We focus on two models; first Merton's model (option price with finite jump activity) and the second is the CGMY model (option price with infinite activity). These models are governed by partial-integro differential equations. We apply an explicit discretization for the differential part and the trapezoidal rule for the integral part. To make these discretizations compatible, the double discretization has been used. The associated error for this technique has been calculated. Moreover, the numerical analysis such as stability and consistency have been studied.

Keywords: Lévy models, numerical analysis, double discretization

1. Introduction

The valuation of option prices based on Black-Scholes model shows inconsistency with its corresponding value in the market. It has become known as "volatility smile" [1]. This is due to the unrealistic assumption that the underlying asset price follows a geometric Brownian motion with constant volatility. The observation of large and sudden price movements has led to the use of stochastic processes with discontinuous jump processes for modeling financial assets. Exponential Lévy models provide a suitable class of models with flexible jumps which allows calibration market with a variety of asymmetric and volatility smiles, [2]. These models have special feature that they provide the price of the option as a solution of integro-partial differential equation (PIDE). These equations contain second order differential operator and a non local integral term that require specific treatment.

*E-mail address: fakharany@aucegypt.edu

Several authors have used finite difference schemes (FD) to solve numerically such PIDEs [3] - [8]. The application of these methods involves several challenges, such as how to approximate the integral term and how to truncate the unbounded domain to keep relevant information such as big jumps. In many cases the kernel of the integrand has singularities that must be treated with care. Moreover, it must be combined correctly the FD approximations of the differential part with the numerical integration of the integral term to produce stable and consistent approximations. The nonlocal character of the integral term involves recurrence systems of equations with dense coefficient matrices. In [8] implicit discretization is used in the time variable and it is proposed a rapidly convergent iterative method for solving the dense matrix problem discussed above. In [3] the authors use an explicit-implicit scheme to obtain numerical approximation for European and barrier options; implicit for the differential part and explicit for the integral part. The stability, consistency, and monotonicity are studied. They assume a particular behavior of the solution outside the truncated domain. Other authors assume a particular behavior of the solution outside the truncated domain. This improvable feature has been observed in other papers [5, 6].

$$\frac{\partial V}{\partial \tau} = \frac{\sigma^2}{2} S^2 \frac{\partial^2 V}{\partial S^2} + rS\frac{\partial V}{\partial S} - rV$$

$$+ \int_{-\infty}^{+\infty} \nu(y)\big[V(Se^y,\tau) - V(S,\tau) - S(e^y - 1)\frac{\partial V}{\partial S}\big]dy, \; S \in (0,\infty), \; \tau \in (0,T], \tag{1}$$

$$V(S,0) = f(S), \qquad S \in (0,\infty), \tag{2}$$

where $V(S,\tau)$ is the option price as a function of the underlying asset S and time to maturity $\tau = T - t$, σ is the volatility, r is the risk-free interest and ν is the measure of the Lévy process. The payoff function $f(S)$ for the vanilla call option is given by

$$f(S) = \max(S - E, 0), \tag{3}$$

where E is the strike price.

A Lévy process is said to be of finite activity if $\int_R \nu(y)dy < \infty$, otherwise it is called of infinite type. In the first case the measure is proportional to a probability density function g, $\nu = \lambda g$. One of the most relevant finite activity models is the jump-diffusion Merton's model [9] that can be written in the following form

$$\frac{\partial V}{\partial t} + \frac{1}{2}\sigma^2 S^2 \frac{\partial^2 V}{\partial S^2} + (r - \lambda K)\, S\, \frac{\partial V}{\partial S} - (r + \lambda)V$$

$$+\lambda \int_0^{\infty} V(S\eta, t)\, g(\eta)\, d\eta = 0, \qquad 0 < S < \infty\,,\; 0 \le t < T\,, \tag{4}$$

where $\eta = e^y$ is the jump amplitude, the expected value for jump size is given by $K = \mathbf{E}[\eta - 1]$ and λ is the jump intensity of the Poisson process. The jump sizes in Merton's model are assumed to be log-normally distributed with mean μ_J and standard deviation σ_J. Its probability density function is given by

$$g(\eta) = \frac{\exp\left(-\frac{1}{2}\left(\frac{\ln(\eta) - \mu_J}{\sigma_J}\right)^2\right)}{\sigma_J\, \eta\, \sqrt{2\pi}}\,. \tag{5}$$

On the other hand, one of the most important Lévy models was proposed by Carr, Geman, Madan and Yor called CGMY model [10]. This model allows diffusions and jumps of finite and infinite activities and its density is given by

$$\nu(y) = \begin{cases} \frac{Ce^{-G|y|}}{|y|^{1+Y}}, & y < 0 \\ \\ \frac{Ce^{-M|y|}}{|y|^{1+Y}}, & y > 0 \end{cases} \tag{6}$$

where $C > 0$, $G \geq 0$, $M \geq 0$, and $Y < 2$. The parameter Y allows to control the fine structure of asset return distribution. For $Y < 0$, the Lévy process is of finite activity. For $0 \leq Y \leq 1$, it is of infinity activity but finite variance, i.e., $\int_{|y|<1} y\nu(y)dy < \infty$. Finally, for $1 < Y < 2$, both the activity and variation are infinite. Note that for $Y = 0$ one gets the well known Variance Gamma process proposed by Madan and Seneta [11] as a particular case.

In this chapter we propose the construction of explicit finite difference schemes for the Merton model PIDE, as an example of finite activity model (2)-(5), and to the PIDE CGMY model, prototype model with infinite activity (1)-(2)-(3)-(6). To cover the whole bounded domain and not lose information of the solution, we use a double spatial discretization with parameters h and δ that provide a uniform scheme in the bounded domain and a non uniform scheme in the remaining unbounded domain. This strategy allows more flexibility for improving the numerical approximation in different parts of the domain as seen in the numerical simulations.

2. The Numerical Scheme for Merton Model

First we apply a change of variables in order to eliminate the convection and reaction terms of the PIDE (4),

$$\left.\begin{array}{l} X = \exp\left((r-\lambda K)(T-t)\right)S, \quad \tau = T-t, \\ U(X,\tau) = \exp\left((r+\lambda)(T-t)\right)V(S,t). \end{array}\right\} \tag{7}$$

The problem (2)-(5) is transformed into

$$\frac{\partial U}{\partial \tau} = \frac{1}{2}\sigma^2 X^2 \frac{\partial^2 U}{\partial X^2} + \lambda \int_0^\infty U(X\eta,\tau)g(\eta)\,d\eta, \tag{8}$$
$$0 < X < \infty\,,\; 0 < \tau \leq T\,,$$

$$U(X,0) = f(X)\,, \qquad 0 < X < \infty\,. \tag{9}$$

To approximate the integral part of the PIDE (8), it is convenient to use the change $\phi = X\eta$. Also by using a parameter $A > 0$, the integral domain is decomposed into two parts, $]0, A]$ and $[A, \infty[$. By substituting $z = \frac{A}{\phi}$ the latter is achieved by an integral expression on the finite interval $]0, 1]$ and the problem (8)-(9) takes the following form

$$\frac{\partial U}{\partial \tau} = \frac{\sigma^2 X^2}{2}\frac{\partial^2 U}{\partial X^2} + \frac{\lambda}{X}(J_1 + J_2), \quad 0 < X < \infty\,,\; 0 < \tau \leq T\,, \tag{10}$$

$$U(X,0) = f(X)\,, \qquad 0 < X < \infty\,, \tag{11}$$

where

$$\left.\begin{aligned} J_1 &= \int_0^A U(\phi,\tau)\, g\left(\tfrac{\phi}{X}\right) d\phi\,; \\ J_2 &= \int_A^\infty U(\phi,\tau)\, g\left(\tfrac{\phi}{X}\right) d\phi \,=\, A\int_0^1 U\!\left(\tfrac{A}{z},\tau\right)\, g\!\left(\tfrac{A}{Xz}\right) \tfrac{1}{z^2}\, dz\,. \end{aligned}\right\} \tag{12}$$

We construct the corresponding scheme to the problem (10)-(12) by FD with time step $k = \frac{\tau}{L}$, and $\tau^l = lk$, $0 \le l \le L$. With respect to the spacial variable X, we construct a uniform mesh in $[0, A]$, with step size $h = \frac{A}{N}$, $X_j = jh$, $0 \le j \le N$. For the other part, a uniform step size δ in $]0,1]$ for the variable z with mesh points $z_j = j\,\delta$, $1 \le j \le M$, $M\delta = 1$, implies a non uniform distribution for the original variable X in $[A,\infty[$,

$$X_j = \frac{A}{z_{N+M-j}} = \frac{A}{1-(j-N)\delta}\,, \quad N \le j \le N+M-1\,.$$

Let us denote the numerical solution by $u_j^l \approx U(X_j,\tau^l)$, and consider the FD approximation forward for the time-derivative and centered for the second spacial-derivative

$$\frac{\partial U}{\partial \tau}(x_i,\tau^l) \approx \frac{u_i^{l+1}-u_i^l}{k} \tag{13}$$

$$\frac{\partial^2 U}{\partial X^2}(X_j,\tau^l) \approx \frac{u_{j+1}^l - 2u_j^l + u_{j-1}^l}{h^2} = \Delta_i^l\,, \tag{14}$$

for the interior points in $[0, A]$, and using $h_j = X_{j+1} - X_j > 0$,

$$\frac{\partial^2 U}{\partial X^2}(X_j,\tau^l) \approx 2\left(\frac{u_{j+1}^l}{h_j(h_j+h_{j-1})} + \frac{u_{j-1}^l}{h_{j-1}(h_j+h_{j-1})} - \frac{u_j^l}{h_j h_{j-1}}\right) = \Delta_i^l\,, \tag{15}$$

for the points X_j, $N \le j \le N+M-2$ in $\left[A, \frac{A}{2\,\delta}\right]$. For the interior points the scheme is given by

$$u_i^l = u_i^l + \frac{k}{2}\sigma^2 X_i^2 \Delta_i^l + \frac{k\,\lambda}{X_i}(J_{1,i}^l + J_{2,i}^l)\,, \quad 1 \le i \le N+M-2\,, \tag{16}$$

where $J_{1,i}^l$ and $J_{2,i}^l$ are the approximations for the integrals (12) using the trapezoidal rule. If we denote $g_{i,j} = g\left(\frac{X_j}{X_i}\right)$, it follows that

$$J_{1,i}^l = h\left(\sum_{j=1}^{N-1} u_j^l\, g_{i,j} + \frac{1}{2}\, u_N^l\, g_{i,N}\right)\,, \quad 1 \le i \le N+M-2, \tag{17}$$

$$J_{2,i}^l = \frac{\delta}{A}\left(\frac{1}{2}\, u_N^l\, g_{i,N}\, X_N^2 + \sum_{j=N+1}^{N+M-1} u_j^l\, g_{i,j}\, X_j^2\right)\,, \quad 1 \le i \le N+M-2. \tag{18}$$

The numerical scheme (16)-(18) must incorporate the initial and boundary conditions

$$u_i^0 = f(X_i) = \max\left(X_i - E, 0\right)\,, \quad 1 \le i \le N+M-1\,, \tag{19}$$

$$u_0^l = 0\,, \qquad u_{N+M-1}^l = u_{N+M-1}^0\,, \qquad 0 \le l \le L\,. \tag{20}$$

This last condition is given assuming linear behavior of the solution for large values of X and corresponds to the original problem [6].

3. The Numerical Scheme for CGMY Model

For this model of infinite activity type, it is required an approximation using Taylor expansion close to the singularity of the integral kernel in (1) such as [4]. We divide the domain of the integral (R) into two regions using a parameter $\varepsilon > 0$; $R_1 = [-\varepsilon, \varepsilon]$ and $R_2 = (-\infty, \varepsilon) \cup (\varepsilon, \infty)$. In R_1 the term $V(Se^y, \tau)$ is expanded by Taylor polynomial in S. This strategy gives approximation of order $\mathcal{O}(\varepsilon^{3-Y})$. Subsequently, the change of variables

$$X = \exp[(r - q - \gamma(\varepsilon))\tau]S, \; U(X, \tau) = \exp[(r + \lambda(\varepsilon))\tau]V(S, \tau), \tag{21}$$

transforms the problem (1)-(2)-(3)-(6) into the following PIDE in which the differential part only remains the diffusion term to avoid numerical oscillations [12].

$$\begin{aligned} \frac{\partial U}{\partial \tau} &= \frac{\hat{\sigma}^2}{2} X^2 \frac{\partial^2 U}{\partial X^2} + J, \; X \in (0, +\infty), \; \tau \in (0, T], \\ U(X, 0) &= f(X), \; X \in (0, +\infty), \end{aligned} \tag{22}$$

where

$$\begin{aligned} J = J(X, \tau, \varepsilon) &= \int_{R_2} \nu(y) U(Xe^y, \tau) dy \\ &= \int_{-\infty}^{-\varepsilon} \nu(y) U(Xe^y, \tau) dy + \int_{\varepsilon}^{\infty} \nu(y) U(Xe^y, \tau) dy. \end{aligned} \tag{23}$$

We introduce the change $\phi = Xe^y$ in order to properly combine the discretization of the differential part with numerical integration. Moreover, J is divided into two integrals, one over a finite interval J_1 and the other on the unbounded domain J_2

$$J = J_1 + J_2 = \int_0^{Xe^{-\varepsilon}} g(X, \phi) U(\phi, \tau) d\phi + \int_{Xe^{\varepsilon}}^{\infty} g(X, \phi) U(\phi, \tau) d\phi, \tag{24}$$

where $g(X, \phi) = \frac{\nu(\ln(\phi/X))}{\phi}$. To evaluate the integrals in the whole domain we introduce the parameter A as in the Merton model. The integral parts J_1 and J_2 corresponding to $\phi > A$ are transformed to integrals over finite domains by $z = \frac{A}{\phi}$.

The numerical scheme for CGMY model (22)-(24) is developed by using the double discretization technique. Note that for each X_i the integral limits $X_i e^{-\varepsilon}$ and $X_i e^{\varepsilon}$ are not necessarily points of the nods, so we split each integral of (24) into the following form

$$\left. \begin{aligned} J_i^l &= J_{i,1}^l + J_{i,2}^l, \\ J_{i,1}^l &= \int_0^{X_{i_1}} g(X_i, \phi) U(\phi, \tau^l) d\phi + \int_{X_{i_1}}^{X_i e^{-\varepsilon}} g(X_i, \phi) U(\phi, \tau^l) d\phi \\ J_{i,2}^l &= \int_{X_{i_2}}^{\infty} g(X_i, \phi) U(\phi, \tau^l) d\phi + \int_{X_i e^{\varepsilon}}^{X_{i_2}} g(X_i, \phi) U(\phi, \tau^l) d\phi. \end{aligned} \right\} \tag{25}$$

where $J_i^l = J(X_i, \tau^l, \varepsilon)$, X_{i_1} is the point of the mesh immediately before $X_i e^{-\varepsilon}$ and X_{i_2} is the point of the mesh immediately after $X_i e^{\varepsilon}$. For the integrals in $(0, X_{i_1}]$ and $[X_{i_{2},\infty})$

we apply the trapezoidal rule. The first mean value theorem has been implemented for the remaining two integrals as follows

$$\int_{X_{i_1}}^{X_i e^{-\varepsilon}} g(X_i,\phi)U(\phi,\tau^l)d\phi \approx \Big(\int_{X_{i_1}}^{X_i e^{-\varepsilon}} g(X_i,\phi)d\phi\Big)u_{i_1}^l, \tag{26}$$

$$\int_{X_i e^{\varepsilon}}^{X_{i_2}} g(X_i,\phi)U(\phi,\tau^l)d\phi \approx \Big(\int_{X_i e^{\varepsilon}}^{X_{i_2}} g(X_i,\phi)d\phi\Big)u_{i_2}^l. \tag{27}$$

Taking into account the previous considerations, the numerical scheme for the PIDE (22) is given by

$$u_i^{l+1} = u_i^l + \frac{k\hat{\sigma}^2}{2}X_i^2\Delta_i^l + k\hat{J}_i^l,\ 1 \le i \le N+M-2, \tag{28}$$

where $\hat{J}_i^l$ is the numerical approximation for the integral term $J_i^l = J(X_i,\tau^l,\varepsilon)$. The initial and boundary conditions for this model are given by

$$u_i^0 = \max(X_i - E, 0),\ \ 1 \le i \le N+M-1, \tag{29}$$

and

$$u_0^l = 0,\ \ 0 \le l \le L, \qquad u_{N+M-1}^l = u_{N+M-1}^0,\ \ 0 \le l \le L-1. \tag{30}$$

4. Positivity, Stability and Consistency

The following lemma provides sufficient conditions that ensure positive solutions of the proposed schemes.

Lemma 1. *Assuming discretization steps* $k = \Delta\tau$, $h = \Delta X$ *in* $[0, A]$ *and* $\delta = \Delta z$ *in* $]0, 1]$, $0 < \delta \le \frac{1}{3}$, *verify:*

$$(C_1)\ \frac{k}{h^2} \le \frac{1}{\sigma^2 A^2}, \qquad (C_2)\ k \le \min\left\{\frac{\delta^2}{\sigma^2(1-2\delta)}, \frac{\delta h}{\sigma^2}\right\}, \tag{31}$$

then the solutions $\{u_i^l\}$ *for the schemes (16)-(20) and (28)-(30) are nonnegative provided that the initial condition* $u_i^0 \ge 0$, $1 \le i \le N+M-1$.

Let us consider $U^l \in R^{N+M-1}$ be the vector containing the numerical solution in all mesh-points at the time τ^l, we say that a numerical scheme is conditionally strongly stable in $\|\cdot\|_\infty$ if the numerical solution U^l remains bounded in $\|\cdot\|_\infty$ with respect to the initial condition for all the time levels regardless of the discretization steps h, δ and k. Here, under the conditions C_1 and C_2 one can show that both schemes (16)-(20) and (28)-(30) are conditionally strongly stable.

With respect to the consistency, using Taylor expansions of the partial derivatives around $\left(X_i,\tau^l\right)$ it can be shown that (16)-(18) is consistent with the PIDE, (10) and (28) is consistent with (22). The local truncation error $T_i^l(U)$ verifies in both cases that

$$T_i^l(U) = \mathcal{O}(h^2) + \mathcal{O}(\delta^2) + \mathcal{O}(k). \tag{32}$$

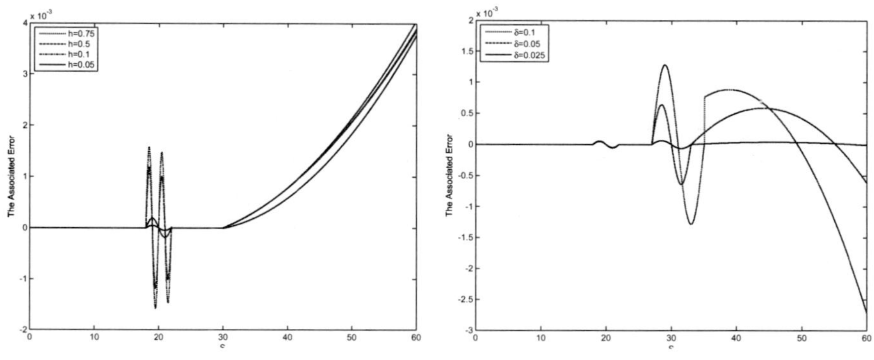

Figure 1. Left: Absolute errors with several values of h and a fixed δ in Example 1. Right: Absolute errors with several values of δ and a fixed h in Example 2.

5. Numerical Results

The following examples illustrate the advantage of the double discretization technique to reduce the error of the numerical solution.

5.1. Example 1

Consider the vanilla call option problem (4)-(5) under Merton jump diffusion model with parameters $T = 1$, $r = 0.05$, $E = 20$, $\sigma = 0.1$, $\mu_J = 0$, $K = 0.005$ and $\lambda = 0.15$. For $A = 60$, $k = 0.001$ and $\delta = 0.05$. Figure 1 (left) shows the reduction of the error in the neighborhood of the strike when the discretization step h decreases while the error remains stationary at points near the boundary numerical A. This coincides with results of other authors [13].

5.2. Example 2

For the same problem in Example 1 by setting $h = 0.25$, Figure 1 (right) shows how the error in the boundary can be reduced with our strategy of double discretization as δ decreases.

The following example examines the error variation of the solution at the strike with respect to parameters h and ε for CGMY model.

5.3. Example 3

Consider the European call option for CGMY process with the following values $C = 1$, $G = M = 5$, $E = 100$, $T = 1, r = 0.1$, $q = 0$, $k = 0.001$, $\delta = 0.1$, $A = 3E$, for several values of Yor parameter $Y = 0.5$, 1.5 and 1.98.

We consider the evaluation of the price option at the strike and $\tau = T$. Table 1 reveals the deviation between our numerical solutions and the reference values used in [14, tables 8-10] for different stepsizes h, and fixed $\varepsilon = 0.12$. Notice that the numerical solution exhibits the expected second order convergence rate ($\mathcal{O}(h^2)$), i.e., α is close to 2.

Table 1. The variation of the error for several values of h.

	$Y = 0.5$			$Y = 1.5$		
h	Absolute error	Relative error	α	Absolute error	Relative error	α
1	7.15×10^{-4}	3.6×10^{-5}	–	4.52×10^{-4}	9.1×10^{-6}	–
0.75	4.12×10^{-4}	2.08×10^{-6}	1.92	2.58×10^{-4}	5.18×10^{-6}	1.95
0.5	1.85×10^{-4}	9.34×10^{-6}	1.95	1.16×10^{-4}	2.33×10^{-6}	1.96
0.25	4.67×10^{-5}	2.36×10^{-6}	1.97	2.9×10^{-5}	5.82×10^{-7}	1.98

Table 2 shows the deviation for different values of ε, when $h = 0.3$.

Table 2. The change of the error due to various values of ε.

	$Y = 0.5$		$Y = 1.5$	
ε	Absolute error	Relative error	Absolute error	Relative error
0.75	1.48×10^{-3}	7.47×10^{-5}	4.17×10^{-4}	8.38×10^{-6}
0.5	8.63×10^{-4}	4.36×10^{-5}	9.37×10^{-5}	1.88×10^{-6}
0.25	2.67×10^{-5}	1.35×10^{-6}	9.48×10^{-6}	1.9×10^{-7}

Conclusion

The PIDEs that govern Merton and CGMY models have been solved numerically using double discretization technique. To apply this technique the domain of the integral part has been split into two regions using a parameter $A > 0$; the first region is bounded while the second is unbounded. In the unbounded region, we use a suitable substitution to convert it into a bounded region, then the discretization has been implemented using explicit discretization for the differential operator and trapezoidal rule for the integral part resulted in uniform discretization for the first region and nonuniform in the second one. The associated error has been calculated for this technique. In light of double discretization technique, we conclude that

1. The error decreases around the strike E and the parameter A as h and δ decrease.

2. The schemes provide a strongly conditional stable solutions.

3. Also, suitable conditions to guarantee positivity of the solutions are shown.

4. The proposed schemes are consistent with the PIDEs.

Acknowledgments

This work has been partially supported by the European Union in the FP7-PEOPLE-2012-ITN program under Grant Agreement Number 304617 (FP7 Marie Curie Action, Project Multi-ITN STRIKE-Novel Methods in Computational Finance).

References

[1] Campbell, J. Y.; Lo, A. W.; MacKinlay, A. C. *The Econometrics of Financial Markets*; Princeton University Press, 1997.

[2] Cont, R.; Tankov, P. *Financial modelling with jump processes*; Chapman and Hall/CRC Press, 2003.

[3] Cont, R.; Voltchkova, E. A finite difference scheme for option pricing in jump diffusion and exponential Lévy models. *SINUM.* 2005, vol. 43, no. 4, 1596-1626.

[4] Wang, I. R.; Wan, J. W. L.; Forsyth, P. A. Robust numerical valuation of European and American options under the CGMY process. *J. Comput. Financ.* 2007, vol. 10 , 31-69.

[5] Almendral, A.; Oosterlee, C.W. Accurate evaluation of european and american options under the CGMY process. *SISC.* 2007, vol. 29, 93-117.

[6] Toivanen, J. Numerical valuation of European and American options under Kou's jump-diffusion model. *SISC.* 2008, vol. 30, no. 4, 1949-1970.

[7] Casabán, M. C.; Company, R.; Jódar, L.; Romero, J. V. Double discretization difference schemes for partial integro-differential option pricing jump diffusion models. *Abstract and Applied Analysis*. 2012, vol. 2012, 1-20.

[8] Tavella, D.; Randall, C. *Pricing Financial Instruments*; Wiley, 2000.

[9] Merton, R.C. Option pricing when the underlying stocks are discontinuous. *JFE.* 1976, vol. 3, no. 1-2, 125-144.

[10] Carr, P.; Geman, H.; Madan, D. B.; Yor, M. The fine structure of asset returns: An empirical investigation. *J. Bus.* 2002, vol. 75, 305-332.

[11] Madan, D. B.; Seneta, E. The Variance Gamma (V.G.) model for share market returns. *J. Bus.* 1990, vol. 63, 511-524.

[12] Sachs, E.W.; Strauss, A.K. Efficient solution of a partial integro-differential equation in finance. *Applied Numerical Mathematics*. 2008, vol. 58, no. 11, 1687-1703.

[13] Almendral, A.; Oosterlee, C.W. Numerical valuation of options with jumps in the underlying. *Appl. Num. Math.* 2005, vol. 53, no. 1, 1-18.

[14] Fang, F.; Oosterlee, C. W. A novel pricing method for European options based on Fourier-cosine series expansions. *SISC.* 2008, vol. 31, no. 2, 826-848.

In: Mathematical Modeling in Social Sciences ...
Editors: J. C. Cortés López et al.
ISBN: 978-1-63117-335-6

Chapter 33

PORTFOLIO COMPOSITION TO REPLICATE STOCK MARKET INDEXES. APPLICATION TO THE SPANISH INDEX *IBEX-35*

***J. C. Cortés*[1,*], *A. Debón*[2,†] *and C. Moreno*[1,‡]**
[1]Instituto Universitario de Matemática Multidisciplinar,
Universitat Politècnica de València, Valencia, Spain
[2]Departamento de Estadística e Investigación Operativa Aplicadas y Calidad,
Universitat Politècnica de València, Valencia, Spain

Abstract

The main goal of this contribution is to provide a methodology to replicate the Spanish stock market index IBEX-35 using the assets of a few companies which perform this index. This will allow us to build up investment portfolios and predictions for this stock market index. The methodology is based in the application of different statistical techniques, namely, linear regression, the principal component analysis, simulation of random variables using Monte Carlo sampling, optimization and stochastic differential equations. In order to determine the weights of the replicating portfolio, a measure of the risk of investment portfolios, usually referred to as Tracking Error Variance, has been used. The period used to apply the proposed methodology is 2 011 and 2 012. After selecting the companies that make up the replicating portfolio, the Log-normal model and sampling Monte Carlo are applied to estimate the value of the shares of each one of these companies. In this way, we are able to estimate the value of the IBEX-35 during the first week of 2 013.

Keywords: Asset Pricing Modeling, Spanish Stock Index, Geometric Brownian Motion, Principal Component Analysis, Tracking Error Variance

---

*E-mail address: jccortes@imm.upv.es
†E-mail address: audeau@eio.upv.es
‡E-mail address: carlotiya20@hotmail.com

1. Introduction and Motivation

The IBEX-35 is the benchmark stock market index of the *Bolsa de Madrid* (Spain), Spain's principal stock exchange [11]. IBEX-35 is a market capitalization weighted index comprising the 35 most liquid Spanish stocks traded in the *Bolsa de Madrid.* This stock market index reflects the economic activity of some of the most important companies operating in Spain. The asset values of these companies indicate their market prices. These values are determined by a large number of random factors which depend on the strategies adopted by rival companies, the international monetary policies, the political stability of the countries where these companies have investments, etc. even, they could be affected by natural disasters. Daily, the investors make their investment decisions taking into account the share prices of the companies trying to anticipate these values. In addition, a wide range of hedging financial products, usually referred to as derivative securities, also depend on the value of the shares traded in the IBEX-35. This motivates the search for appropriate methodologies to replicate the IBEX-35 in order to forecast this financial index which allow us to design effective investment strategies. Currently this task turns out to be very difficult due to the high volatilities rates that affect the financial markets. In this chapter, we will provide a methodology to replicate and, therefore to forecast the IBEX-35. The study will be based on both stochastic differential equations of Itô-type and statistic techniques [6].

The Spanish stock market index IBEX-35 is computed by the following formula [11]:

$$\text{IBEX-35}(t) = \text{IBEX-35}(t-1) \times \frac{\sum_{i=1}^{35} \text{Cap}_i(t)}{\sum_{i=1}^{35} \text{Cap}_i(t-1) + J}, \tag{1}$$

where:

- t denotes the time instant where the IBEX-35 is computed.
- $S_i(t)$ is the number of assets of the company i, $1 \le i \le 35$, at the time instant t.
- $P_i(t)$ is the price of share of the company i, $1 \le i \le 35$, at the time instant t.
- $\text{Cap}_i(t)$ is the capital of the company i, $1 \le i \le 35$, at the time instant t. This value is the product of $S_i(t)$ by $P_i(t)$: $\text{Cap}_i(t) = S_i(t) \times P_i(t)$.
- J is an adjustment coefficient which value is set by the members of the IBEX-35 Administrator Committee according to The Technical Rules to the Computation and Performance of IBEX-35 Index, [12]. Unless exceptional events affecting the trade market happen, J takes the null value, $J = 0$.

Therefore, in order to provide predictions of the IBEX-35, the value of $P(t)$ is required. In practice, this value must be forecast using a mixture of analytic and statistical techniques. In this chapter, the lognormal model and Monte Carlo sampling will be used to construct predictions of $P(t)$ and then, using (1) the IBEX-35 will be predicted, [6, 4].

2. Methodology to Replicate and Predict a Stock Index

The methodology to replicate a stock index that will be used in this chapter consists of three steps [5]:

1. Determine the number of companies to be used in the replicating portfolio. This will be based on both regression analysis and Principal Component Analysis, [9].

2. Calculate the weights corresponding to each company selected in the Step 1. It will be done applying an optimization technique usually referred to as Tracking Error Variance, [1, 7, 8].

3. For each one of the companies selected in the Step 1, predict its value. This will be done using the Log-normal model, [2].

Principal Component Analysis (PCA) is a mathematical procedure that uses orthogonal transformation to convert a set of observations of possibly correlated variables into a set of values of linearly uncorrelated variables called *Principal Components*. The number of principal components is less or equal than the number of original variables. This transformation is defined in such a way that the first principal component has the largest possible variance (that is, accounts for as much of the variability in the data as possible), and each succeeding component in turn has the highest variance possible under the constraint that it be orthogonal to (i.e., uncorrelated with) the preceding components, [9]. Let us explain in detail how PCA method has been applied in our study. First, we point out that during the period where our study has been performed, $2\,011 - 2\,012$, just 31 companies were traded in the IBEX-35 uninterruptedly over the whole period.

- For each one of these 31 companies, let us denote by $V_i(t)$ the week average value of the company i, $1 \le i \le 31$ at the week t, where $t = 0, 1, \ldots, 104$. Notice that two years have approximately 104 weeks.

- For each one of these companies, we consider its relative profitability:

$$PR_i(t) = \frac{V_i(t) - V_i(t-1)}{V_i(t-1)}, \quad t = 1, \ldots, 104, \quad 1 \le i \le 31. \tag{2}$$

- Denoting $\overline{PR}_i$ and $\sigma(PR_i)$ the average and standard deviation of relative profitability of company i, $1 \le i \le 31$, respectively, we introduce its standard relative profitability at the time instant t:

$$\overline{PR}_i(t) = \frac{V_i(t) - \overline{PR}_i}{\sigma(PR_i)}, \quad t = 1, \ldots, 104, \quad 1 \le i \le 31. \tag{3}$$

- For each one of the 31 companies, we perform a linear regression of $\overline{PR}_i(t)$ against the standard relative profitability of IBEX-35 and denote by $\mathbf{u}_i$, $1 \le i \le 31$, the vector of residuals for each one of the 31 companies. Notice that $\mathbf{u}_i$ is a column vector of dimension 104. This generates a matrix of residuals of size 104×31. Next, we transform each entry of this matrix according to the formula

$$\mathbf{v}_i = \left(\left(\mathbf{u}_i^{\mathsf{T}} \mathbf{u}_i \right)^{-1} - 1 \right)^{1/2} \mathbf{u}_i, \quad 1 \leq i \leq 31, \tag{4}$$

the super-index ^{T} stands for the transpose operator for vectors or matrices.

- Then, we apply PCA method to the matrix of size 104×31 whose columns are the vectors $\mathbf{v}_i$.

Notice that if the PCA would have applied to the profitability of the companies, then the companies would have been clusterized according to their yields. However, when the PCA is applied on the residuals obtained after a simple regression has been performed, it implies that two assets which have similar PCA coefficient have a strong relationship with respect to the unexplained part of the IBEX-35. The same can be said when the coefficients are different.

The Tracking Error Variance is the error that appears when the index is replicated by a portfolio made up of, say, N assets [7]. It is given by

$$\text{TEV} = (\mathbf{q}_p - \mathbf{q}_b)^{\mathsf{T}} V (\mathbf{q}_p - \mathbf{q}_b) = \mathbf{x}^{\mathsf{T}} V \mathbf{x}, \tag{5}$$

where V is the covariance matrix of size N of the assets, $\mathbf{q}_p$ and $\mathbf{q}_b$ are vectors whose N components are the weights of the portfolio and the benchmark index, respectively. Then, according to (5), the k-th component of the vector $\mathbf{x}$, $1 \leq k \leq N$, is the variation (difference) between the weight of k-th asset of the portfolio and the benchmark index. The quadratic function that minimizes the TEV

$$\left.\begin{array}{ll} \text{Min.} & \displaystyle\sum_{i=1}^{T} (\mathbf{w}^{\mathsf{T}} \mathbf{r}_i - \mathbf{R}_i)^2, \\ & \\ \text{s.t.} & \mathbf{w}^{\mathsf{T}} \mathbf{1} = 1, \end{array}\right\} \tag{6}$$

where $\mathbf{1} = (1, \ldots, 1)^{\mathsf{T}}$, $\mathbf{w}$ is the unknown vector whose components are the weights of the N assets of the replicating portfolio, and $\mathbf{r}_i$ and $\mathbf{R}_i$ are the vectors of standard profitability of each of the N assets and benchmark stock index, respectively, the during the period i, [8]. Notice that the value of N has previously determined by PCA method.

As we pointed out previously, once the replicating portfolio has been determined, modelling the dynamic evolution of the price $P(t)$ for each one of the underlying in the portfolio is required. This will be made using the *Log-normal model* based on the following Itô-type differential equation

$$dP(t) = \mu P(t)\, \mathrm{d}t + \sigma P(t)\, \mathrm{d}W(t), \quad P(0) = P_0, \tag{7}$$

where $\mu \in R$ and $\sigma > 0$ are parameters that denote the drift and volatility of $P(t)$, respectively, and $W(t)$ is the standard Wiener process or Brownian motion. Using the Itô's Lemma, the stochastic differential equation (7) can be solved. It leads to:

$$P(t) = P_0 \exp\left(\left(\mu - \frac{1}{2}\sigma^2 \right) t + \sigma W(t) \right), \quad t \geq 0, \tag{8}$$

which is usually referred to as the *Geometric Brownian Motion* [6]. Parameters μ and $\sigma > 0$ need to be determined. In this paper the calibration of this two parameters has been done using both moment and maximum likelihood methods. With aim it is more convenient to take logarithms in (8) and handle log-returns, $\ln(P(t))$ rather than prices $P(t)$. In fact, this facilitates the determination of the probabilistic distribution $\ln(P(t))$ since, by definition the Wiener process is Gaussian: $W(t) \sim \mathrm{N}\left(0; \sigma\sqrt{t}\right)$. Therefore, from the (8) one gets

$$\ln(P(t)) - \ln(P_0) = \left(\mu - \frac{1}{2}\sigma^2\right) t + \sigma W(t) \sim \mathrm{N}\left(\left(\mu - \frac{1}{2}\sigma^2\right) t; \sigma\sqrt{t}\right), \quad t \geq 0. \quad (9)$$

In this way, the application of moment and maximum likelihood methods to calibrate μ and σ is easier. Once this has been made, the prices $P(t)$ of the assets of each one of the companies selected by PCA method to replicate the IBEX-35 index can be predicted directly by applying (8). To complete these predictions, we will also compute these values using Monte Carlo simulation. This will be made sampling values of Wiener process taking into account the following identity $W(t) \stackrel{\mathrm{d}}{=} \sqrt{t}\, Z$, $Z \sim \mathrm{N}(0;1)$, where d stands for distribution identity.

Once the model has been set, we will measure the quality of its predictions $\{\hat{S}_i : 1 \leq i \leq K\}$ with respect the data $\{S_i : 1 \leq i \leq K\}$ using the mean square error (MSE) and the mean absolute percentage error (MAPE) given, respectively, by

$$\mathrm{MSE} = \sqrt{\frac{\sum_{i=1}^{K}(S_i - \hat{S}_i)^2}{K}}, \quad \mathrm{MAPE} = \frac{100}{K}\sum_{i=1}^{K}\frac{|S_i - \hat{S}_i|}{S_i}. \quad (10)$$

3. Application of the Proposed Methodology

In this section we will apply the methodology introduced in the previous section to predict the IBEX-35. The period chosen to perform the study corresponds to the whole years 2 011 and 2 012 and, predictions will be performed for the first week of 2 013. According to the IBEX-35 regulations, if a company does not fulfil specific conditions, it can leave this Spanish stock market index. In this case, a new company is invited to take part of this index. As a consequence, the number of companies that were trading in the IBEX-35 index during the whole period 2 011–2 012 were not 35 but 31. This entails that we will perform a non–complete replication of the IBEX-35, i.e., the TEV will be different from zero.

According to procedure developed in the previous section, for every of these 31 companies, in the first step, we have performed a simple linear regression of $\overline{PR}_i(t)$ (see (3)) against the standard relative profitability of IBEX-35. The residuals of these regressions have been transformed according to the formula (4). Then, a PCA upon the obtained covariance matrix has been performed. From this analysis, 5 principal components with an eigenvalue greater or equal than 1 have been computed. This permits to explain 68.30% of the variability of the original data. In order to obtain the variables which have a greater associated coefficient, a Varimax rotation has been carried out, [5]. The obtained results are collected in Table 1.

In our context, the application of PCA method to replicate the stock market index IBEX-35 means the identification of the companies to be included in the replicating portfolio in

Table 1. Selection of the companies belonging to the IBEX-35 using PCA with 5 components.

Residuals	Component 1	Component 2	Component 3	Component 4	Component 5
ABE	0.609849	0.233442	0.120727	0.0982592	-0.217647
ABG	0.5127	0.681366	0.0959179	-0.0428052	0.0413134
ACS	0.217702	0.69292	0.296148	0.0814548	-0.00584972
ACX	0.618928	0.295785	0,256561	0.404563	0.142418
AMS	0.76331	0.418522	0.205378	0.124788	0.170492
ANA	0.464501	0.43441	0.336398	0.155075	0.0377348
BBVA	0.0153832	0.0899231	0.0059288	0.0971096	**0.883105**
BKT	0.0366953	0.701121	0.041704	0.245449	0.264954
BME	0.527849	0.469765	-0.330067	0.379111	0.152217
CABK	0.471537	0.590644	0.0377869	0.145969	0.188546
ELE	0.646503	0.080045	-0.336167	0.241721	0.0970999
ENG	0.739641	0.203219	0.04658	0.0560641	-0.0503325
FCC	0.345131	0.571481	0.229819	0.192017	-0.142756
FER	0.422495	0.574466	0.0904471	0.276229	-0.145659
GAS	0.381419	0.146611	0.642183	0.00419437	0.056171
GRF	0.704649	0.508733	0.145918	0.20843	0.185448
IBE	0.0433486	0.222361	**0.87691**	-0.0787693	-0.0735303
IDR	0.387339	0.443269	0.320782	0.421594	0.214985
ITX	0.662614	0.44929	0.065448	0.0853688	0.0203353
MAP	0.287116	0.394315	-0.349104	0.600941	-0.0209808
MTS	0.622967	0.146017	0.294421	0.486542	0.159389
OHL	0.673266	0.0882317	0.244336	0.304199	-0.0909408
POP	0.178069	**0.825872**	0.0541848	0.100903	0.17543
REE	**0.807375**	0.276547	0.21314	-0.0703676	-0.0387846
REP	0.208571	0.307679	0.242265	**0.601543**	-0.11764
SAB	0.309588	0.76297	-0.0674907	0.167853	0.3352
SAN	0.0141658	0.222165	-0.0817586	-0.139401	0.859701
SYV	0.446835	0.535111	0.215502	0.372488	0.00580241
TEF	0.113078	0.0229173	0.669893	0.295581	-0.0427899
TL5	0.563414	0.220373	0.0601854	0.381146	0.356435
TRE	0.774803	0.223208	0.0881508	0.248906	0.116526

such a way that the explained variance is maximum. Additionally, it is also desirable that the mathematical results agree with the economic situation. This is not a simple issue and often the results provided by PCA can become somewhat subjective. In our case, the selected shares to build the replicating portfolio belong to three Spanish economic sectors, namely, banking (**BBVA** and **POP**ULAR), electricity (**IBE**RDROLA and **R**ed **E**lectrica **E**spañola) and energy (**REP**SOL).

In Table 2, we show the weights corresponding to each of the 5 selected companies. These figures have been computed solving the optimization problem (6) using the *Solver* tool by Excel.

Table 2. Determination of the weights of each company selected by PCA with 5 components.

Company	BBVA	IBE	POP	REE	REP
Weight	49.02%	23.79%	5.34%	9.22%	12.63%

Note that the companies selected by PCA technique do not correspond to ones which have the highest weight in the IBEX-35. This confirms us that appropriate statistic tools are required to perform a replicating portfolio. It can be explained because an analysis of the covariances of all the shares needs to be performed to build an adequate replicating portfolio.

Table 3. Predictions of the shares of each selected company that perform the replicating portfolio.

Company	Observation	Prediction	Difference	s^2
BBVA	7.09	6.98	0.11	0.002090
IBE	4.11	4.15	0.04	0.002745
POP	3.27	3.00	0.27	0.006524
REE	39.05	37.76	1.29	0.000739
REP	15.56	15.44	0.12	0.002452

Table 4. Predictions of the IBEX-35 in the week 105 corresponding to the first week of 2 013.

Company	Number of shares	Prediction	Capitalization	IBEX-35
BBVA	5 448 000	6.98	18 641 651.77	
IBE	6 139 000	4.15	6 061 024.476	
POP	8 409 000	3.00	13 346 803.277	
REE	135 000	37.76	470 211.048	
REP	1 256 000	15.44	24 448 948.653	
TOTAL			28 968 639.230	
PREDICTION				8 440.55

Table 5. MSE and MAPE associated to the predictions of each company performing the replicating portfolio using the PCA method with 5 components. Weighted MSE and MAPE associated to the IBEX-35 prediction.

Company	BBVA	IBE	POP	REE	REP	IBEX-35
Weight	49.02%	23.79%	5.34%	9.22%	12.63%	–
MSE	0.75	0.70	0.99	3.4	2.82	**1.26**
MAPE	8.65%	14.39%	20.27%	6.85%	25.37%	**11.32%**

So far, we have proposed 5 companies that will perform the replicating portfolio. As we pointed out previously, predictions for each one of these 5 shares need to be constructed.

Table 6. Predictions of the IBEX-35 in different weeks during 2 013.

PREDICTION 2013	**PCA** (5 components)	**PCA** (8 components)
20 weeks		
Absolute difference w.r.t. the IBEX-35	4.75	3.55
% w.r.t. the IBEX-35	0.056%	0.04%
12 weeks		
Absolute difference w.r.t. the IBEX-35	120.83	257.77
% w.r.t. the IBEX-35	1.45%	3.15%

Table 7. Predictions of the IBEX-35 in different weeks during 2 012.

PREDICTION 2012	**PCA** (5 components)	**PCA** (8 components)
20 weeks		
Absolute difference w.r.t. the IBEX-35	24.65	5.07
% w.r.t. the IBEX-35	0.29%	0.06%
12 weeks		
Absolute difference w.r.t. the IBEX-35	27.90	51.39
% w.r.t. the IBEX-35	0.33%	0.60%

This has been made applying the Log–normal model and Monte Carlo sampling for two different periods: the last 12 weeks (3 months) and the last 20 weeks (5 months) of 2 012. The goal is to predict the IBEX 35 for the first week of 2 013. Table 3 collects the obtained results using the Log–normal model for 20 weeks which correspond to the best results among the predictions constructed by the four combinations of methods and data set.

Table 4 shows the capitalization of the 5 selected companies in the week 105 (first week of 2 013). These values have been computed by multiplying the predicted value of the shares of each company (see column 3 of Table 3) by the number of assets corresponding to each company according to the referred data (see column 2 in Table 4). From the values of these capitalizations and considering the weights obtained by TEV (see Table 2), the value 8 440.55 points has been forecast to the IBEX-35. The real data of the IBEX-35 for this date was 8 435.80. Hence, there is a difference of 4.75 points between the real value and the prediction of the IBEX-35, i.e., the prediction has 0.056% relative error.

In Table 5, we collect both the MSE and MAPE associated to the 5 companies performing the replicating portfolio when PCA method is applied with 5 components. Once, these errors have been carried out, an average value of the MSE and MAPE associated to the IBEX-35 prediction is determined according to the weights of each one of the 5 companies (see file 2 in Table 5). The associated errors are acceptable since neither of them is greater than 15%.

Conclusion

In this chapter we have presented different methodologies to predict the value of the Spanish stock market IBEX-35. The obtained results show that, in general, all the methods provide satisfactory results. Considering the different scenarios analysed, it can be stated the application of PCA technique achieved better results when we imposed the condition that eigenvalues associated to the corresponding variance-covariance matrix were greater than 0.8 rather than 1 since it entails further explanation of the variability of the original data. In this way, the difference between the real value of the IBEX-35 at the first week of 2 013 and the forecast value was 3.55 points which represents 0.04% of relative error (see Table 6). Similar comments can be done for the predictions in different weeks of 2 012 (see Table 7). Intuitively, it agrees with the fact that the IBEX-35 forecast value depends on the values of the companies which perform the replicating portfolio. Therefore, the most explanation of the variability of the replicating portfolio, the most accuracy of the IBEX-35 forecast value. The explained variability increases as the number of companies included in the replicating portfolio does, however this is got at expense of higher computational cost. Hence, an adequate number of companies must be considered to reach a balanced solution. I our case, we considered two acceptable solutions, 5 and 8 companies out of 31 achieving better results with 8 companies. After selecting the companies that have performed the replicating portfolio, we have applied the TEV method to determine the the weight of each selected company. These weights influence directly on the forecast value of the IBEX-35. The BBVA company had a weight of almost 50% in the portfolio, so its accurate prediction has further influence on the estimated value of the IBEX-35. We checked that the methods used to predict the value of this company, namely, Log-normal model and Monte Carlo sampling, have been very accurate using for this end measures of goodness-of-fit. It explains the quality obtained results.

References

[1] Beasley J.E., Meade N.Y., Chang T.-J. (2003): *An evolutionary heuristic for the index tracking problem*, *European Journal of Operational Research* 148, 621–643.

[2] Back K. (2005): A Course of Derivative Securities: *Introduction to Theory and Computation, Series: Springer Finance*, Springer, Berlin.

[3] Black F., Scholes M. (1973): *The pricing of options and corporate liabilities*, *Journal of Political Economy* 81, 637–659.

[4] Gedam S.G., Beaudet S.T. (2000): *Monte Carlo Simulation using Excel Spreadsheet for Predicting Reliability of a Complex System*, *Proceedings Annual Reliability and Maintainability Symposium*, 188–193.

[5] Guijarro F., Moya I. (2008): *Propuesta metodológica para la selección de acciones en la réplica de índices*, *Revista de Economía Financiera* 16, 26–51 (in Spanish).

[6] Øksendahl B. (1998): *Stochastic Differential Equations*, Springer, Berlin.

[7] Roll R. (1992): *A mean/variance analysis of tracking error*, *The Journal of Portfolio Management* 18, 13–22.

[8] Rudolf M., Wolter H.-J., Zimmermann H. (1999): *A linear model for tracking error minimization*, *Journal of Banking & Finance* 23, 85–103.

[9] Shlens J. (2009): *A Tutorial on Principal Component Analysis*, Center for Neural Science, New York University New York City, NY 10003-6603 and Systems Neurobiology Laboratory, Salk Insitute for Biological Studies La Jolla, CA 92037 (Document on-line: September 30, 2013; Version 3.01).

[10] Sociedad de Bolsas: www.sbolsas.es, www.ibex35.com, (document on-line) (in Spanish).

[11] www.ibex35.com (Accessed at: October 20, 2013).

[12] www.sbolsas.es (Accessed at: October 20, 2013).

INDEX

#

A

B

C

D

E

F

G

H

I

K

L

M

T

U

V

W